LACAME 2012

C.A. Barrero Meneses • E. Passamani Caetano
C.E. Rodríguez Torres • C. Pizarro
L.E. Zamora Alfonso
Editors

LACAME 2012

Proceedings of the 13th Latin American Conference on the Applications of the Mössbauer Effect, (LACAME 2012) held in Medellín, Colombia, November 11–16, 2012

Previously published in *Hyperfine Interactions*
Volume 224, 2014

Springer

Editors
César Augusto Barrero Meneses
Instituto de Física
Universidad de Antioquia
Medellín, Colombia

Edson Passamani Caetano
Centro de Ciências Exatas
Departamento de Física
Universidade Federal do Espírito Santo
Vitória ES, Brazil

Claudia Elena Rodríguez Torres
Departamento de Física y Ciencias Exactas
Universidad Nacional de La Plata
La Plata, Argentina

Carmen Pizarro
Facultad de Química y Biología
Universidad de Santiago de Chile
Santiago, Chile

Ligia Edith Zamora Alfonso
Departamento de Física
Universidad del Valle
Cali, Colombia

ISBN 978-94-007-6481-1 ISBN 978-94-007-6482-8 (eBook)
DOI 10.1007/978-94-007-6482-8
Springer Dordrecht Heidelberg New York London

Library of Congress Control Number: 2014935866

Printed on acid-free paper

Springer is part of Springer Science+Business Media (www.springer.com)

Preface

Published online: 28 March 2013

Mössbauer Spectroscopy is an experimental technique based on the nuclear resonant emission and absorption of gamma rays. It was discovered in 1958 by Professor Rudolf Mössbauer who received the Nobel Prize for Physics in 1961. Due to its high sensitivity to the local physical properties of the systems under study, the technique has proven to be very useful in several areas of scientific and technological research.

The "Latin American Conference on the Applications of the Mössbauer Effect-LACAME" is a series of conferences started in 1988 in Rio de Janeiro and has grown steadily since then, changing every 2 years the venue from different nations in which Mössbauer research laboratories exist. The purposes of LACAME are to bring together the Latin American and the International community, to promote activities of basic and applied research, to encourage participation and interaction among students, young and senior researchers, and to strengthen international cooperation around the field. LACAME is the best opportunity to share experiences gained in the process by each of the different countries as well as a good opportunity for the students of Latin American countries to discuss their researches with specialists of Latin America and abroad that work with the same experimental characterization technique. Topics addressed in these conferences include Advances in Experimentation and Data Processing, Amorphous, Nanocrystals and Nanoparticles, Applications in Soils, Mineralogy, Geology and Archaeology, Biological and Medical Applications, Catalysis, Corrosion and Environment, Chemical Applications, Structure and Bonding, Industrial Applications, Magnetism and Magnetic Materials, Multilayers, Thin Films and Artificially Structured Materials, Physical Metallurgy and Materials Science, and others related with the use of Mössbauer spectroscopy.

In 2012, the Institute of Physics at the Universidad de Antioquia (Medellín-Colombia) had the privilege to lead the planning and development of this important academic event in cooperation with Planetario de Medellín "Jesús Emilio Ramírez González" and Parque Explora. In order to ensure the continuity of the Latin American Mössbauer Community, in LACAME-2012 different activities were programmed encouraging the creation of the new generation for Mössbauer spectroscopy in Latin America. Therefore, the "International School on Recent Advances in Mössbauer Spectroscopy", the "Rudolf Mössbauer Latin American Thesis Award", and the "Best Presentation Award" were specially organized to promote active participation of young scientists. Additionally, internationally recognized experts of the highest scientific level as well as young prominent researchers from Latin America were invited as Speakers and also as Tutors of the International School.

The Organizing Committee decided to honor the outstanding scientific contributions of Prof. R. Mössbauer, who passed away in 2011, through the "Rudolf Mössbauer Latin American Thesis Award-2012". Also, several academic sessions were devoted to honor the memory of our friends and colleagues Profs. Fernando González-Jiménez (Venezuela), I. P. Suzdalev (Russia), A. Vértes (Hungary) and H. Rechenberg (Brazil), who regretfully passed away in the last years. Soon after finishing the Conference, we received the sad news, that our colleague and friend Dra. Judith Dessimoni (Argentina) died on November 26 of 2012. We certainly shall miss all of them.

Medellín, known as the 'city of the eternal spring', welcomed the Mössbauer community. Nowadays, this city has been chosen as the most innovative city in the world by the Urban Land Institute, Citigroup and The Wall Street Journal. It is becoming the destination of choice for many new and seasoned visitors. It is a vibrant, vital and rapidly developing place that offers an excellent platform for varied activity as tourism, conventions and academic conferences.

We are convinced that the readers will enjoy the scientific contributions published in the present volume, which give an overview of the science currently done around the Applications of the Mössbauer Effect. We would like to express our deep gratitude to all the institutions supporting LACAME 2012, and thank all the presenters, participants, and members of the Organizing, Latin American and Scientific Committees for making this conference possible. The conference counted with the presence of 131 participants, mostly students, including leading researchers in the field coming from 16 countries of Latin America, Europe, Asia, and North America. Over 120 works were presented both domestic and from abroad; all of them with excellent scientific quality and valuable contributions to the field. In this proceeding volume, the papers accepted for publication in the Journal of Hyperfine Interactions were submitted through the online system via editorial management newly provided by Springer.

C.A. Barrero Meneses and J. Mazo-Zuluaga (Chairs)

XIII Latin American Conference on the Applications of the Mössbauer Effect LACAME-2012
Universidad de Antioquia - Planetario de Medellín - Colombia
11-16 November 2012

Contents

Chairmen and committees

Published online: 28 March 2013

Committees

Chairpersons
C.A. Barrero Meneses (UdeA)
J. Mazo-Zuluaga (UdeA)

Local Organizing Committee

K.E. García Téllez (UdeA)
J.A. Osorio (UdeA)
A.L. Morales Aramburo (UdeA)
J. Restrepo Cárdenas (UdeA)
O.L. Arnache Olmos (UdeA)
J.E. Tobón (UdeA)
A.A. Velásquez Torres (EAFIT)
F.R. Pérez (UPB)
G.A. Pérez Alcázar (UV)
L.E. Zamora Alfonso (UV)
Y.A. Rojas Martínez (UT)
D. Oyola Lozano (UT)
H. Bustos Rodríguez (UT)
B.C. Londoño (UTP)
R. Rodríguez (UAO)

Latin American Committee

E.M. Baggio Saitovitch (CBPF-Brazil)
F. González Jiménez (UCV-Venezuela)
G.A. Pérez Alcázar (UV-Colombia)
J.A. Jaén (UP-Panamá)

N.R. Furet Bridón (CNIC-Cuba)
N. Nava (IMP-México)
R.C. Mercader (UNLP-Argentina)
V.A. Peña Rodríguez (UNMSM-Perú)
C. Pizarro (USACH-Chile)

Scientific Committee

E.M. Baggio Saitovitch (CBPF-Brazil)
R.C. Mercader (UNLP-Argentina)
F.H. Sánchez (UNLP-Argentina)
C. Saragovi (CNEA-Argentina)
C. Partiti (USP-Brazil)
C. Larica (UFES-Brazil)
E.C. Passamani (UFES-Brazil)
C.A. Barrero Meneses (UdeA-Colombia)
N.R. Furet Bridón (CNIC-Cuba)
J.A. Jaén (UP-Panamá)
A. Bustamante Domínguez (UNMSM-Perú)
J.R. Gancedo (IQFR-Spain)

SPONSORS

Universidad de Antioquia
Planetario de Medellín "Jesús Emilio Ramírez González"
Parque Explora
COLCIENCIAS
IOP Publishing
Universidad del Tolima
Universidad EAFIT
Universidad Tecnológica de Pereira
Medellín Convention and Visitors Bureau
Alcaldía de Medellín
Centro Latinoamericano de Física-CLAF
WissEL GmbH
Embajada de Francia en Colombia
Sociedad Colombiana de Física
Universidad Pontificia Bolivariana
Universidad Autónoma de Occidente-Cali
SENA
Springer Science, Business Media–The Netherlands

DOI 10.1007/978-94-007-6482-8_1
Reprinted from *Hyperfine Interactions*, DOI 10.1007/s10751-013-0819-4

Mössbauer investigation of the reaction of ferrate(VI) with sulfamethoxazole and aniline in alkaline medium

Virender K. Sharma · Zoltan Homonnay · Karolina Siskova · Libor Machala · Radek Zboril

Published online: 13 February 2013

Abstract Mechanisms on the oxidation of sulfamethoxazole (SMX) and aniline by ferrate(VI) ($Fe^{VI}O_4^{2-}$, Fe(VI)) in alkaline medium suggested the formation of Fe(VI)-SMX or Fe(VI)-aniline intermediates, respectively. Fe(V) and Fe(IV) as other intermediate iron species have also been proposed in the mechanism. In this paper, rapid freeze Mössbauer spectroscopy was applied in rapidly frozen samples to explore intermediate iron species in the reactions of SMX and aniline with Fe(VI). In both reactions, Fe(VI)-SMX and Fe(VI)-aniline intermediates were not seen in second-minute time scale. Fe(V) and Fe(IV) were also not observed. Fe(III) was the only final species of the reactions.

Keywords Ferrate · High-valent iron species · Fe(II) · Fe(III) · Antibiotics · Electron-transfer · Rapid-freeze technique · Mössbauer spectroscopy

Thirteenth Latin American Conference on the Applications of the Mössbauer Effect, LACAME 2012, Medellín, Columbia, 11–16 November 2012.

V. K. Sharma (✉)
Center of Ferrate Excellence and Chemistry Department,
Florida Institute of Technology, 150 West University Boulevard,
Melbourne, FL 32901, USA
e-mail: vsharma@fit.edu

Z. Homonnay
Laboratory of Nuclear Chemistry, Institute of Chemistry,
Eotvos Lorand University, Pazmany P.s. 1/A, Budapest, Hungary

K. Siskova · L. Machala · R. Zboril
Regional Center of Advanced Materials and Technologies,
Palacky University in Olomouc, 17. listopadu 1192/12,
771 46 Olomouc, Czech Republic

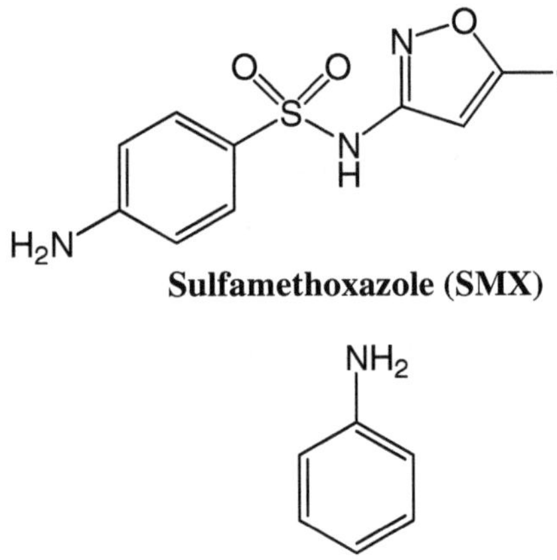

Fig. 1 Structures of sulfamethoxazole (SMX) and aniline

1 Introduction

The application of potassium ferrate(VI) (K_2FeO_4, Fe(VI)) in water and wastewater treatment is attracting increasing interest [1–8]. Fe(VI) is a strong oxidizing agent with a standard reduction potential of 2.20 V under acidic conditions, however it is relatively mild oxidant in the basic conditions with a standard reduction potential of 0.72 V [9]. Like ozone, Fe(VI) does not produce carcinogenic bromate ion in treatment of Br^- containing water [7]. Because of growing concern of chlorine disinfection by-products (DBPs), Fe(VI) shows promise as an alternative disinfectant [10]. Fe(VI) can also be an effective oxidant for removal of organic micropollutants such as estrogens and certain pharmaceuticals [11–17]. No antibacterial activity against *E. coli* after complete removal of antibiotic, trimethoprim by Fe(VI) was observed [14].

The understanding of the mechanisms of oxidation reactions carried out by Fe(VI) is in progress [18, 19]. In our laboratory, we are attempting to advance knowledge on the mechanism of oxidation of antibiotics by Fe(VI) [14, 20]. The present paper deals with antibiotic, sulfamethoxazole (SMX), which consists of two moieties, an aniline ring and a five-membered heterocyclic aromatic group, connected to both sides of the sulfonamide linkage (-NH-SO_2-) (Fig. 1). In the proposed mechanism of Fe(VI) with SMX, intermediate iron in +4 oxidation state has been suggested [20]. However, no direct evidence was provided to confirm this intermediate iron species. Furthermore, a recent study proposed Fe(VI)-SMX intermediate species and analyzed their kinetics results based on this assumption [15]. Moreover, studies on aniline, a substructure molecule of SMX (see Fig. 1) have also been performed [21–23]. Under condition of using excess molar amounts of Fe(VI) to aniline, nitrobenzene was the product of the reaction (1) while azobenzene was formed when excess aniline was used (2).

$$FeO_4^{2-} + C_5H_5NH_2 \rightarrow Fe\,(II) + C_5H_5NO_2 \quad (1)$$

$$FeO_4^{2-} + C_5H_5NH_2 \rightarrow Fe\,(II) + C_5H_5N = NC_5H_5 \quad (2)$$

A free radical mechanism for the reaction, based on EPR measurements was proposed when excess amounts of aniline to Fe(VI) was used [21]. In this free radical mechanism, FeO_4^{3-} (Fe(V)) was also the intermediate high-valent iron species. However, a study under similar experimental conditions contradicted the free radical based mechanism, and the imidoiron(VI) intermediate as the intermediate was suggested before the formation of final product, *cis*-azobenzene (3) [21].

$$FeO_4^{2-} \xrightarrow[\text{milliseconds-seconds}]{C_6H_5NH_2} \text{imidoiron (VI) intermediate} \xrightarrow[\text{seconds-minutes}]{C_6H_5NH_2} \textit{cis}\text{-azobenzene} + \text{Fe (II)} \quad (3)$$

In the last few years, we are attempting to explore mechanism of the oxidation reactions of Fe(VI) using the Mössbauer spectroscopy in order to establish any intermediate iron(VI)-substrate species or to distinguish one-electron steps (Fe(VI) → Fe(IV) → Fe(III)) and two-electron steps (Fe(VI) → Fe(IV) → Fe(II)) [5, 24, 25]. The oxidation states of iron can be determined unequivocally by the Mössbauer technique [26] and therefore, any intermediate iron species of the reaction may be known. In the current paper, we have applied Mössbauer technique to the reactions of Fe(VI) with SMX and aniline to learn about the mechanism of the reactions.

2 Experimental methods

The samples for Mössbauer spectroscopy were prepared by mixing solutions of 0.2 M Fe(VI) with either SMX or aniline in 10 M NaOH and in 0.01 M phosphate buffer (pH 11.0). The experiments at pH 9.0 for the reaction of Fe(VI) with aniline were performed under anaerobic solutions in which 0.2 M Fe(VI) (a ferrate(VI) sample enriched in ^{57}Fe isotope) and 0.2 M aniline were mixed to achieve the molar ratio 1:1. Anaerobic conditions were obtained by degassing with a stream of nitrogen to remove oxygen prior to mixing them together and the volume used for each solution was 100 μL. The mixing of solutions was performed in a Mössbauer sample holder (i.e., in a tiny bowl) within second time scale. The sample holder was then immersed into liquid nitrogen (78 K). The cooling rate provided by this method preserves the original structure of the solution and iron species in the liquid state [25].

The transmission ^{57}Fe Mössbauer spectra were recorded at $T = 100$ K using a Mössbauer spectrometer in a constant acceleration mode with a $^{57}Co(Rh)$ source and velocity range from −5 to 5 mm/s. The isomer shift values are given relative to metallic alpha iron at room temperature. The Mössbauer spectra were evaluated assuming Lorentzian line shapes using the least squares method with the help of the MossWinn^R code. It was expected that the effects of non-ideal absorber thickness and variable recoil-free fractions for iron atoms in non-equivalent structural sites are all within experimental error.

3 Results and discussion

Initially, solid SMX was added into 0.1 M Fe(VI) solution in 10 M NaOH solution. It was expected that such high concentration of hydroxide would stabilize any intermediate Fe(V)/Fe(IV) species [27]. As shown in Fig. 2, the Mössbauer

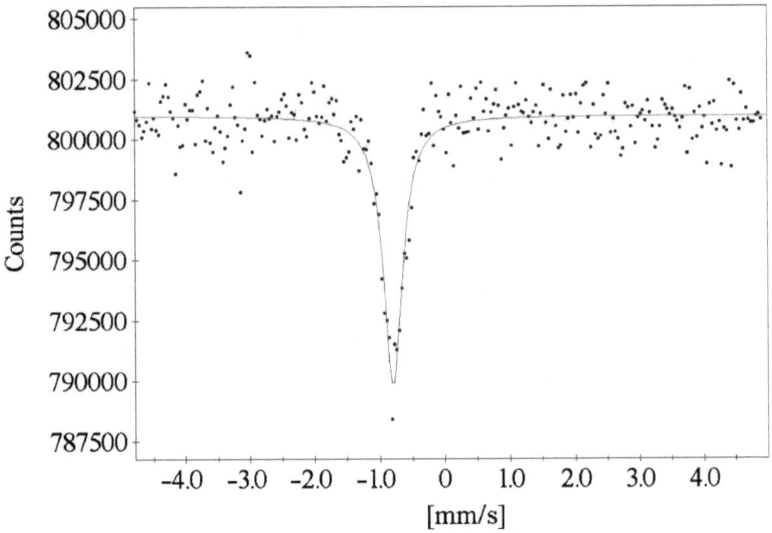

Fig. 2 Mössbauer spectrum of the frozen solution of the mixture of 0.1 M Fe(VI) and solid SMX

spectrum of the frozen solution showed a single component having isomer shift of −0.82 mm/s and this represents non reacted Fe(VI). The Fe(V) and Fe(IV) species as intermediates of the reaction between Fe(VI) and SMX were not seen. Furthermore, no reaction between Fe(VI) and SMX was observed because no Mössbauer spectra of Fe(III)/Fe(III) as the final product of Fe(VI) reduction were found.

Next, the experiments between Fe(VI) and SMX were extended to pH 11 in phosphate buffer to increase the rate of the reaction. Lowering the pH generally increases the rate of the reactions in the alkaline medium [13, 28, 29]. In this set-up, two solutions were mixed at pH 11 and were frozen in ~10 s to 180 s time intervals. Mössbauer spectra of these frozen solutions, shown in Fig. 3, had one singlet and a doublet, which represent Fe(VI) and ferrihydrite, respectively. The identification of ferrihydrite is based on Mössbauer parameters of the doublet. The doublet component increase with the reaction time (see Fig. 3a and b). The final Fe(III) species is consistent with the earlier study on the reaction of Fe(VI) with SMX [20]. The intermediate high-valent iron species were again elusive even at pH 11. It is possible that the intermediate iron species, Fe(V) and Fe(IV), formed during the reaction, were either reacting with SMX or self-decomposing to give Fe(II)/Fe(III) within the time scale (~10 s) of the freezing of the mixed sample.

Similarly, the study of Fe(VI) with aniline was carried out in 10 M NaOH and at pH 11. In 10 M NaOH solution, no reaction of Fe(VI) with aniline was observed and Mössbauer spectrum had a single component of Fe(VI). Lowering the reaction solution to pH 11 showed the decay of Fe(VI) with the concomitant growth of the doublet component. Again, no intermediate iron species could be observed. It seems that if Fe(VI)-aniline intermediate formed during the reaction, it must have a life of less than 10 s i.e. time needed to mix the solution, followed by freezing. The doublet component corresponds to Fe(III). This contradicts with the earlier work in which Fe(II) was the final product [22]. It is very likely that Fe(II) produced from the

Fig. 3 Mössbauer spectra of the frozen solutions of the mixture of 0.1 M Fe(VI) and 0.01 M SMX in phosphate buffer at pH 11.0. **a** 10 s; **b** 180 s

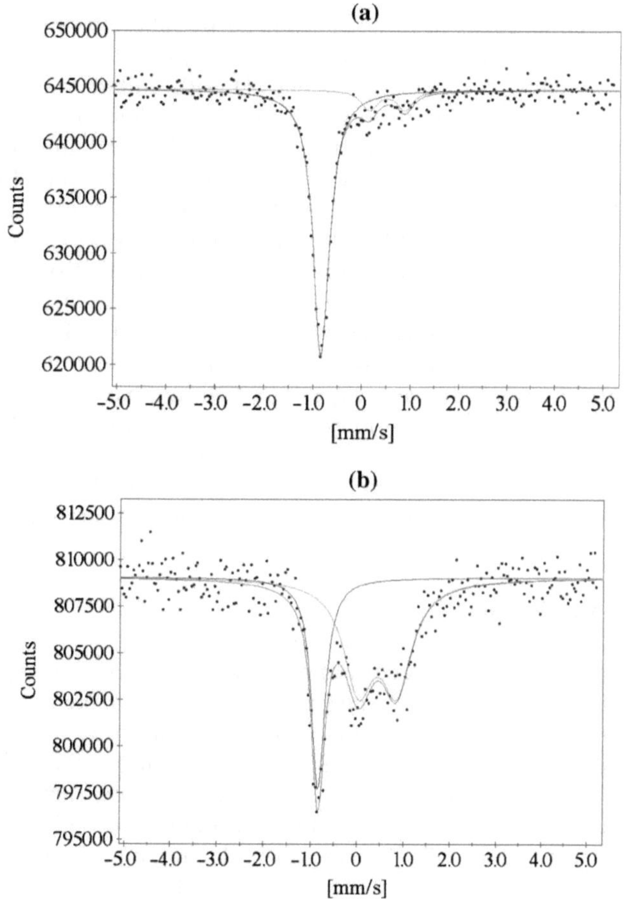

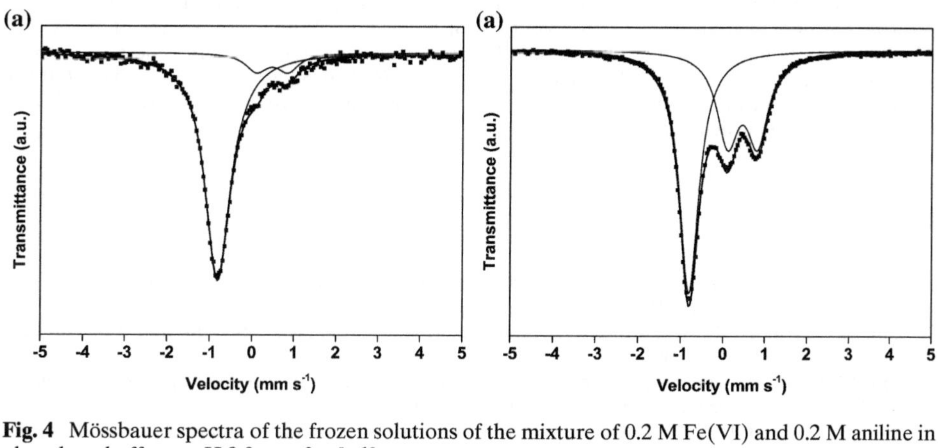

Fig. 4 Mössbauer spectra of the frozen solutions of the mixture of 0.2 M Fe(VI) and 0.2 M aniline in phosphate buffer at pH 9.0. **a** ~ 3 s; **b** 60 s

reaction might have been oxidized by inherent oxygen in the reaction solution. It is known that the oxidation of Fe(II) at such high pH is rapid [30].

Finally, the reaction of Fe(VI) with aniline was studied at pH 9.0, but under anaerobic conditions to make sure no oxygen present in the solution. The reaction was studied at a molar ratio of 1:1 ([Fe(VI)];[aniline]). The Mössbauer spectra, collected after immediate freezing of the solution (~3 s) and 60 s, are presented in Fig. 4. The results were not different from the reaction conducted at pH 11. Besides not observing any of high-valent iron species, expected Fe(II) was also not seen. Because of anaerobic condition, the oxygenation of Fe(II) to form Fe(III) could be ruled out. This suggest the possibility of the reaction between Fe(VI) and Fe(II) under studied conditions, which is fast ($k \geq 10^6$ M^{-1} s^{-1}) and yielded Fe(III) [31].

It is noteworthy that the Mössbauer parameters of the resultant Fe(III) species are the same in every experiment within experimental error, and are close to that of ferrihydrite. This shows that even at not too high pH (i.e., pH 9.0) the phosphate buffer could not compete with OH^- ions in coordinating to iron and the formation of hydroxides/oxyhydroxides were preferred. This process can also stabilize Fe(III) and therefore accelerates decomposition of intermediate oxidation states of iron.

4 Conclusions

Mössbauer spectroscopy approach was shown as an experimental technique to study the mechanism of oxidation of sulfamethoxazole and aniline by Fe(VI). However, freezing of samples of the reaction within three seconds was not sufficient in identifying any of the possible intermediate iron species, Fe(VI)-SMX/Fe(VI)-aniline, Fe(V) and Fe(IV), during the reaction. Further improvement in freezing of reaction samples on the time scale of milliseconds is needed to establish the intermediate iron species, which will allow better understanding of role of high-valent iron, Fe(V) and Fe(IV) species in oxidation reactions of Fe(VI).

Acknowledgements V.K. Sharma acknowledges the partial support of United States National Science Foundation (CBET 1236331) for ferrate research. K. Siskova thanks to the Grant Agency of the Czech Republic for financial support (grant no. P108/11/P657). The authors also gratefully acknowledge the support by the Technology Agency of the Czech Republic "Competence Centres" (project TE01010218), the support by the Innovations—European Regional Development Fund (project CZ.1.05/2.1.00/03.0058 of the Ministry of Education, Youth and Sports of the Czech Republic) and funding from the Operational Program Education for Competitiveness—European Social Fund (projects CZ.1.07/2.3.00/20.0056, CZ.1.07/2.3.00/20.0058 and CZ.1.07/2.3.00/20.0017 of the Ministry of Education, Youth and Sports of the Czech Republic). We thank anonymous reviewers for their comments which improved the paper.

References

1. Yang, B., Ying, G.G., Zhao, J., Liu, S., Zhou, L.J., Chen, F.: Water Res. **46**, 2194 (2012)
2. Lee, Y., von Gunten, U.: Water Res. **44**, 555 (2010)
3. Hu, L., Martin, H.M., Arce-bulted, O., Sugihara, M.N., Keating, K.A., Strathmann, T.J.: Environ. Sci. Technol. **43**, 509 (2009)
4. Hu, L., Page, M., Marinas, B., Shisler, J.L., Strathmann, T.J.: In: Proceedings—Water Quality Technology Conference and Exposition, hul1/1 (2010)
5. Filip, J., Yngard, R.A., Siskova, K., Marusak, Z., Ettler, V., Sajdl, P., Sharma, V.K., Zboril, R.: Chem. Eur. J. **17**, 10097 (2011)

6. Sharma, V.K.: J. Environ. Sci. Health. Part A, Environ. Sci. Eng. Toxic Hazard. Substance Control **45**, 645 (2010)
7. Sharma, V.K.: Environ. Sci. Technol. **44**, 5148 (2010)
8. Zboril, R., Andrle, M., Oplustil, F., Machala, L., Tucek, J., Filip, J., Marusak, Z., Sharma, V.K.: J. Hazard. Mater. **211–212**, 126 (2012)
9. Wood, R.H.: J. Am. Chem. Soc. **80**, 2038 (1958)
10. Sharma, V.K.: Water Sci. Technol. **55**, 225 (2007)
11. Sharma, V.K., Li, X.Z., Graham, N., Doong, R.A.: J. Water Supply: Res. Technol., AQUA **57**, 419 (2008)
12. Sharma, V.K.: Chemosphere **73**, 1379 (2008)
13. Anquandah, G., Ray, M.B., Ray, A.K., Al-Abduly, A.J., Sharma, V.K.: Environ. Technol. **32**, 261 (2011)
14. Anquandah, G.A.K., Sharma, V.K., Knight, D.A., Batchu, S.R., Gardinali, P.R.: Environ. Sci. Technol. **45**, 10575 (2011)
15. Lee, Y., Gunten, U.V.: Water Res. **44**, 555 (2010)
16. Lee, Y., Zimmermann, S.G., Kieu, A.T., Gunten, G.V.: Environ. Sci. Technol. **43**, 3831 (2009)
17. Anquandah, G.A.K., Sharma, V.K.: J. Environ. Sci. Health. Part A, Environ. Sci. Eng. Toxic Hazard. Substance Control **44**, 62 (2009)
18. Sharma, V.K., Sohn, M., Anquandah, G., Nesnas, N.: Chemosphere **87**, 644 (2012)
19. Sharma, V.K.: Coord. Chem. Rev. **257**, 495 (2013)
20. Sharma, V.K., Mishra, S.K., Nesnas, N.: Environ. Sci. Technol. **40**, 7222 (2006)
21. Johnson, M.D., Hornstein, B.J., Wischnewsky, J.: ACS Symp. Ser. **985**(Ferrates), 177 (2008)
22. Johnson, M.D., Hornstein, B.J.: Chem. Commun. **0**, 965 (1996)
23. Huang, H., Sommerfeld, D., Dunn, B.C., Lloyd, C.R., Eyring, E.M.: J. Chem. Soc. Dalton Trans. **0**, 1301 (2001)
24. Sharma, V.K., Siskova, K., Machala, L., Zboril, R.: AIP Conf. Proc. **1489**, 139 (2012)
25. Homonnay, Z., Noorhasan, N.N., Sharma, V.K., Szilagyi, P.A., Kuzmann, E.: ACS Symp. Ser. **985**, 257 (2008)
26. Perfiliev, Y.D., Sharma, V.K.: ACS Symp. Ser. **985**(Ferrates), 112 (2008)
27. Cabelli, D.E., Sharma, V.K.: ACS Symp. Ser. **985**(Ferrates), 158 (2008)
28. Sharma, V.K.: Chemosphere **73**, 1379 (2008)
29. Sharma, V.K.: J. Environ. Manag. **92**, 1051 (2011)
30. Trapp, J.M., Millero, F.J.: J. Sol. Chem. **36**, 1479 (2007)
31. Sharma, V.K., Bielski, B.H.J.: Inorg. Chem. **30**, 4306 (1991)

DOI 10.1007/978-94-007-6482-8_2
Reprinted from *Hyperfine Interactions*, DOI 10.1007/s10751-013-0843-4

Analysis of broadened Mössbauer spectra using simple mathematical functions

Analysis of broadened Mössbauer spectra

A. Cabral-Prieto

Published online: 20 March 2013

Abstract Simulated and experimental broadened Mössbauer spectra are analyzed using several distribution functions. The resolution Hesse and Rübartsch data are reproduced in order to analyze the origin of the oscillations appearing in the recovered distribution function. The lined triangular distribution is used and some of its properties are described. The no implicit *n*th-nomial distribution function $P(x) = (a\text{Cos}(\pi x) + b\,\text{Sin}(\pi x))^n$ is introduced, complementing the Window and Hesse and Rübartasch no implicit distribution functions. This new no implicit distribution function gives similar results of those of Window's method. In addition, the Window method has also been modified by inserting a smoothing factor λ_C. For $0 < \lambda_C < 1$ a hyperfine distribution with low resolution may be obtained; for $\lambda_C > 1$, the opposite is obtained. The Levenberg-Marquardt algorithm is used to solve the involved Fredholm integral equation rather than the typical second order regularized algorithm. From the extracted hyperfine field distribution functions of the Mössbauer spectra of the amorphous and crystallized $Fe_{70}Cr_2Si_5B_{16}$ magnetic alloy the short range atomic order for the amorphous state of this alloy can be inferred.

Keywords Mössbauer spectroscopy · No implicit and implicit hyperfine distributions · Amorphous alloys · Short range atomic order

1 Introduction

Broad Mössbauer spectra are frequently obtained from materials where slightly different iron sites are present; examples such as amorphous magnetic alloys [1–10];

Proceedings of the Thirteenth Latin American Conference on the Applications of the Mössbauer Effect, (LACAME 2012), Medellín, Colombia, 11–16 November 2012.

A. Cabral-Prieto (✉)
Depto. de Química, Instituto Nacional de Investigaciones Nucleares,
Apdo. Postal 18-1027, Col. Escandón,
Deleg. M. Hidalgo, C. P. 11801, México D. F., México
e-mail: agustin.cabral@inin.gob.mx

nanometric materials with crystal size distribution(s), like goethite [8], hidrotalcites containing iron, stainless steels, etc. [1–10]. The hyperfine parameters of the Mössbauer spectra of these materials (isomer shift δ, quadrupole splitting Δ, and the hyperfine magnetic field H) are usually evaluated as distribution functions ($P(x)$ with $x = \delta$, or Δ, or H) rather than as discrete values. Several methods have appeared in the literature where a sum of magnetic patterns (1) or an implicit distribution like the Gaussian line are used [1, 7, 8, 11]. Others methods assume no implicit expressions for the $P(x)$ distribution. Whereas Window [2] describes $P(x)$ in terms of a Fourier series expansion (FSE), Hesse and Rübartsch [3] use a discrete step function. According to this line of thought, the nth nomial distribution function $P(x) = (a\mathrm{Cos}(\pi x) + b\,\mathrm{Sin}(\pi x))^n$ assumes neither a previous knowledge of the $P(x)$ distribution, and is introduced in this study. The resulting distributions from this new function are practically similar to those obtained from the Window method [3]. The implicit lined triangle distribution function is also introduced in this study. In addition to these new proposed no-implicit and implicit distribution functions the Window method has been modified by inserting a smoothing factor λ_C which can take the values $0 > \lambda_C \geq 1$. For a highly resolved distribution function a $\lambda_C > 1$ is required. This smoothing factor multiplies the trigonometric argument of the Cosine or Sine function appearing in the Window method [2].

Generally speaking, one can get practically the same results using anyone of these three no-implicit distributions functions (Window, Hesse and Rübartsch, and present methods). Particularly, the Hesse and Rübartsch method is faster than the others but it is basically limited to one parameter, the number of points of the distribution (NSUB), to get a well resolved distribution. The Window method originally had the number of terms of the FSE in order to obtain a well resolved distribution. The introduction of the smoothing factor λ_C in the present work gives an additional parameter in order to optimize a solution. Similarly the new nth-nomial distribution function depends on the number of terms $(n + 1)$ of the distribution. In addition to this, the later no implicit distribution handles even and odd functions simultaneously. This property would allow studying them in a future work in order to see if it is possible to reduce the oscillations that appear in any solution of these no implicit methods.

2 Experimental

The resolution Hesse and Rübartsch data consisting of a magnetic sextet modified by two Gaussian distribution functions centered at 17.5 and 20 T of standard deviation $\sigma = 0.85$ T each are reproduced [3]. The above simulated spectrum and those of the amorphous and crystallized soft magnetic $Fe_{77}Cr_2\,Si_5B_{16}$ alloy (S3A alloy from Metglas) are analyzed by using the unmodified and modified Window methods [2] and the new proposed distribution function.

3 Methodology

Present analysis assumes the thin absorber approximation where the resonance line is a Lorentzian distribution of FWHM $\Gamma = 0.194$ mm/s. In the transmission

Mössbauer set up, a broadened magnetic Mössbauer spectrum can be represented by the Fredholm integral of the first kind as

$$S_{calc} = \int_c^d P(B)\, L_6(v, B)\, dB, \tag{1}$$

where $S_{calc}(v)$ is the calculated spectrum as a function of the relative velocity between the absorber and Mössbauer source. $L_6(v, B)$ stands for the magnetic sextet of normalized Lorentzian lines and $P(B)$ is the unknown probability distribution function of the hyperfine magnetic field B. The condition $\int_c^d P(B)dB = 1$ has to be met if quantitative information is required from the measured spectrum $S_m(v)$.

The solution of integral equation (1) needs a carful numerical treatment because the resulting hyperfine distribution(s) and the fitted spectrum may posses an oscillatory behavior due to statistical fluctuations on the experimental data set $S_m(v)$ or rounding errors (3, 7, 9). In order to solve this problem several regularized algorithms are at hand. Singular Value Decomposition (SVD), Truncated Singular Value Decomposition (TSVD), Generalized SVD (GSVD), the Tikhonov-Arsenine regularization or the Philips-Twomey regularization methods are some of them. The Tikhonov regularization methods are classified as zero, first and second order regularization algorithms. Any of these algorithms regularizes the ill-posed problem derived from the solution of integral equation (1). From the author's perspective, the preferred algorithm used in Mössbauer spectroscopy is the Philips-Twomey or Tikhonov second order regularization method after proposing it by Hesse and Rübartsch [3]. The zero and first order regularization algorithms, or the SVD, TSVD GSVD have been little explored. The Levenberg-Marquadt (LM) algorithm, which is the standard routine to solve non-linear problems, may also be considered as the Tikhonov zero order regularization method, and is used in the present work. The resulting normal equations of this algorithm may be written as

$$\left(J^T J + \lambda_{LM} I\right) \Delta x = J^T b, \tag{2}$$

where J represents the Jacobian matrix and J^T its transpose, λ_{LM} is the Levenberg-Marquardt factor that controls the searching route to obtain the "best solution" between the Gauss-Newton and the Steepest-Descent methods, I is the identity matrix; λ_{LM} is an internal parameter of the computing program and one has not to worry about it; b represents the weighted residuals between the measured data $S_m(v)$ and the calculated $S_{calc}(v)$ ones. Expression (2) is practically the same as that of the standard form of Tikhonov zero order regularization algorithm

$$\left(J^T J + \lambda_T I\right) \Delta x_T = J^T b \tag{3}$$

The goal of the Tikhonov λ_T factor in (2) is to provide a "smooth solution" of the desired hyperfine $P(x)$ distribution and of the fitted spectrum $S_c(v)$. In this case, the Tikhonov λ_T factor is user defined. Technically, the effect of both the LM and Tikhonov factors, λ_{LM}, λ_T in (2) and (3), respectively, is to dampen the contributions of small singular values in the solution of the $J^T J$ matrix.

Thus the Levenberg-Marquadt algorithm can, under certain conditions, be considered the same to the Tikhonov zero order regularization. Thus, a first introspection of the solutions that can provide the LM algorithm is presented using several distribution functions.

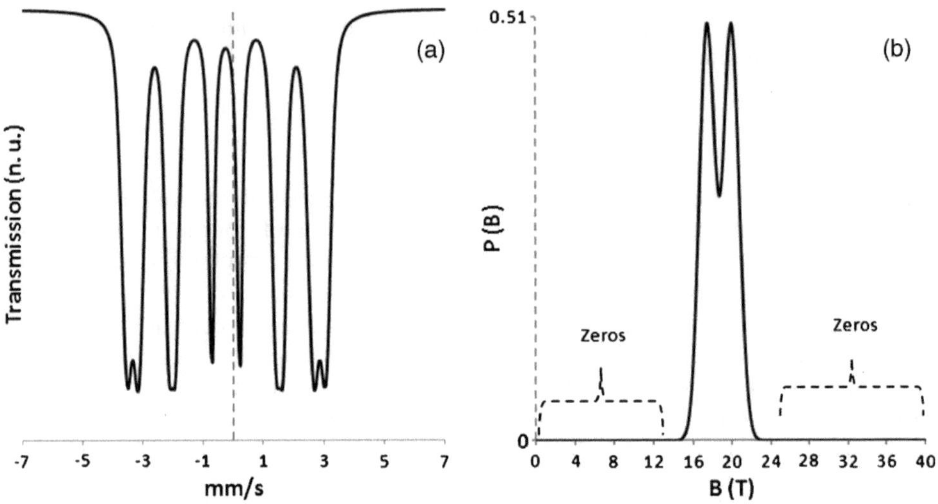

Fig. 1 Hesse and Rübartsch's data [3]. (**a**) A distributed sextet by (**b**) two Gaussian lines, centered at 17.5 and 20 T

4 Results

4.1 Hesse and Rübartsch's data

The resolution test data of Hesse and Rübartsch's paper is reanalyzed. Figure 1 shows the corresponding data of example 3. (ii) from reference [3], where a magnetic sextet is distributed by a two Gaussian shaped functions ($P(B)$), centered at the magnetic field values B = 17.5 and 20 T and having a standard deviation $\sigma_H = 0.85\ T$ each, as shown in Fig. 1a and b. A first observation must be made from the two Gaussian distribution function, Fig. 1b: the plotted distribution function $P(B)$ has two regions of zeros as indicated by the keykets.

These zeros appear from 0.0 to 12.5 T and from 25 to 40 T. That is, the integration limits of integral equation (1) should be $c = 12.51$ and $d = 25.1$ T only.

The recovered distribution function $P(B)$ from Fig. 1a when applying the unmodified Window method [3] as suggested in reference [3], using the limits $c = 0$ and $d = 40$ in the integral equation (1), and using 15, 25 and 35 terms of the FSE, is shown in Fig. 2. The resulting distributions have oscillations as those shown in reference (3). The best fit is obtained when using 35 terms of the FSE but still small oscillations are present in the resulting distribution, Fig. 2c'. A little higher oscillations appear if the Lorentzian width of the sextet is fixed to its actual value ($W_L = 0.1926$ mm/s [3]). Such oscillations are removed if the c and d integral limits are, for example 12.5 and 25 T, respectively, or if they are calculated by the program as shown in Fig. 3a', respectively. Two presentations of the resulting distribution can be made, the typical superimposed distribution (Fig. 3a' and the separated one, Fig. 3b'. The squared chi fitting criterion is improved by two orders of magnitude if the integral limits are guessed and the distributions are separated, Fig. 3b'. In the last two cases the line width of the sextet was also evaluated and was recalculated with zero error as shown in the Fig. 3b' case.

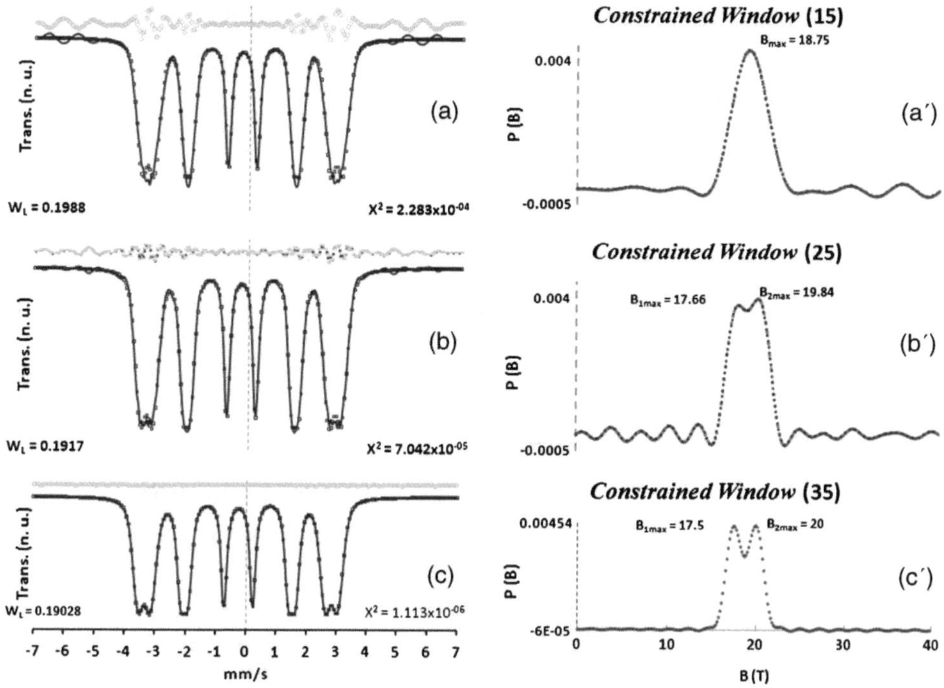

Fig. 2 Hyperfine field distributions by Hesse and Rübartsch [3] using the Window method [2]. On the top of each left figure the un-normalized residuals are plotted. The Lorentzian width W_L was evaluated and was not recovered

The case for which the Gaussian distribution function is replaced by a triangular distribution function is shown in Fig. 3c'. Firstly, note that in this case a single parameter (the number shown in parenthesis in these figures) is only required to define a triangular distribution function. In the present case, the computing program also requires a single parameter to produce both triangles as shown in Fig. 3c'. Thus this triangular distribution speeds up any calculation as compared to those using the Window or other methods. Secondly, it can be noticed from the small residuals that this triangular distribution function is not a bad approximation to the original Gaussian distribution function $P(B)$. The integration limits are also evaluated by the program and in this case the possibilities of oscillations in the fitted spectrum and distribution are also removed.

4.2 S3A magnetic alloy

Following the above strategy, i. e. optimizing the integral limits to prevent any spurious oscillations, the CXMS and transmission spectra of the amorphous soft magnetic alloy S3A, treated at 50C/20' and 430C/20', respectively, are next analyzed. Figure 4 shows tow typical hyperfine field distributions (HFD) from the amorphous state of the alloy when using the unmodified [2] and modified Window methods with 6 terms of the FSE (and $\lambda_c = 1.0$) and 12 terms of the FSE (and $\lambda_c = 1.1109$), either evaluating or fixing the intensity ratios of lines two and five (I25) of the magnetic sextet.

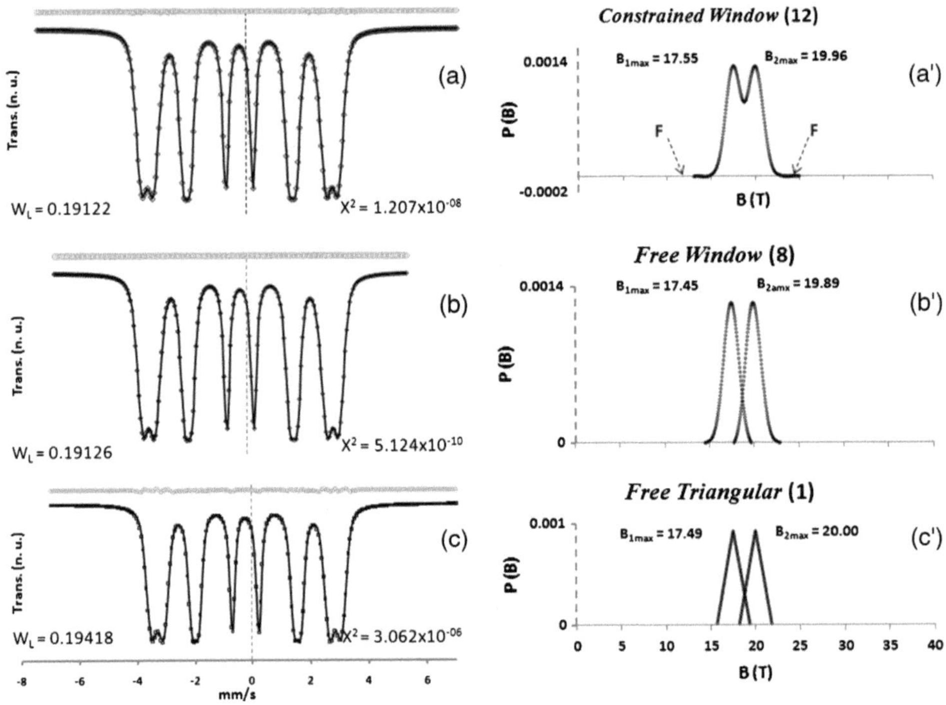

Fig. 3 Oscillations free hyperfine distributions obtained by (**a**) fixing the *c* and *d* integral limits, using 12 terms of the FSE, or (**b**) guessed using 8 terms of the FS only. The distribution can be obtained as (**a**) superimposed or (**b** and **c**) separated one

A single Gaussian distribution is suggested from Fig. 4a'. A more complex distribution is obtained if 12 terms of the FSE and λ_c is evaluated. The latter distribution will be named as the low resolved hyperfine field distribution (LR-HFD). Whereas Fig. 4a' may suggest a single phase, i. e., the amorphous phase of the alloy, Fig. 4b' might suggest a set of Fe-Cr-Si-B crystalline phases.

From the Mössbauer spectroscopy, and particularly from the LR-HFD, still remains ambiguous if the experimental Mössbauer spectrum of metallic glasses, like the S3A alloy, indeed originates from the amorphous phase (Fig. 4a') or not (Fig. 4b').

A similar LR-HFD was reported by Le Caer and Dubois in 1979 arising from the amorphous alloy $Fe_{79.5}Si_{1.5}B_{19}$ [7]. As a result of this, a great deal of research has been carried out to elucidate the local atomic structure around the iron, boron and phosphorus atoms of these metallic glasses in order to be able to interpret the spectral features of Fig. 4b' suggesting the presence of crystalline material [7–14]. Nowadays is generally believed that the local structure around the boron atoms in metallic glasses possess short-range atomic-order (SRAO) rather than short-range atomic-disorder (SRAD). Research work in this area is still in progress to define such a matter (13, 14). The SRAO perception came after studying ^{10}B NMR spectroscopy in some metallic glasses containing ^{10}B, where sharp NMR peaks arise from different supposedly amorphous Fe-B phases [8, 13].

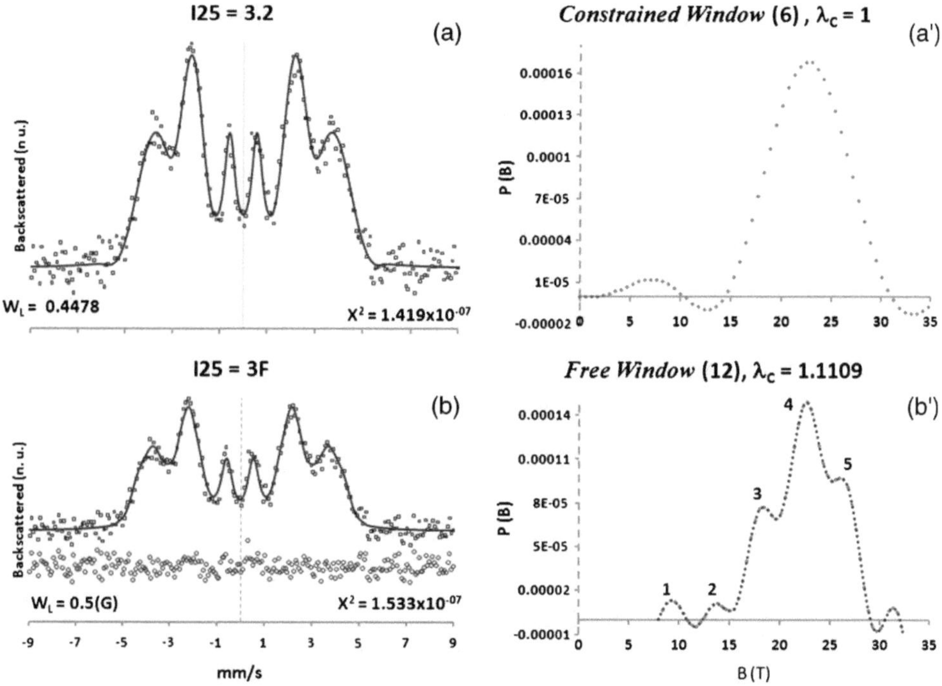

Fig. 4 (**a**) Fitted CXMS spectrum of the amorphous alloy using the unmodified Window's method ($\lambda C = 1$), (**a'**) an unresolved hyperfine field distribution is obtained when using six terms of the FSE. (**b**) Fitted CXMS spectrum of the amorphous alloy using the modified Window's method ($\lambda C = 1.1109$), (**b'**) a low resolved hyperfine field distribution is obtained when using 12 terms of the FSE

The question that arose at Le Caer and Dubois's time (1979) was: Are real the spectral features of the LR-HFD as shown in Fig. 4b'? or they are the product of mathematical artifacts? As far as the author understands, these spectral features appearing in Fig. 4b' have not been answered quite satisfactorily.

In the present case, the mathematical artifacts (spurious oscillations) can mostly be discharged from Fig. 4b' due to the fact that the integral limits are optimized and minimal oscillations would be expected. The statistical nature of the present experimental data may also produce additional oscillations. However, this also can be ruled out because the statistical quality of the Mössbauer data of Le Caer and Dubois's paper [7] is better than in the present case, and in spite of this the obtained distributions and the fitted spectrum are practically the same as those reported by Le Caer and Dubois [7]. In addition to this, no matter if a zero order or a second order regularized algorithm is used. Thus, in order to give, from the Mössbauer point of view, support to the LR-HFD structure of Fig. 4b', the transmission Mössbauer spectrum (TMS) of the heat treated S3A alloy at 530C/20' was obtained, which is shown in Fig. 5b.

This TMS was fitted using the trigonometric nth-nomial distribution function $P(x) = (a\mathrm{Cos}(\pi x) + b\,\mathrm{Sin}(\pi x))^n$ with $n = 29$. The resulting coefficients of this trigonometric distribution function are evaluated as simple parameters by the program. Take the example case for $n = 2$ where the resulting coefficients are: a^2, $2ab$,

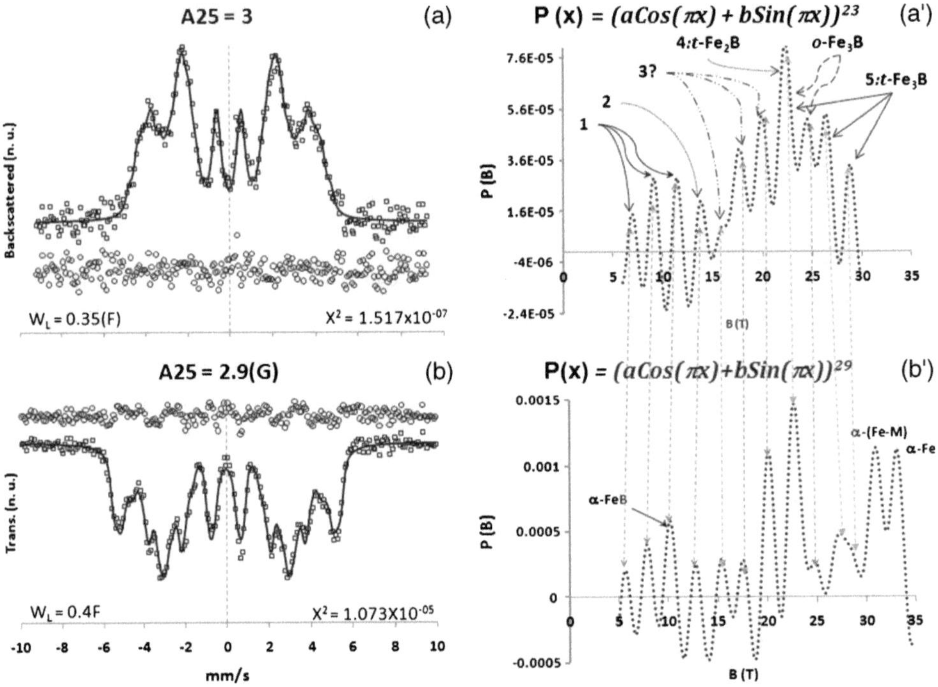

Fig. 5 (**a**) CXMS Mössbauer spectrum of the amorphous alloy fitted using the trigonometric sum with an exponent $n = 23$; (**b**) Transmission Mössbauer spectrum of the crystallized alloy fitted using the nth-binomial expansion distribution with $n = 29$

b^2 and the corresponding numerical values are: 1, 2, 1, respectively, with $a = 1$ and $b = 1$. The evaluated coefficients generally differ from the preceding whole numbers.

For comparison purposes, the fitted Mössbauer spectrum and the resulting HFD of the amorphous alloys are also presented in Fig. 5a and a'; in this case a 23th-nomial trigonometric function was used and a high resolved HFD (HR-HFD) is now shown for both these cases, Fig. 5a' and b' having practically the same peak structure for fields lower than 30 T. A small shift is observed between these peaks only. Is it a coincidence such a peak matching? Or is it an evidence of the short-range atomic structure of the amorphous state of the alloy?

Large oscillations with negative values appearing in Figs. 4 and 5b's should not be considered, they were enhanced in order to obtain HR distributions containing the typical information of these alloys [14, 15]. Negative values can be eliminated but LR distributions are obtained which are not useful for the present discussion.

Thus the LR-HFD in Fig. 4b' may be rationalized as follows:

(i) The small broad peak at 9.25 T, peak 1 shown in the LR-HFD of Fig. 4b', and a first group of three peaks, gathered by curved arrows as shown in the HR-HFD of Fig. 5b', located at 5.47, 7.81 and 10.16 from which an average peak value of 7.81 T is obtained. This average value deviates −15.53 % the peak 1 position in Fig. 4b'. If this average value is taken from the HR-HFD of the amorphous material such deviation is less than 1 %. Broken lines between peaks of Figs. 4 and 5b' show the corresponding shifts.

(ii) The next smaller peak at 13.83 T in Fig. 4b' (peak 2) practically matches its position with the peak at 13.8 T from the HR-HFD in Fig. 5a'. From Fig. 5b', i. e. from the crystallized material this peak is at 12.73 T which deviates a −7.7 %.
(iii) Peak 3 from the LR-HFD in Fig. 4b' is located at 18.4 T. The next gathered three peaks from the HR-HFD signalized by a **(3?)** in Fig. 5a' are located at 15.97, 17.58 and 19.65 T, making them an average value of 17.7 T, which deviates a −3.6 % from peak 3 in Fig. 4b'. The corresponding average value from these peaks in Fig. 5b' is 17.58 %, i. e., −4.47 %
(iv) The highest peak of the LR-HFD in Fig. 4b' (peak number 4) is at 22.58 T. The corresponding highest peak of the HR-HFD in Fig. 5a' signalized by **(4, *t*−Fe_2B)** is located at 21.95 T, which deviates −2.7 % from peak 4 in Fig. 4b. The central peak position from Fig. 5b' is 22.58 T, i .e., the same as in the amorphous state. This peak actually arises from a superposition of several ones as next explained and supported by references [16, 17].
(v) The next peak to follow from the LR-HFD in Fig. 4b' is peak number 5 which is located at 25.92 T. The gathered three peaks from the HR-HFD identified by **5: *o*-Fe_3B** in Fig. 5a' are located at 2.95, 26.09 and 28.38 T, making an average value of 25.47 T, which deviates −1.71 % from peak 5 in Fig. 4b'. The corresponding average value of from Fig. 5b' is 25.85 T, with less than 1 % deviation.
(vi) Peak number 6 in Fig. 4b' was not matched with none of Fig. 5a' and b'.

It can be inferred from the above analysis that the LR-HFD distribution of this amorphous material arises from a large superposition of Mössbauer patterns arising from different iron sites.

The peak structure shown in Fig. 5b' is very well known in the literature [14, 15] in such a way that every peak has practically been identified for some particular Fe-phases as indicated in Fig. 5a' and b'. Those peaks identified by **(3?)** in Fig. 5b' may have different origin depending on the alloy under study [14, 15]. The main point to enhance here is that every peak belongs to a magnetic material in its crystallized form (16, 17).

Based on this analysis, the LR-HFD peaks arising from the Mössbauer spectrum of the amorphous alloy may be interpreted as a result of average values that arise from the corresponding hyperfine magnetic fields of the involved crystallized Fe-B phases for each soft magnetic alloy (8, 13, 14, 15). Thus the present amorphous alloy would consist of a mixture of Fe-B phases having SRAO [7–17], and it can be concluded that one is able to infer unambiguously from the LR- and HR-HFD's of the Mössbauer spectra of the amorphous alloy the presence of SRAO.

5 Conclusions

The Hesse and Rübartsch data were reanalyzed using the unmodified Window method [3] and the Levenberg-Marquardt algorithm. It was shown from this analysis that, the oscillations shown in the Hesse and Rüartsch paper [4] arise form rounding errors mainly and they can be removed completely by taking care of the integration limits (*c* and *d*) of integral equation (1). In addition to this the triangular function was introduced in this work as an approximation to the Gaussian distribution function giving satisfactory results. On the other hand, experimental Mössbauer spectra of the amorphous and crystallized magnetic S3A alloy were analyzed using the new no implicit distribution function $P(x) = (a\mathrm{Cos}(\pi x) + b\,\mathrm{Sin}(\pi x))^n$. The main result of this analysis indicates that the low and high resolved structures appearing in the HFD of the amorphous material may be real. The LR-HFD of the amorphous alloy can be interpreted as arising from average values due to the different Fe-B phases as detected in the high resolved HFD of the crystallized alloy. Similarly, from the comparison between the HR-HFD of the amorphous and crystallized alloy a direct comparison can be made. Practically the same number of peaks appears in both these highly resolved distributions, suggesting therefore the presence of SRAO in the amorphous S3A alloy. The shift between these peaks may be associated to the particular properties of the SRAO and LRAO solids. That is, the CXMS spectrum comes from the surface and the TMS spectrum from the bulk. Such a shift may be associated with different hyperfine parameters between the Mössbauer spectra of the surface and bulk material.

Acknowledgement We thank to Dr. R. Hasegawa from Metglass Inc. Co. USA for his donation of a long ribbon of the 2605 S3A amorphous alloy.

References

1. Varret, F., Gerard, A., Imbert, P.: Phis. Stat. Solidi B **43**, 423 (1971)
2. Window, B.: J. Phys. E: Sci. Intrum. **4**, 401 (1971)
3. Hesse, J., Rübartsch, A.: J. Phys. E: Sci. Instrum. **7**, 526 (1974)
4. Gonser, U., Ghafari, M., Wagner, H.G.: J. Magn. Magn. Mater. **8**, 175 (1978)
5. Vincze, I., Babic, E.: Solid State Commun. **25**, 689 (1978)
6. Vincze, I., Babic, E.: Solid State Commun. **27**, 1425 (1978)
7. Le Caer, G., Dubois, J.M.: J. Phys. E: Sci. Intrum. **12**, 1083 (1979)
8. Ford, J.C., Budnick, J.I., Hines, W.A., Hasegawa, R.: J. Appl. Phys. **55**, 2268 (1984)
9. Dubois, J.M., Le Caer, G.: J. Physique Colloque **43**, C9(12), C9–67 (1982)
10. Kopcewicz, M., Wagner, H.G., Gonser, U.: J. Physique Colloque **46**, C8(12), C8–151 (1985)
11. Eibschüts, M, Lines, M.E. Chen, H.S., Matsumoto, T: J. Phys. F. Met. Phys. **14**, 505 (1984)
12. Ideczak, R., Konieczny, R., Chojcan, J: Hyperfine Interact **208**, 1 (2012). doi:10.1007/s10751-011-0458-6
13. Pokatilov, V.S.: Phys. Solid State **49**(12), 2217 (2007)
14. Pokatilov, V.S., Dimitrieva, T.G., Pokatilov, V.V., D'yakonova, N.B.: Phys. Solid State **54**(9), 1790 (2012)
15. Barinov, V.A., Tsurin, V.A., Voronin, V.I., Surikov, V.T.: Phys. Met. Metallogr. **108**(1), 50 (2009)
16. Frank, H., Rsenberg, M.: J. Magn. Magn. Mater. **7**, 168 (1978)
17. Sánchez, F.H., Zhang, Y.D., Budnick J.I., Hasegawa R.: J. Appl. Phys. **66**(4), 1671 (1989)

DOI 10.1007/978-94-007-6482-8_3
Reprinted from *Hyperfine Interactions*, DOI 10.1007/s10751-013-0863-0

Mössbauer and vibrational DOS studies of diluted magnetic tin oxides and nano iron oxides

K. Nomura · A. I. Rykov · A. M. Mudarra Navarro · C. E. Rodriguez Torres · L. A. Errico · C. A. Barrero · Y. Yoda

Published online: 18 April 2013

Abstract The magnetic properties and Mössbauer results for SnO_2 doped with ^{57}Fe are reviewed, and the values of isomer shift and quadrupole splitting are compared with the results obtained by *ab initio* calculations. It is concluded that the exchange interactions between oxygen defects and magnetic atoms are responsible for long range magnetic interactions of dilute Fe ions dispersed in SnO_2. Fe atom precipitated clusters may be formed in highly Fe doped SnO_2 samples by annealing at relatively high temperatures for several hours. The reduction of the particle size to nano-scale dimensions induces magnetization, which can be associated with oxygen defects. We have measured the nuclear inelastic scattering (NIS) spectra of Fe oxides, and ^{57}Fe and (Co or Mn) doped SnO_2 synthesized mainly by sol–gel methods and we have derived the vibration density of states (VDOS). The local phonons are sensitive to the presence of precipitated clusters.

Proceedings of the Thirteenth Latin American Conference on the Applications of the Mössbauer Effect (LACAME 2012), Medellín, Colombia, 11–16 November 2012.

K. Nomura (✉) · A. I. Rykov
School of Engineering, The University of Tokyo, Tokyo, Japan
e-mail: k-nomura@t-adm.t.u-tokyo.ac.jp, dqf10204@nifty.com

A. I. Rykov
Mössbauer Effect Data Centre, Dalian, China

A. M. Mudarra Navarro · C. E. Rodriguez Torres · L. A. Errico
Universidad Nacional de La Plata, La Plata, Argentina

C. A. Barrero
University of Antioquia, Madellín, Colombia

Y. Yoda
Japan Synchrotron Radiation Research Institute, Hyogo, Japan

Keywords Vibration DOS · Iron doped tin oxides · Diluted magnetic oxides · Nuclear inelastic scattering · Mössbauer · $MnFe_2O_4$ · $CoFe_2O_4$ · Mn and Fe co-doped SnO_2 · Co and Fe co-doped SnO_2

1 Introduction

^{57}Fe atoms are magnetic and can be used as Mössbauer nuclear probes. Recently, oxide semiconductors doped with magnetic atoms have excited much interest in issues of both applied spintronics and basic physics of diluted magnetic atoms acquiring a larger magnetic moment with decreasing magnetic atom concentration. Dielt et al. predicted that transparent semiconductors with wide bandgap such as ZnO and GaN become ferromagnetic at room temperature (RT) by doping with dilute magnetic atoms [1]. Chemically synthesized powders of SnO_2 doped with Co and Fe have been studied by Punnoose et al. [2, 3], and Hays et al. [4]. The authors suggested a close relationship between the structural and magnetic properties in Fe-doped SnO_2 nano-scale particles. In all of these works, the observed ferromagnetism has been attributed to interaction between the magnetic impurities, but the main source of the observed ferromagnetism has never been associated with magnetically ordered defects yet. It was reported that when oxygen vacancies near Fe atom are introduced, the magnetic moment increases [5]. Several kinds of Fe states were identified by synthesizing SnO_2 doped with ^{57}Fe using a sol gel method [6].

The ferromagnetism of SnO_2 doped with Fe is presumed to be induced either by the spin ordering of magnetic atoms through localized spins trapped at oxygen defects or by the formation of precipitated magnetic clusters [6]. It is necessary to reveal how the ferromagnetic atoms, either dispersed or clustered, are incorporated inside the SnO_2 semiconductor, and how Fe ions substitutionally located at Sn site interact with the oxygen vacancies. Magnetism in diluted magnetic semiconductors (DMS) has been considered by some theories to be induced by the RKKY interaction, or carrier-induced interaction between magnetic ions [7–9], or by impurity band exchange interaction, which implies that the magnetic moments are hybridized under the conduction band [10], or by the large polaron model of magnetism mediated by defects [11].

Sb and Fe co-doping of SnO_2 increases its electrical conductivity and induces magnetism. The doping by Sb ions induces a higher solubility of Fe ions in SnO_2, and magnetization of cassiterite based materials increases with Fe concentration [12]. The doping effects of non magnetic Zn ions were also studied for Fe and Sb doped SnO_2. It was found that the saturation magnetization increases with the addition of less than 10 % of non-magnetic Zn ions [13], and that Fe and Co [14, 15] (Fe and Mn [16], or Fe and Ni [17]) co-doped SnO_2 gave rise to enhanced magnetization as compared with the single atom doped SnO_2. In the case of diluted ^{57}Fe substituted at the Sn site in SnO_2 obtained by a sol–gel method, the experimental Mössbauer parameters were compared with *ab initio* calculations [18] to determine the local structure around iron in SnO_2 rutile structure.

On the other hand, a nuclear resonance inelastic scattering (NIS) spectrum can be obtained due to the annihilation and creation of vibration energy quanta at both sides of the nuclear resonance elastic scattering peak. As a supplementary analysis, it is important to measure the NIS spectra of various iron oxides, which may be precipitated by synthesis and annealing.

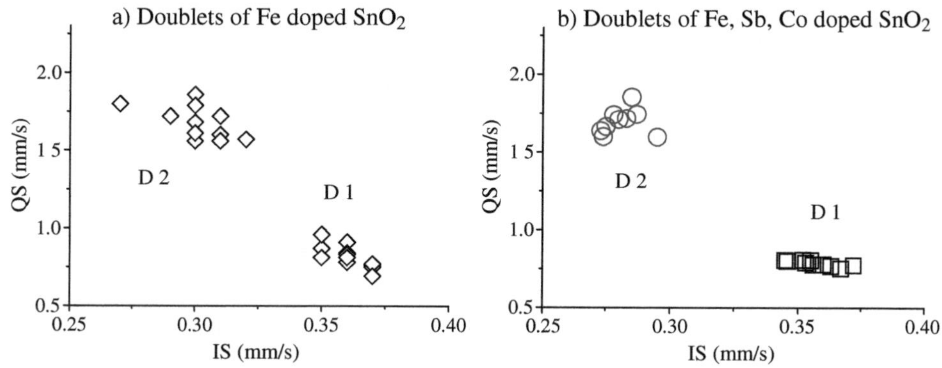

Fig. 1 Relationship between IS and QS values on Fe doped SnO_2 [6] and Fe, Co, Sb co-doped SnO_2 [20]

2 Experimental

^{57}Fe doped $SnO_{2-\delta}$, Fe oxides, (Fe, Mn) oxides, and (Fe, Co) oxides were obtained by a sol–gel method [6]. Only γ-Fe_2O_3 was donated by Toda Kogyo Corporation. Mössbauer spectra were measured by a conventional transmission method with velocity calibration by α-Fe. NIS spectra were measured using the BL09 beam with energy resolution of 2.5 meV in the synchrotron facility of SPring8. While the incidence energy of 14.41 keV was modulated from −80 meV to +80 meV, the NIS spectra were measured several times by detecting the scattered resonant X-rays after Mössbauer effect with APD detectors. Vibration densities of states (VDOS) were derived from the NIS spectra [19].

3 Results

3.1 Mössbauer parameters and oxygen vacancy

Figure 1 shows the relationship between isomer shift (IS) and quadrupole splitting (QS) values in two doublets, observed in the Mössbauer spectra of Fe doped SnO_2 [6] and (Fe, Co and Sb) doped SnO_2 [20], respectively. Fe-doped SnO_2 has a major doublet 1 (D1) with IS values of 0.34–0.37 mm/s and QS values of 0.5–0.9 mm/s, and a minor doublet 2 (D2) with IS = 0.29–0.34 mm/s and QS = 1.5–1.8 mm/s. (Fe, Co, Sb) co-doped SnO_2 has D1 with IS = 0.34–0.37 mm/s and QS = 0.7–0.8 mm/s, and D2 with IS = 0.27–0.29 mm/s and QS = 1.6–1.8 mm/s. Both doublets correspond to Fe (III). The IS values of D2 for the group of (Fe, Co, Sb) doped SnO_2 are relatively smaller than those of D2 for SnO_2 only doped with Fe, which show a relatively large distribution. On the other hand, the magnetic hysteresis for (Fe, Co, and Sb) doped SnO_2 is broader than that of Fe-doped SnO_2. The relative fraction of D2 decreases with the increase of Sb doping rate, whereas the proportion of the broad sextet increases with the increase of Fe concentration [20].

According to our *ab initio* calculations [18] the hyperfine parameters for D2 are close to the calculated ones for the model of Fe substituting Sn in the rutile structure

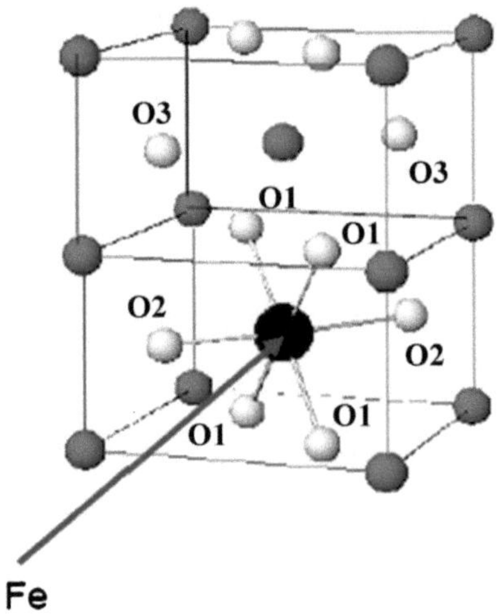

Fig. 2 The structure of SnO_2 with the substituted Fe and different types of oxygen vacancies as explained in the text

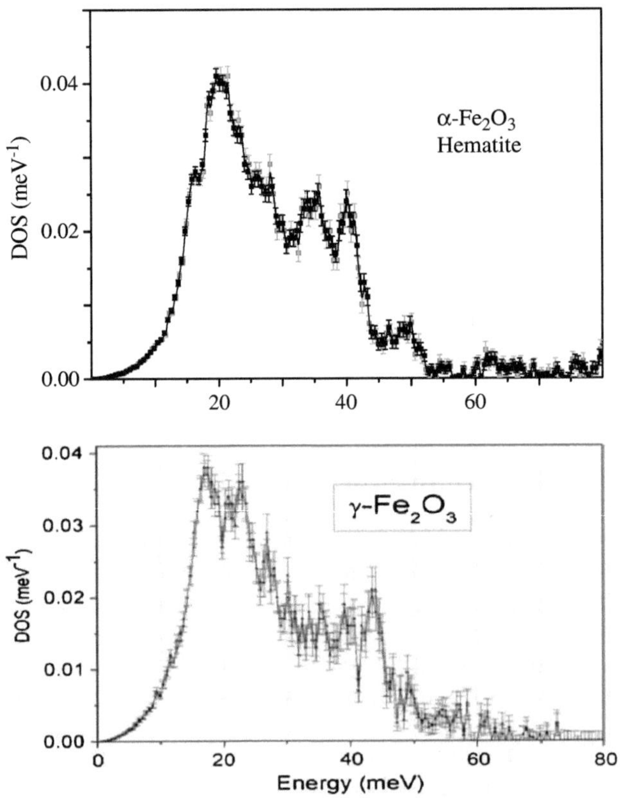

Fig. 3 Vibration DOS of hematite (α-Fe_2O_3), and maghemite (γ-Fe_2O_3)

with an oxygen vacancy at O2 (apical oxygen of the octahedron, see Fig. 2). In this case the calculated values are: IS = 0.32 and QS = 1.43 mm/s. The experimental values are also in agreement with the model where two iron atoms substitute tin atoms, one in the centre and the other in a corner of unit cell sharing an oxygen

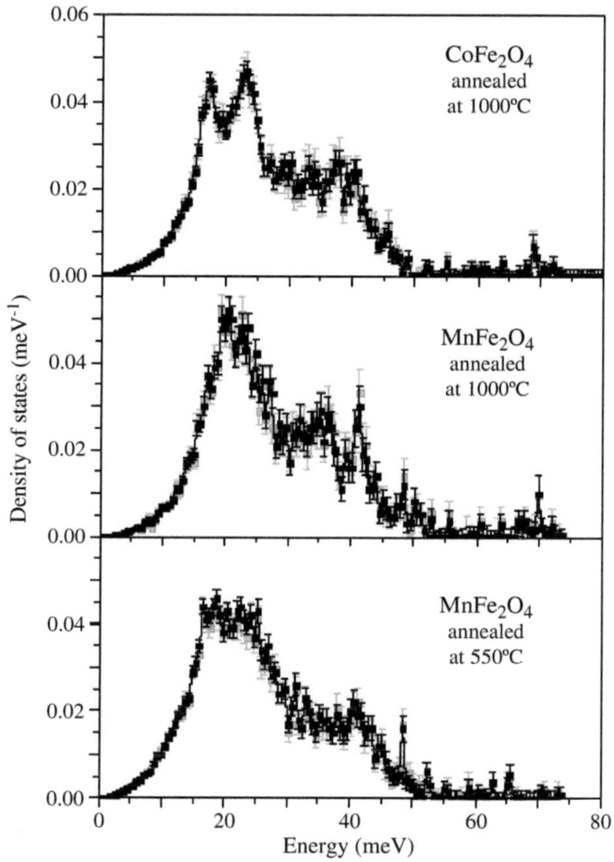

Fig. 4 Vibration DOS of Co ferrite and Mn ferrite: **a** $CoFe_2O_4$ annealed at 1,000 °C, **b** $MnFe_2O_4$, annealed at 1,000 °C, and $MnFe_2O_4$ annealed at 550 °C

vacancy. In this case the calculation predicts the existence of two non-equivalent iron atoms with hyperfine parameters: $IS_1 = 0.32$ and $QS_1 = 1.30$ mm/s and $IS_2 = 0.31$ and $QS_2 = 2.03$ mm/s, respectively.

In the case of doublet D1, the experimental values are between the calculated ones for the model of iron substituting Sn with an oxygen vacancy at O1 (basal plane of the octahedron, see Fig. 2) and the model of two iron substituting tin central atom of the unit cell along the c-axis that share an oxygen vacancy at O1. The calculated hyperfine parameters for these configurations are: IS = 0.48 mm/s and QS = 0.92 mm/s (for one iron with vacancy at O1) and IS = 0.35 mm/s and QS = 0.51 mm/s (for two irons sharing an oxygen vacancy at O1).

Interstitial Fe sites cannot be responsible for the observed hyperfine interaction data because the calculated IS values (between 0.5 and 0.7 mm/s) are larger than the experimental ones.

We can therefore conclude that the experimental values of hyperfine parameters are consistent with iron substituting tin in the rutile structure associated with oxygen vacancies near iron impurities.

A magnetic sextet with IS = 0.37 mm/s and H = 51 T was also observed, and it can be associated with hematite precipitated in the SnO_2 matrix. The hematite contains Sn atoms because the Morin transition was not observed at low temperature [12]. In

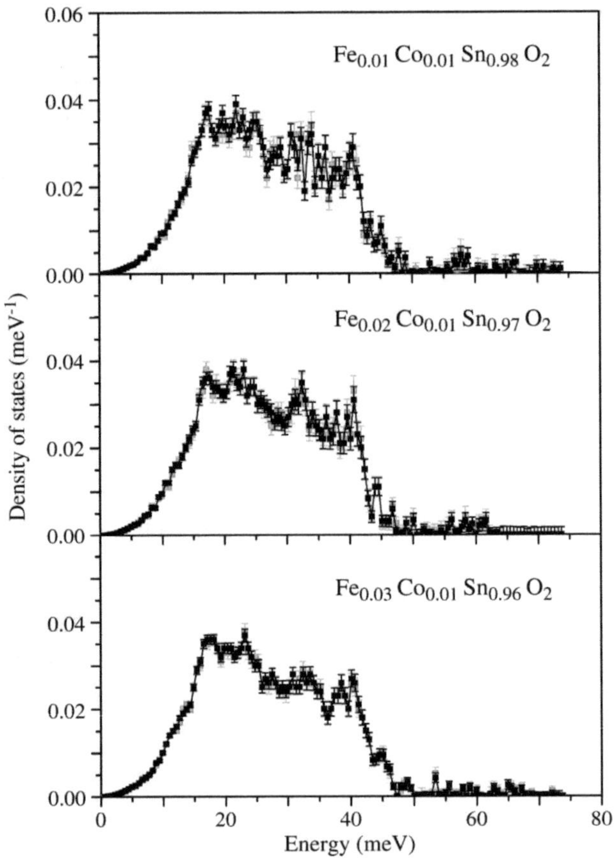

Fig. 5 VDOS of SnO_2 doped with Fe and Co. Average crystallite size: 46 nm, 43 nm, and 36 nm of SnO_2 for 1 %Fe + 1 %Co doped SnO_2, 2 %Fe + 1 %Co doped SnO_2, and 3 %Fe + 1 %Co doped SnO_2, respectively

addition to the sextet with 51 T, another broad sextet with H = 49 T was observed. We are uncertain whether or not this sextet is caused by precipitates of maghemite.

In the case of highly Fe doped SnO_2, and two TMs co-doped SnO_2, it can be considered that the magnetism originates not only from the impurity clustering but also from double exchange interactions of dispersed Fe(III) and TM(II) mediated by oxygen defects produced at the interface of a cluster or at a grain boundary. It is known that point defects like *F* centers, which consist of two adjacent singly occupied vacancies, can also be ferromagnetically ordered [21]. Now, it was observed that the shape of the low temperature Mössbauer spectra did not noticeably change with respect to the room temperature spectrum. These data suggest that the relaxation can be originating from weakly interacting magnetic polarons or small size polaron clusters. Bergqvist et al. [22] have proposed that the magnetic ordering in DMS materials is greatly affected by magnetic percolation, and that the main source of ferromagnetism comes from the short range interatomic exchange interactions that are strongly localized in real space. Moreover, it is also possible that magnetic defects can be also contributing to the magnetization value. Therefore, the interface defects between clusters and host matrix oxide are considered to be an important source of magnetism.

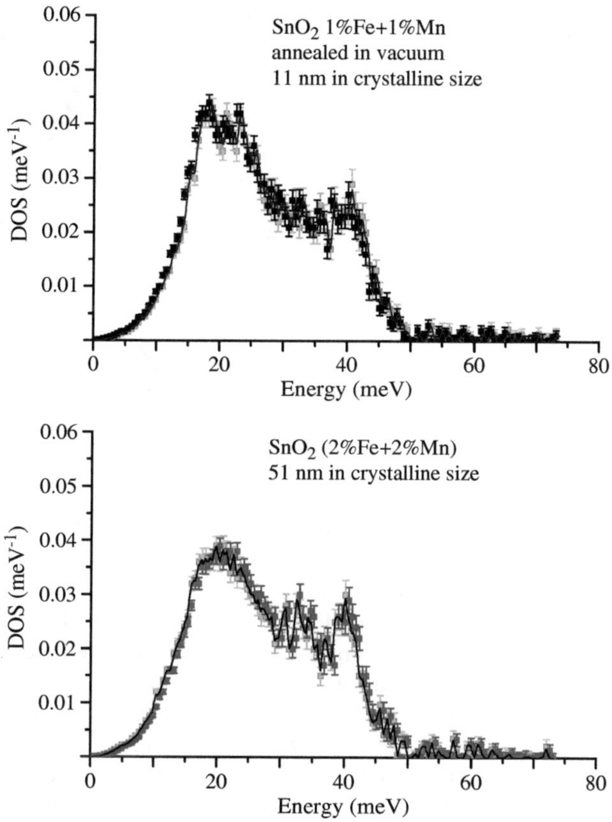

Fig. 6 VDOS of SnO_2 doped with Fe and Mn (Doping rate: 1 % Fe + 1 % Mn, 2 % Fe + 2 % Mn)

3.2 VDOS of Fe doped SnO_2 and ferrite oxides

The oxides SnO_2(Fe), γ-Fe_2O_3, α-Fe_2O_3, $CoFe_2O_4$, and $MnFe_2O_4$ were measured by NIS. Figure 3 shows VDOS of α-Fe_2O_3 and γ-Fe_2O_3. The corundum structure of α-Fe_2O_3 showed particular phonon peaks at 16, 21, 27, 35, 40 and 49 meV, whereas the inverse spinel structure of γ-Fe_2O_3 shows particular peaks at 18, 23, 27, 35, 39, 44, and 50 meV.

VDOS of the inverse spinel structures of $CoFc_2O_3$, and $MnFe_2O_3$, prepared by a sol–gel method and treated at different annealing temperatures, are shown in Fig. 4. The main peaks are around 17, 23, 38, and 40 meV for $CoFe_2O_3$, whereas the peaks are at 20, 23, 35, and 42 meV for $MnFe_2O_3$ annealed at 1,000 °C, and at 18, 23, 43 meV for $MnFe_2O_3$ annealed at 550 °C. Phonon softening is observed for $MnFe_2O_3$ annealed at low temperatures. It is believed to be due to the low degree of crystallinity. The inverse spinel type of ferrites show similar VDOS'es, which may be somewhat influenced by the amount of Fe(III) ions occupied at tetrahedral and octahedral sites.

Figure 5 shows the VDOS of SnO_2 doped with 1 %, 2 % or 3 %Fe and 1 %Co. With increasing Fe contents, a shoulder appears at the energies lower than 15 meV. The typical shapes at larger energies are similar to those of a ferrite, although VDOS of SnO_2 also shows a somewhat similar shape [23]. It may suggest that some spinel-type of Fe and Co compounds cannot be excluded. In contrast, from the VDOS

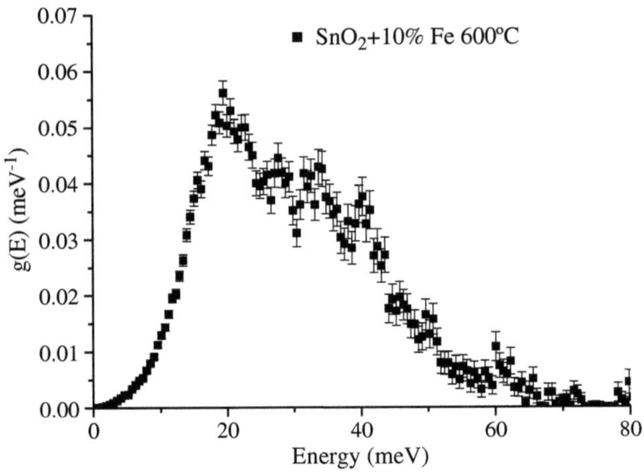

Fig. 7 VDOS of 10 % Fe doped SnO_2 annealed at 600 °C

results, pure α-Fe_2O_3 is excluded. It has a peak at 49 meV, which is not observed elsewhere. The VDOS distribution of SnO_2 doped with Fe and Co are observed below 47 meV as shown in Fig. 5.

The VDOS of SnO_2 doped with Fe and Mn looks more similar to that of SnO_2 doped with Fe as shown in Fig. 6. If the small clusters of Fe oxides are precipitated in SnO_2, the distribution of VDOS spreads to low energy and high energy regions as shown in Fig. 7. In this case, the crystallites of SnO_2 are characterized by a small size, giving rise to the VDOS enhancement. The VDOS in lower energy regions is enhanced for (Fe, Co) doped SnO_2 and (Fe, Mn) co-doped SnO_2.

The precipitated Fe compound cannot be clearly identified but the composition of the small clusters might be derived from the VDOS. There is a possibility that small clusters such as $CoFe_2O_4$ and $MnFe_2O_4$ are formed in several 10 nm crystalline SnO_2, although XRD could not detect those compounds.

4 Summary

Two doublets and sextets with sharp and broad relaxation peaks have been observed in Mössbauer spectra of Fe doped and two transition metal co-doped SnO_2. It is clear from *ab initio* calculations that when Fe(III) is located at Sn site of SnO_2, one oxygen vacancy is created. One doublet (D1) is due to Fe (III) with one oxygen vacancy in the basal plane of the octahedron. The other doublet (D2) is due to Fe(III) with an oxygen vacancy in the vertex of the octahedron, or with two substitutional Fe(III), between which one oxygen is vacant. The sextet with sharp peaks is due to α-Fe_2O_3 with H = 51 T. The saturation magnetization is enhanced for (Fe, Co) co-doped SnO_2 [14] or (Fe, Mn) co-doped SnO_2 [16], in contrast to single metal ion doped SnO_2. The broad sextet with 47–49 T cannot be assigned uniquely although it may be identified either with a large magnetic polaron or with γ-Fe_2O_3.

In this paper, (Co, Fe) and (Mn, Fe) ferrite oxides and dilute (Fe, Co) or (Fe, Mn) co-doped SnO_2 were prepared by a sol–gel method and measured by NIS to obtain the VDOS. It is concluded from VDOS that clusters may be formed in (Fe, Co) or (Fe, Mn) co-doped SnO_2 with highly doping.

References

1. Dielt, T., Ohno, H., Matsukura, F.: Phys. Rev. B **63**, 195205 (2001)
2. Punnoose, A., Hays, J., Gopal, V., Shutthanandan, V.: Appl. Phys. Lett. **85**, 1559 (2004)
3. Punnoose, A., Hays, J., Thurber, A., Engelhard, M.H., Kukkadapu, R.K., Wang, C., Shutthanandan, V., Thevuthasan, S.: Phys. Rev. B **72**, 54402 (2005)
4. Hays, J., Punnoose, A., Baldner, R., Engelhard, M.H., Peloquin, J., Reddy, K.M.: Phys. Rev. B **72**, 75203 (2005)
5. Rodriguez Torres, C.E., Errico, L., Golmar, F., Mudarra Navarro, A.M., Cabrera, A.F., Duhalde, S., Sanchez, F.H., Weissmann, M.: J. Magn. Magn. Mater. **316**, e219 (2007)
6. Nomura, K., Barrero, C., Sakuma, J., Takeda, M.: Phy. Rev. B **75**, 184411 (2007)
7. Ruderman, M.A., Kittel, C.: Phys. Rev. **96**, 99 (1954)
8. Kasuya, T.: Prog. Theor. Phys. **16**, 45 (1956)
9. Yosida, K.: Phys. Rev. **106**, 893 (1957)
10. Coey, J.M., Douvalis, A.P., Fitzgerald, C.B., Venkatesan, M.: Appl. Phys. Lett. **84**, 1332 (2004)
11. Coey, J.M., Venkatesan, M., Stamenov, P., Fitzgerald, C.B., Dorneles, L.S.: Phys. Rev. B **72**, 24450 (2005)
12. Nomura, K., Barrero, C.A., Kuwano, K., Yamada, Y., Saito, T., Kuzmann, E.: Hyperfine Interact. **191**, 25 (2009)
13. Nomura, K., Rykov, A., Nemeth, Z., Yoda, Y.: Hyperfine Interact. **205**, 129 (2012)
14. Nomura, K., Okabayashi, J., Okamura, Yamada, Y.: J. Appl. Phys. **110**, 83901 (2011)
15. Kono, S., Nomura, K., Yamada, Y., Okabayashi, J.: Hyperfine Interact. **205**, 105 (2012)
16. Okabayashi, J., Nomura, K., Kano, S., Yamada, Y.: Jpn. J. Appl. Phys. **51**, 23003 (2012)
17. Okabayashi, J., Kono, S., Yamada, Y., Nomura, K.: J. Appl. Phys. **112**, 73917 (2012)
18. Mudarra Navarro, A.M., Rodriguez Torres, C.E., Errico, L.A.: in private communication
19. Rykov, A.I., Nomura, K., Mitsui, T., Seto, M.: Physica B **350**, 287 (2004)
20. Nomura, K., Shima, A.: AIP Conf. Proc. (MSMS2012) **1489**, 13 (2012)
21. Stoneham, A.M.: Theory of Defects in Solids, chap. 14. Clarendon Press, Oxford (1975)
22. Bergqvist, L., Eriksson, O., Kudrnovsky, J., Drchal, V., Korzhavyi, P., Turek, I.: Phys. Rev. Lett. **93**, 137202 (2004)
23. Rykov, A.I., Nomura, K., Sakuma, J., Barrero, C., Yoda, Y., Mitsui, T.: Phys. Rev. B **77**, 014302 (2008)

DOI 10.1007/978-94-007-6482-8_4
Reprinted from *Hyperfine Interactions*, DOI 10.1007/s10751-013-0870-1

Effect of sintering conditions on the magnetic and structural properties of $Fe_{0.6}Mn_{0.1}Al_{0.3}$ synthesized by mechanical alloying

J. M. Marín · Y. A. Rojas · G. A. Pérez Alcázar · B. Cruz · M. H. Medina Barreto

Published online: 26 June 2013

Abstract In order to study the effect of sintering condition on the structural and magnetic behavior of prealloyed metallic powders of $Fe_{0.6}Mn_{0.1}Al_{0.3}$ system, two different thermal treatments were employed. All samples were previously milled and then compacted. Later, the sintering process was carried out in two cycles. For the first one, a sintering time of 2 h was followed by a cooling process governed by the inertia of the furnace. In the second treatment, a sintering time of 0.17 h with a controlled slow ramp of 1 °C/min between 500 °C and 250 °C. All samples were characterized by X-ray diffraction and Mössbauer spectroscopy. It was found that the sintering time improves the crystallinity while the magnetic behavior was modified by the changes in the cooling rate.

Keywords FeMnAl · Thermal treatment · X-ray diffraction · Mössbauer spectroscopy

Proceedings of the Thirteenth Latin American Conference on the Applications of the Mössbauer Effect, (LACAME 2012), Medellín, Colombia, 11–16 November 2012.

J. M. Marín · M. H. Medina Barreto (✉) · B. Cruz
Grupo de Investigación en Propiedades Magnéticas y Magnetoópticas de Nuevos Materiales, Departamento de Física, Universidad Tecnológica de Pereira, A. A. 097 Vereda La Julita, Pereira, Colombia
e-mail: mmedina@utp.edu.co

Y. Rojas
Grupo de Ciencia de Materiales y Tecnología del Plasma, Universidad del Tolima, A.A. 546, Ibagué, Colombia

G. A. Pérez Alcázar
Departamento de Física, Universidad del Valle, A. A. 25360, Cali, Colombia

1 Introduction

Fe-Mn-Al alloys have technical and scientific interest because of the presence of different properties depending on concentration and thermal treatment. It has been observed that these alloys can be prepared in several ways such as casting and mechanical alloying. In addition, if high temperature treatment takes place dramatic changes could be detected by X-ray diffraction and Mössbauer spectrometry due to ordering processes appearing at temperatures above 427 °C [1]. Mechanical alloying (MA) is a simple process based on high-energy ball milling to produce non-equilibrium powdered materials which cannot always be obtained by conventional methods, leading to nanostructured powders with different and new properties in comparison to other processes [2, 3]. Rebolledo et al. [4] show that the $Fe_{0.6}Mn_{0.1}Al_{0.3}$ system produced by MA is a soft magnetic material and this character is improved by increasing the milling time. However, the subsequent sintering leads the system to steady state with paramagnetic behavior at room temperature. Therefore, the softening during milling is lost. Furthermore, the sublimation of Mn above 527 °C affect the distribution of Mn atoms in the alloy as reported in the FeMnC system [5].

We have synthesized $Fe_{0.6}Mn_{0.1}Al_{0.3}$ steels by the complete pulvimetallurgical processes starting with prealloyed powders that were compacted and then sintered. In the process, the final structural and magnetic properties were studied using X-ray diffraction and Mössbauer spectroscopy. The milled and compacted powders were sintered up to maximum temperature of 1100 °C and then cooled in two different ways. The crystallite sizes and the mean magnetic hyperfine fields were compared against each other.

2 Experimental method

Using the powder metallurgy via, $Fe_{0.6}Mn_{0.1}Al_{0.3}$ powders were mechanically alloyed in a high energy planetary ball mill with a ball mass-to-powder mass ratio of 9:1 for 3, 9 and 15 h. Samples of 1,5 g were uniaxially compacted in a rigid die with a compaction force of 150 kN obtaining cylindrical pellets of 12 mm diameter and 2 mm height. Green samples were sintered in a tubular furnace by using a heating rate of 10 °C/min until was reached a maximum temperature of 1100 °C in argon atmosphere to prevent oxidation. The cooling of the test tubes was performed using two different cycles. Cycle I (Fig. 1a) has a sintering time of 2 h and then cooled at the rate of the inertia of the oven (cooling of the resistance). In the Cycle II (Fig. 1b) the same sintering temperature was considered, but the sintering time was 0.17 h. In this case, the cooling was performed using the furnace inertia rate until 500 °C and then with a programmed ramp until 250 °C, lowering the temperature at a rate of 1 °C/min. X-ray patterns were taken at room temperature in a Bruker D8 Advance difractometer. Mössbauer spectra were collected using a ^{57}Fe source in standard transmission geometry. Data from the difractograms was refined with the Rietveld method by GSAS (General Structure Analysis System) software [6], and Mössbauer spectra were fitted with MOSFIT program.

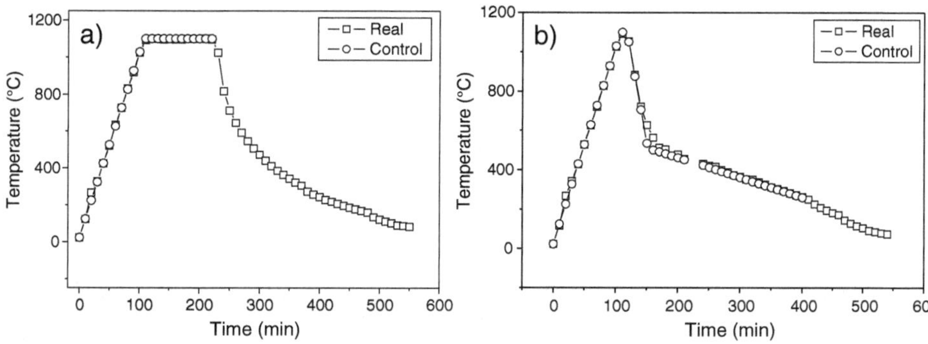

Fig. 1 Sintering cycles (**a**) Cycle I and (**b**) Cycle II

3 Results and discussion

Figure 2 shows some of the X-ray patterns acquired from powders milled (3 and 9 h), compacted and sintered for both cycles. All the studied samples exhibit the presence of a bcc structural phase, showing characteristic diffraction peaks of the planes (110), (200) and (211). The presence of the bcc phase suggests that the Mn and Al atoms are distributed inside the Fe cell. For short milling times, the alloy is not completely consolidated, but after sintering a full diffusivity of the minor elements is achieved and the reached lattice parameter is around 0,2902 nm (±0,0001 nm), in agreement with previous results obtained by milling [4, 7]. This value is independent of the sintering method employed, slightly above the typical lattice parameter of pure iron (0,2870 nm). This enhancement of the lattice parameter is due to the presence of Al which presents a bigger atomic size compared to that of the Fe atom, and is in according with previous results reported for melted and ordered Fe-Al alloys [8, 9] as well as grinded and heat treated Fe-Al alloys [10–12].

The crystallite size in the $Fe_{0.6}Mn_{0.1}Al_{0.3}$ consolidated samples was obtained from the refinement made by GSAS. Figure 3 shows that the crystallite size values increase for the sintered samples in the Cycle I, in comparison to the values in Cycle II. Therefore, an improvement in the crystallinity is found. A similar behavior has been reported in magnetic materials [13, 14] as well as in non-magnetic materials [15].

Results acquired from Transmission Mössbauer Spectroscopy (TMS) at room temperature for 3, 9 and 15 h in the Cycle I, are shown in Fig. 4a. The spectra display the formation of a magnetically disordered phase and a broad paramagnetic line near the central region. The fitting realized with the MOSFIT program exhibits the presence of two components: a broad singlet and a hyperfine field distribution (HFD) between 60 kOe and 300 kOe. Figure 4b shows the probability function for the hyperfine field distribution. This type of fit, and remembering the XRD results which show only the presence of a bcc Fe-Mn-Al lattice, indicates that this lattice is disordered and composed by different ferromagnetic Fe sites, which explain the HFD, and paramagnetic Fe sites, which explain the paramagnetic line. The ferromagnetic Fe sites are those rich in Fe and the paramagnetic ones are those rich in Al, as it has been found in previous reports of these type of samples [4, 16].

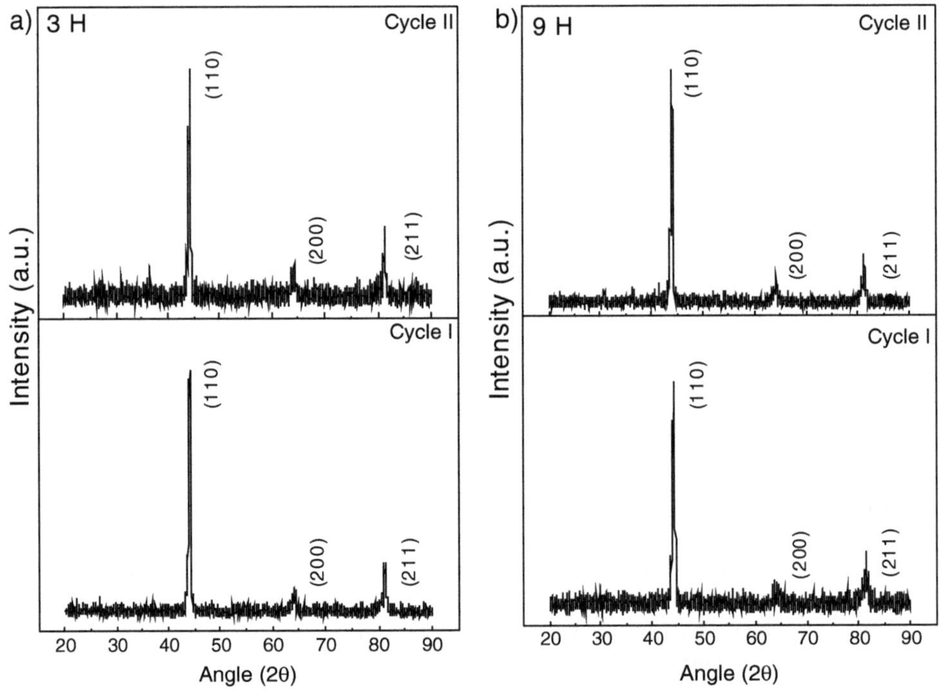

Fig. 2 X-Ray difractograms from both cycles for 3 and 9 h of milling

Fig. 3 Crystallite size evolution over milling time and sintering cycle

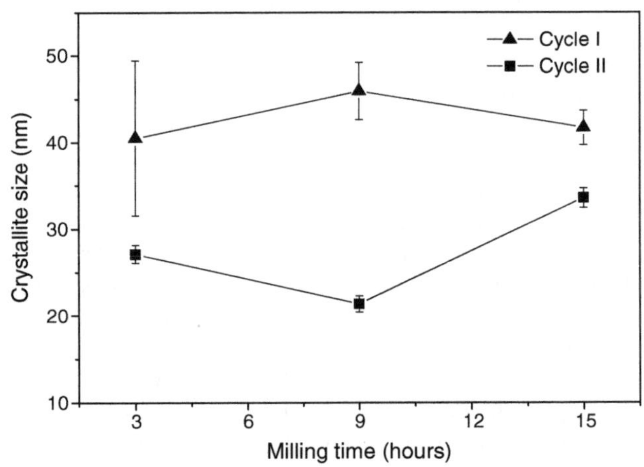

Figure 5a shows the Mössbauer spectra and their corresponding HFDs of the $Fe_{0.6}Mn_{0.1}Al_{0.3}$ samples sintered by using Cycle II. The fitting process was similar to those obtained with the cycle I, with the presence of a singlet and a hyperfine field distribution between 30 kOe and 300 kOe. Figure 5b shows the respective probability function for the hyperfine field distribution. However, the relative percentual amount of each component shows variations for each cycle, as can be observed

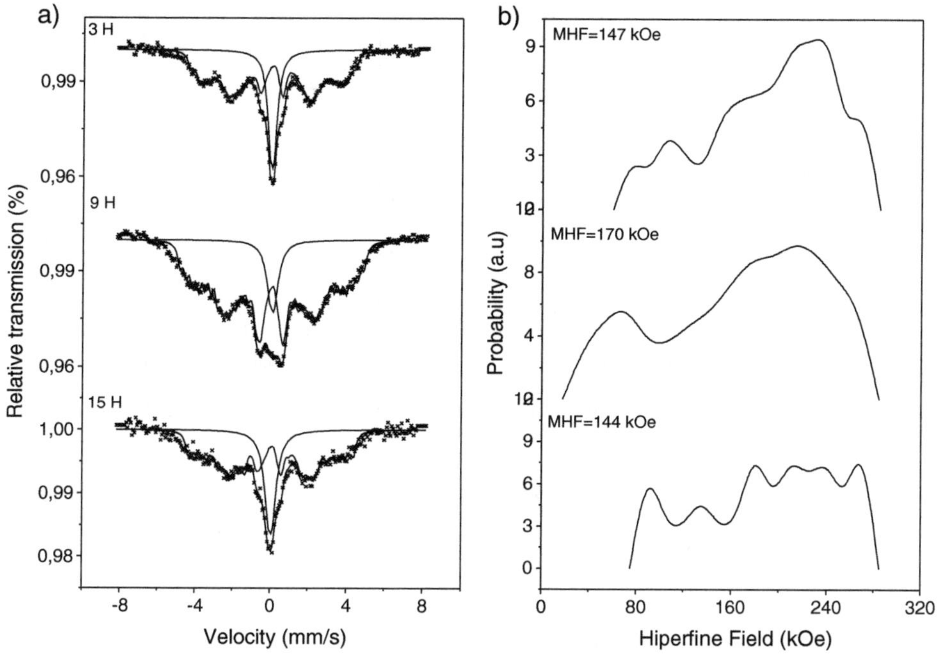

Fig. 4 Room temperature Mössbauer spectra (**a**) fitted by MOSFIT for Cycle I against milling time and (**b**) probability function for the hyperfine field distribution

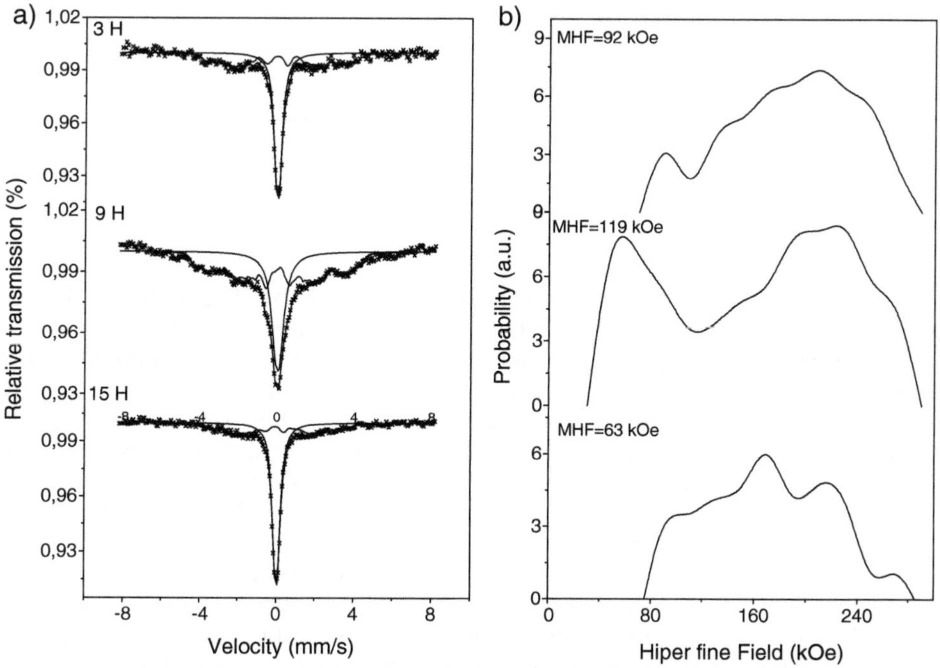

Fig. 5 Room temperature Mössbauer spectra (**a**) fitted by MOSFIT for Cycle II against milling time and (**b**) probability function for the hyperfine field distribution

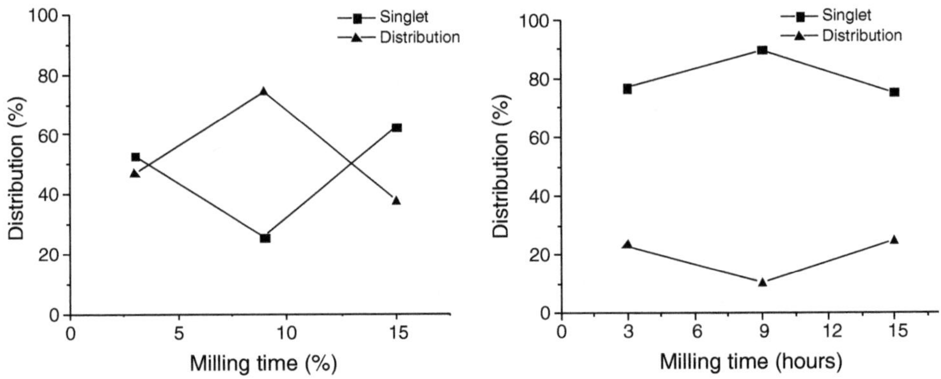

Fig. 6 Distribution percentages for both sintering cycles: Cycle I (*left*) and Cycle II (*right*)

Fig. 7 Mean hyperfine field variation against milling time and sintering cycle

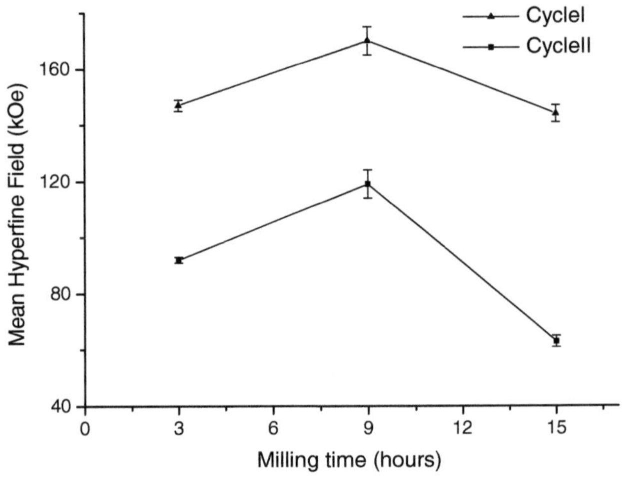

in Fig. 6. For Cycle II there is a greater presence of the singlet (about 80 %). Conversely, there is not a clear predominant component in Cycle I. This behavior may be related to configurational changes that occur in the neighborhood of the iron atoms. A sintering time of 2 h combined with a faster cooling (≈2,4 °C/min between 500 °C and 250 °C) in Cycle I allowed to retain part of the magnetic disorder which was present at high temperature. With the slower cooling of Cycle II (≈1 °C/min between 500 °C and 250 °C) and also shorter sintering time, the magnetic accommodation of the atoms inside the alloy reveals a steady state more stable (in our case, a paramagnetic one) and a decrease of the disorder degree. These results are in agreement with reports in $Fe_{60}Al_{40}$ alloys where these systems are paramagnetic when they are ordered and ferromagnetic when they are disordered [17, 18]. Regarding the relationship of the cooling rate with variations on magnetic properties, Thota et al. [19] have found that changes in the neighborhood of an iron atom and the quantity of the relative phases presents a clear dependance respecting cooling conditions.

Finally, Fig. 7 shows the mean hyperfine field for the samples treated with both cycles. The sintered systems obtained in the Cycle I has a higher mean hyperfine fields then that in the Cycle II. This behavior can be properly related to variations in the crystallite size derived from the diffractograms refinement; thereby, crystallite sizes of the order of 40 nm to 45 nm generate a mean hyperfine field near to 160 kOe, while crystallite sizes between 20 nm and 30 nm show approximate 38 % decrease in the mean hyperfine field, independent from milling time. Thus, variations in the mean hyperfine field exhibit an increase in the long-range order of the crystals benefited from rapid cooling, favoring the ferromagnetic interactions of Fe atoms with its neighborhood.

4 Conclusions

In conclusion, according to the established parameters during the thermal treatment, changes on the structural and magnetic properties $Fe_{0.6}Mn_{0.1}Al_{0.3}$ disordered alloy obtained via powder metallurgy were achieved. Independently of the thermal treatment employed, samples exhibited a bcc structure and the coexistence of two phases: a paramagnetic component and a disordered ferromagnetic component. Nevertheless, in the diffraction patterns a relation between crystallite size and sintering time was revealed. Additionally, an increment on the long-range order of the crystals reached by a faster cooling and a longer sintering time, favors the presence of a disordered magnetic phase.

Acknowledgements Authors would like to thank Universidad Tecnológica de Pereira for its financial support under project **3-12-3**. J.M. Marín is grateful with Colciencias for the scholarship in the "Jóvenes Investigadores e Innovadores" program.

References

1. Bremers, H., Frick, M., Hesse, J.: Hyperfine Interact. **94**(1), 1855–1859 (1994)
2. Suryanarayana, C.: Prog. Mater. Sci. **46**, 97–103 (2001)
3. Song, H., Wu, Y., Tangm, C., Yuan, S., Gong, Q., Liang, J.: Tsinghua Sci. Technol. **14**(3), 300–306 (2009)
4. Rebolledo, A., Romero, J., Cuadrado, R., Gonzĺez, J., Pigazo, F., Palomares, F., Medina, M.H.: J. Magn. Magn. Mater. **316**, 418–421(2007)
5. Hryha, E., Dudrova, E., Tadashi, M. (eds.): ISBN: 978-953-307-980-6, InTech, Available from: http://www.intechopen.com/books/application-of-thermodynamics-to-biological-and-materials-science/the-sintering-behaviour-of-fe-mn-c-powder-system-correlation-between-thermodynamics-and-sintering-pr (2011)
6. Larson, A., Dreele, R.V.: General structure analysis system (GSAS). Los Alamos National Laboratory Report LAUR, pp. 86–748 (2000)
7. Pérez Alcázar, G.: Rev. Acad. Colomb. Cienc. **28**(107), 265–274 (2004)
8. Taylor, A., Jones, R.M.: J. Phys. Chem. Solids **6**, 16 (1958)
9. Pérez Alcázar, G.A., Galvão da Silva, E.: J. Phys. F, Met. Phys. **17**, 2323 (1987)
10. Apianiz, E., Plazaola, F., Garitaonandia, J.S., Martin, D., Jiménez, J.A.: J. Appl. Phys. **93**(10), 7649–7651 (2003)
11. Martin Rodríguez, D., Apianiz, E., Plazaola, F., Garataoandia, J.S., Jiménez, J.A., Schmool, D.S., Cuello, G.J.: Phys. Rev. B **71**, 212408 (2005)
12. Nogués, J., Apianiz, E., Sort, J., Amboage, M., d'Astuto, M., Mathon, O., Puzniak, R., Fita, I., Garitaonadia, J.S., Suriñac, S., Muñoz, J.S., Baró, M.D., Plazaola, F., Baudelet, F.: Phys. Rev. B **74**, 024407 (2006)

13. Singh, J.P., Payal, R.S., Srivastava, R.C., Agrawal, H.M., Chand, P., Tripathi, A., Tripathi, R.P.: J. Phys. Conf. Ser. **217**, 1–4 (2010)
14. Abdoli, H., Asgharzadeh, H., Salahi, E.: J. Alloy. Compd. **473**, 116–122 (2009)
15. Garcia, S., Portelles, J., Martinez, F., Font, R., Quiñones, J.R., Siqueiros, J.M.: Rev. Mex. Fís. **49**(1), 15–19 (2003)
16. Valderruten, J., Pérez, G., Civale, L.: J. Mater. Sci. **39**, 5445–5449 (2004)
17. Das, G.P., Rao, B.K., Jena, P., Deevi, S.C.: Phys. Rev. B **66**, 184203 (2002)
18. Gialanella, S., Amils, X., Baró, M.D., Delcroix, P., Le Caer, G., Lutterotti, L., Suriñach, S.: Acta Mater. **46**, 3305 (1998)
19. Thota, H., Garg, A., Pandey, B., Verma, H.C.: Hyperfine Interact. **187**, 1–184 (2001)

DOI 10.1007/978-94-007-6482-8_5
Reprinted from *Hyperfine Interactions*, DOI 10.1007/s10751-013-0814-9

Design of an Auger Electron Mössbauer Spectrometer (AEMS) using a modified cylindrical mirror analyzer

F. Moutinho · C. Rojas · L. D'Onofrio

Received: 3 December 2012 / Accepted: 21 January 2013 / Published online: 19 February 2013

Abstract We have designed and built an Auger Electron Mössbauer Spectrometer (AEMS) for the detection of resonant ^{57}Fe Auger electrons using a modified commercial cylindrical mirror analyzer (CMA). The CMA final aperture was modified intentionally in order to increase electron transmission at the expense of reducing its energy resolution, from an original value of 0.5 % to a value of 11 % after the modification. The Channeltron detector electronics and the pre-amplifier were also modified in order to increase the counting efficiency. The electron energy analyzer is selective in energy in the 30 eV–3000 eV range, so the spectrometer can be used to detect MNN (45 eV) and LMM (600–700 eV) Fe Auger signals, what gives it a high surface sensitivity for Fe containing samples. We have used it to acquire the Fe LMM Auger signals generated from the de-excitation process after γ-Ray resonant nuclear absorption. The spectrometer can be used to study samples non-enriched in ^{57}Fe, with acquisition times from 5 to 7 days, what is a big advantage. From electron trajectory Monte Carlo simulations in metallic iron, the mean-escape-depth of the detected Auger signals has been estimated in approximately 1 nm. Fe K conversion electrons and KLL Auger electrons with mean escape depths of 129 nm and 78 nm respectively also contribute to the detected signal although in a lesser proportion.

Keywords AEMS · Surface sensitivity · Energy resolution · Mean escape depth

Proceedings of the Thirteenth Latin American Conference on the Applications of the Mössbauer Effect, (LACAME 2012), Medellín, Columbia, 11–16 November 2012.

F. Moutinho (✉) · C. Rojas · L. D'Onofrio
Centro de Física Experimental del Sólido, Facultad de Ciencias,
Universidad Central de Venezuela, Caracas, Venezuela
e-mail: fmoutinho@fisica.ciens.ucv.ve

1 Introduction

Mössbauer Spectroscopy analysis can be carried out in either of two modes: transmission or backscattering; being the electron backscattering mode [1, 3] the best choice for the study of phenomena associated to the surface region like, for instance, the oxidation and corrosion of steels. In the backscattering mode where Mössbauer resonant electrons are detected, there are different techniques or variants that can be applied, which differ mainly in their depth resolution or sensitivity to the surface region. Techniques like; Depth Selective Conversion Electron Mössbauer Spectroscopy (DCEMS) and Auger Electron Mössbauer Spectroscopy (AEMS) are selective in energy by the use of electron analyzers tuned around an energy center value (E) with a given energy resolution (R). By choosing to detect electrons with low enough characteristic kinetic energies, one can guarantee that the majority of the detected signal comes from the surface region of the sample. In the present work the AEMS technique is used.

Mössbauer surface analysis of metallic samples using AEMS can provide results with a very high surface sensitivity due to the shorter inelastic mean free path (IMFP) of the 600–700 eV Fe LMM Auger electrons (1–2 nm), compared with the IMFP of the 7.3 keV Fe K conversion electrons (5 nm) [1]. In AEMS the mean escape depth of the detected signal can be close to the IMFP [1] of the electron signal while in integral CEMS (ICEMS) it can reach values as high as 80 nm [3]. However the use of the AEMS technique is not so common due to the complexity and high cost that implies the use of an electron energy analyzer, an ultra-high vacuum chamber, high activity sources and ^{57}Fe-enriched samples. Surface sensitivity refers to the relative proportion of surface and bulk signals but even if a technique has a high surface sensitivity it can give a very small signal due to the small absolute number of atoms characteristic of the surface region. In order to diminish acquisition times ^{57}Fe sample enrichment and/or high activity sources are usually required to increase the intensity of resonant signal available.

Some authors [4, 5] were able to perform AEMS spectroscopy using a gamma source (^{57}Co) of 20–50 mCi, high transmission electron analyzers and samples enriched in ^{57}Fe. Our contribution here is to construct an instrument able to perform Mössbauer surface analysis with the big advantage of using a conventional 50 mCi gamma source to study non enriched ^{57}Fe samples and with acceptable acquisition times from 5 to 7 days. We used an Auger Electron Spectrometer installed in an ultra-high vacuum chamber and converted it to an Auger Electron Mössbauer Spectrometer with a very low cost, modifying an old Cylindrical Mirror Analyzer (CMA) [2] specifically for this Mössbauer application.

2 Design and building of the spectrometer

2.1 The source-sample-detector system

The first part of this project was to design the source-sample-detector system. In order to locate the gamma source close to the sample, which is at the center of the chamber, a stainless steel tube was connected to a 70 mm chamber port. The tube was lead lined and finishes in an aluminized Mylar window. The final pressure achieved

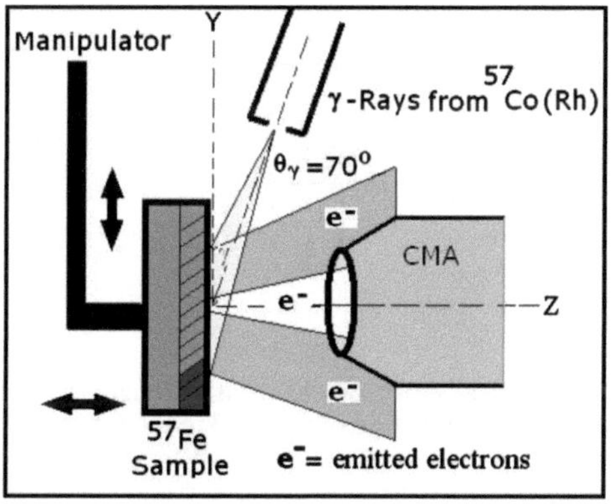

Fig. 1 Source-sample-detector array in the spectrometer

with this condition is 10^{-6} mbar. For ultra-high vacuum condition a welded beryllium window is required.

The lead collimator tube has one side attached to a lead disk of 3 mm thick with a 5 mm diameter centered hole. Also a Plexiglas filter of 1 mm thick has been added to reduce the 6.3 keV X-rays from the source. An aluminum rod, joined to the motor fixed in the outside of the chamber allows the gamma source to be very close of the sample without breaking the vacuum. The collimator design let us to illuminate just a centered region of the sample at the working distance and minimizes the noise from the X-rays. The distance from source to sample is ~3.5 cm and from sample to face of the CMA~3 cm. The source-sample-detector configuration is shown the Fig. 1.

2.2 The modified CMA

The second part consisted in modifying a small commercial CMA electron energy analyzer in order to increase its transmission at the expense of reducing its resolution, what was possible by altering its optics. This CMA had an annular entrance window that defines an angle of acceptance for the electrons between 40° and 60°. This electron energy analyzer is selective in energy in the 30 eV–3000 eV range. After studying its operation it was found that the original resolution for a focused electron beam (0.5 %) is controlled by a third slit (slit 3) placed at the back of this analyzer. This slit acts as an energy filter of the dispersed electrons on their path from the inner cylinder of the CMA to the Channeltron detector. This slit was completely removed and the position of the detector was adjusted. The resolution was measured, using a broad electron beam in order to simulate the area shone by the gamma source. It was found that the resolution changes from 0.5 % to 4 % on defocussing the electron beam and then to 11 % on removing slit 3, what is convenient for our purposes since in that way the three Fe LMM Auger peaks, at approximately 600 eV, 650 eV and 700 eV, can be merged into a single, more intense, peak. Figure 2 shows the schematics for the function of slit 3 and a picture of it.

Figure 3 shows the elastic peak spectra used to determine the analyzer resolution, defined as the ratio of $\Delta E/E$ where E is the peak energy and ΔE is the full width at half maximum, obtained with the defocussed electron beam: 3(a) with the slit in position and 3(b) without the slit.

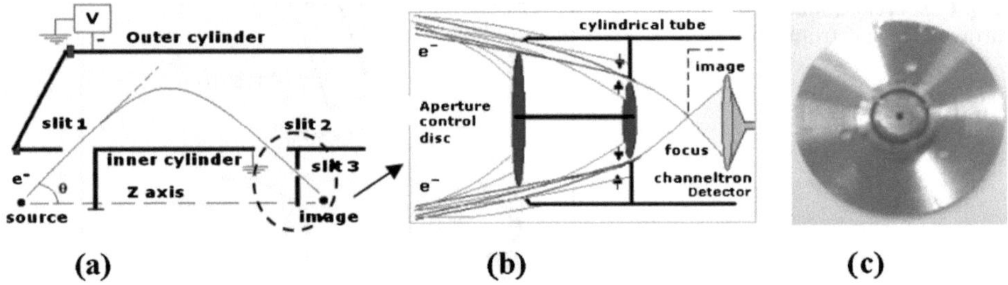

Fig. 2 **a** Main trajectory of electrons through the inner cylinder and the slit 3 of the CMA. **b** Schematic for the function of the slit 3 in the CMA. **c** Picture of the Slit 3

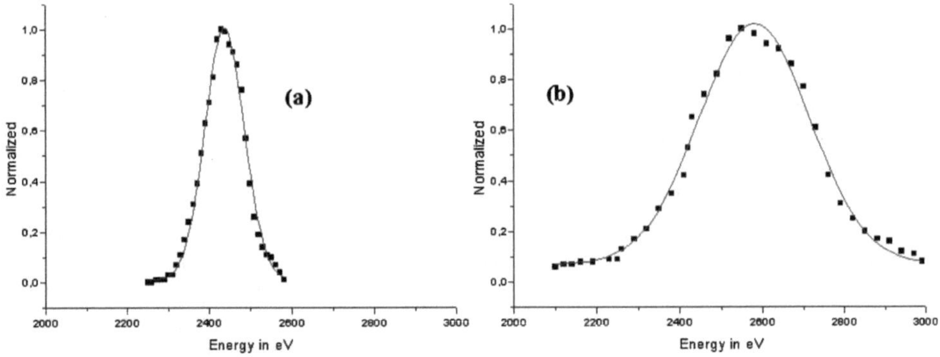

Fig. 3 **a** Resolution with the slit in position (4 %). **b** Resolution with the removed slit (11 %)

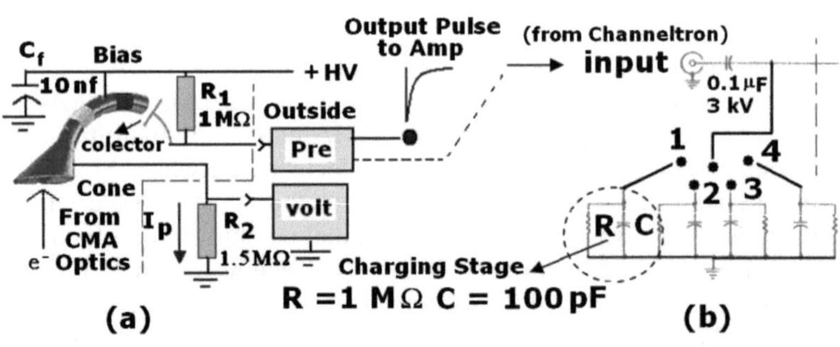

Fig. 4 **a** Electronic Arrangement made in the Channeltron detector connected to the pre-amplifier and multimeter. **b** Modified section in the Pre-amplifier ORTEC-113

The analyzer was reassembled positioning the Channeltron detector closer to the CMA body in order to maximize the signal at the detector.

2.3 Detector and modified electronics

The electronics associated to detection was also modified to enhance its response to pulse counting, especially for low pulse count rates. Changes were made both

Fig. 5 **a** Internal view. **b** External part showing motor coupling

in the Channeltron detector and in the pre-amplifier (ORTEC-113), increasing the sensitivity of the analyzer in a noticeable fashion. Figure 4 shows these arrangements.

Figure 5a and b show the internal and external view of the AEMS spectrometer that has been constructed.

3 Experimental results

Initially we tested the equipment with a very low activity (1 mCi) ^{57}Co/Rh source, and an enriched (>90 %) ^{57}Fe sample using an acquisition time of 7 days. We also changed the center of the analyzer pass energy in order to observe possible changes in the sensitivity to the detected signal. The results are shown in Fig. 6. In this experiment it was observed that the pass energy in the range of 580–650 eV produces the highest sensitivity, so for the rest of the experiments the pass energy was fixed at 650 eV, what coincides with the central peak of the Fe LMM Auger spectrum. This Fe LMM Auger spectrum consists of three peaks of comparable intensity located approximately at 600 eV, 650 eV and 700 eV. With the pass energy centered at 650 eV and the quoted 11 % resolution, the three Auger peaks located from 590 eV to 710 eV can be merged into a single peak, what increases the amount of detected signal without varying much its escape depth.

The second experiment was performed using a 50 mCi ^{57}Co/Rh source and a non-enriched iron metallic foil as the sample. Figure 7 shows the spectrum obtained after a 7-day acquisition time.

The spectrum of Fig. 7 shows an emission of 0.6 %. Despite that the effect measured is very low, the spectrum can be well detected, and confirm the possibility of using iron samples with a natural isotopic composition.

4 Monte Carlo simulations

We used David Joy Monte Carlo simulation programs [6] for electron trajectories in solids to estimate the mean escape depth of the signals detected by our spectrometer when an iron metal sample is analyzed. For this purpose we obtained the energy distribution of the electrons emerging from a face of a thin iron film when a beam of

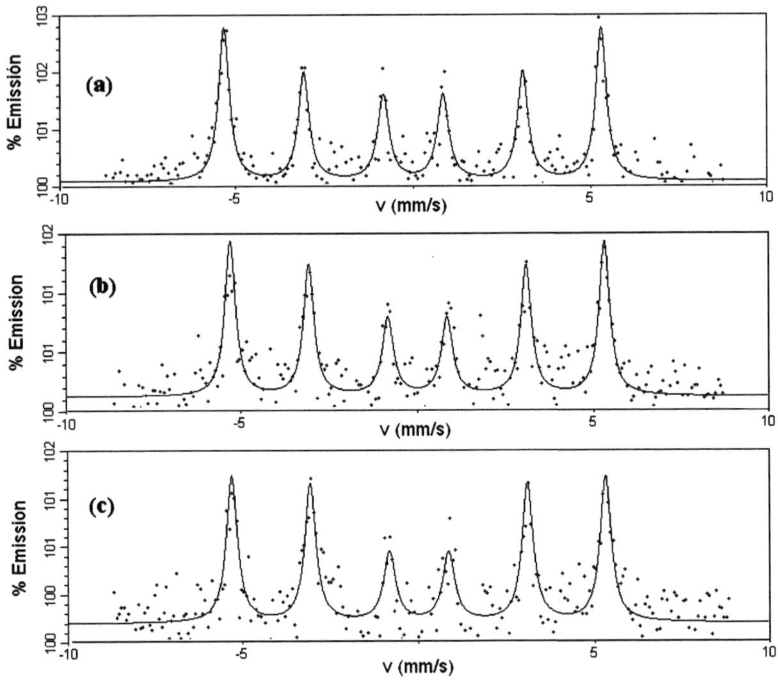

Fig. 6 AEMS spectra of an enriched ^{57}Fe sample with a 1 mCi $^{57}Co/Rh$ source acquired in 7 days. **a** Pass energy at 580 eV, 3 % of emission, **b** 650 eV, 2 %, **c** 700 eV, 1.5 %

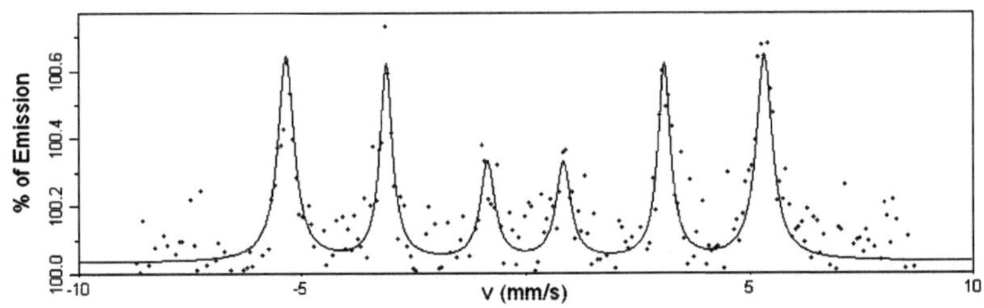

Fig. 7 AEMS spectrum of a non-enriched metallic iron sample, using a 50 mCi $^{57}Co/Rh$ source acquired in 7 days. Pass energy at 650 eV, 0.6 % of emission

10^5 electrons impinges on the other face. Film thicknesses were varied from 0.25 nm to 230 nm and beam energies of 7.3 keV, 5.4 keV, 600 eV, 650 eV and 700 eV were used with an incidence angle corresponding to the acceptance window of the CMA analyzer. These energy values correspond respectively to the K conversion electrons, to the KLL Auger electrons and to the triplet LMM Auger electrons. The combined LMM triplet, with a low 9 % emission probability, was chosen because of its high surface sensitivity. The more energetic K and KLL electronic emissions were included in the simulation because of their high emission probabilities that amount to 81 % and 43 % respectively. Figure 8a shows the electron energy distributions obtained when a 7.3 keV conversion electron beam traverses iron films of 50 nm,

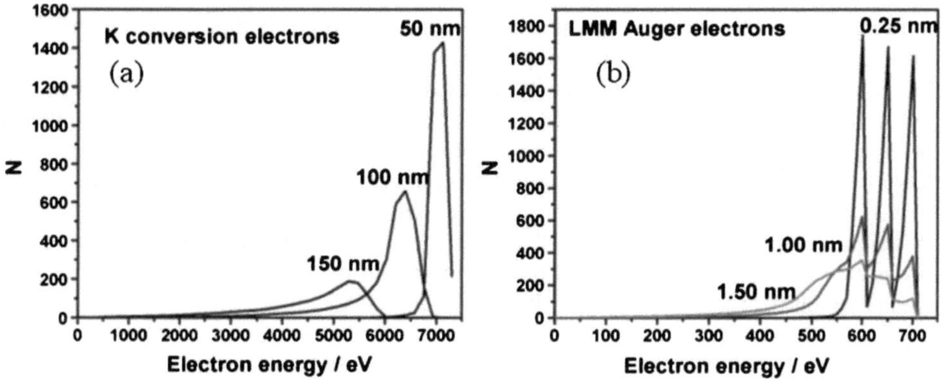

Fig. 8 Electron energy distributions as a function of iron slab thickness for **a** 7.3 keV K conversion electrons. **b** Triplet (600, 650, 700) eV LMM Auger electrons

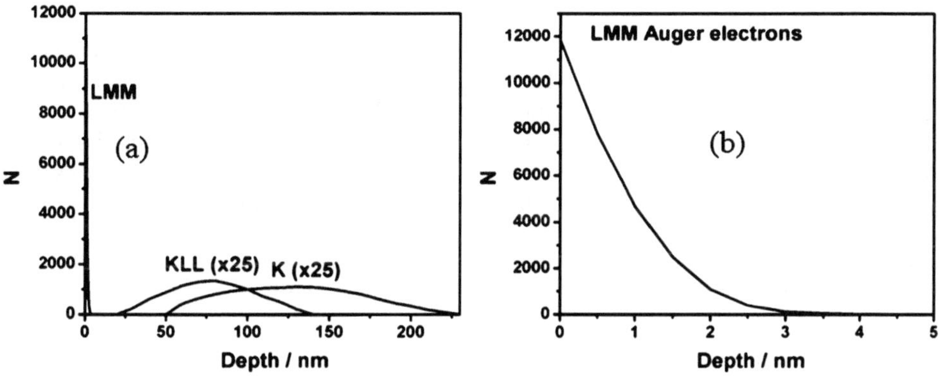

Fig. 9 **a** Depth distribution for the detected LMM, KLL and K electron signals from metallic iron. **b** LMM depth distribution zoomed from **a**

100 nm and 150 nm. As can be observed, the thicker the film the smaller the peak maximum and the broader the tail toward lower energies. After suffering inelastic scattering a significant number of these resonant electrons can reduce its energy to reach the (590 eV–710 eV) energy window defined by the CMA. The number of K, KLL and KLM electrons within this energy window, weighted by their corresponding emission probabilities, were evaluated using the results of the Monte Carlo simulation. Figure 8b shows the electron energy distributions corresponding to the triplet LMM Auger signal, obtained after traversing very thin iron slabs. This time, the reduction in intensity within the (590 eV–710 eV) CMA energy window occurs more rapidly with increasing iron thickness, an indication of the high surface sensitivity of these signals.

By counting the number of K, KLL and LMM electrons within the (590 eV–710 eV) CMA energy window, weighted by their emission probabilities, one can obtain the depth distribution functions for these three kinds of signals, detected by the analyzer in AEMS. Figure 9a shows the three separated contributions. The detected LMM signals are originated in a very thin surface layer that can be better appreciated in Fig. 9b, where one observes a monotonic decrease with depth that disappears after 5 nm. From this depth distribution one obtains a value of 0.75 nm, less than 3 iron monolayers, for the mean-escape-depth of the Fe LMM signals, what amounts for the 62 % of the total Mössbauer electron signal. The depth distributions for the K and KLL contributions can be observed in Fig. 9a, where their intensities have been amplified 25 times to be able to appreciate them. The KLL distribution forms a broad band from 20 nm to 140 nm, giving a mean-escape-depth of 78 nm for this signal, which represents a 16 % of the total. The K distribution forms an even broader band, from 50 nm to 230 nm, giving a 129 nm mean-escape-depth for this signal, which represents a 22 % of the total.

5 Conclusions

We have constructed an AEMS spectrometer which offers the advantage of acquiring Mössbauer spectra from non-enriched samples, using 50–25 mCi sources and acquisition times from 5 to 7 days.

Monte Carlo electron trajectory simulations allowed us to calculate the mean-escape-depth of the detected Fe LMM Auger signals in metallic iron, taking into account the energy resolution of the analyzer. The main Mössbauer ^{57}Fe electron signals that are being detected in this spectrometer are the LMM Auger signal, the KLL Auger signal and the K conversion electron signal, being the LMM Auger signal the one which contributes to the surface sensitivity of the technique.

Acknowledgements Financial support from CDCH-UCV under Project PG 03-8216-2011/1 and from FONACIT under Project PEII 3896 are gratefully acknowledged.

References

1. Gancedo, J.R., Gracia, M., Marco, J.F., Tabares, J.A.: H iperfine Interact. **111**, 83–92 (1998)
2. Briggs, D., Seah, M.P.: Practical Surface Analysis: Auger and Photoelectron Spectroscopy. Wiley (1990)
3. Moutinho, F., et al.: Design of a conversión electron Mössbauer spectrometer based on an electron multiplier. Evaluation of the mean-escape-depth of the detected signals. Hiperfine Interact. **195**, 141–147 (2010)
4. Klingelhöfer, G., Meisel, W.: Study of very thin oxide layers by conversion and Auger electrons. Hiperfine Interact. **57**, 1911–1918 (1990)
5. Klingelhöfer, G., Imkeller, U., Kankeleit, E., Stahl, B.: Surface sensitivity of low energy electron Mössbauer spectroscopy (LEEMS) using ^{57}Fe. Hiperfine Interact. **69**, 819–822 (1991)
6. Joy, D.: Monte Carlo Modeling for Electron Microscopy and Microanalysis. Oxford University Press, Oxford (1995)

DOI 10.1007/978-94-007-6482-8_6
Reprinted from *Hyperfine Interactions*, DOI 10.1007/s10751-013-0769-x

^{57}FE Mössbauer spectroscopy studies of Tektites from Khon Kaen, Ne Thailand

B. F. O. Costa · G. Klingelhöfer · E. I. Alves

Published online: 7 February 2013

Abstract Room temperature ^{57}Fe Mössbauer effect spectroscopy has been used to investigate the local Fe environment in a set of indochinite tektites from Thailand. A MIMOS II spectrometer in backscattering geometry has been used in the study, so that no sample preparation at all was required. The spectra have been analysed in terms of discrete spectral components using Voigt functions. The results are similar to those obtained with analysis done with quadrupole splitting distributions. In all cases the Mössbauer spectra show a broadened asymmetric quadrupole split doublet. The Fe^{2+} sites have been distinguished in Fe with octahedral and tetrahedral coordination. The Fe^{3+}/Fe^{2+} and Fe^{2+} tetrahedral/octahedral ratios have been determined from the relative areas of the various spectral components. Their values have been discussed in the light of actual literature.

Keywords Tektites · Natural glasses · Iron oxidation states

Thirteenth Latin American Conference on the Applications of the Mössbauer Effect, LACAME 2012, Medellín, Columbia, 11–16 November 2012.

B. F. O. Costa (✉)
CEMDRX, Physics Department, University of Coimbra,
Rua Larga, 3004-516 Coimbra, Portugal
e-mail: benilde@ci.uc.pt

B. F. O. Costa · G. Klingelhöfer
Institut für Anorganische und Analytische Chemie,
Joh. Gutenberg-Universität Mainz, 55099 Mainz, Germany

E. I. Alves
Centre for Geophysics and Department of Earth Sciences,
Geophysical Institute, University of Coimbra, 3000-134 Coimbra, Portugal

1 Introduction

Tektites are glass beads, usually black or very dark, rarely more than a few cm in length, showing peculiar aerodynamic shapes and small pits which give hints to their formation. These glasses were for some time believed to may have originated from high-velocity volcanic material emitted from the Moon, escaping its gravitational field, and eventually landed on Earth [1]. Nowadays there is consensus that they are natural glasses originated from fused material spilled during a meteorite impact [2] a hypothesis which had already been defended by Faul [3], without the proper attention. Tektites found on land fall into three general morphological categories; splash-form tektites, aerodynamically shaped tektites and Muong Nong type tektites. The Muong Nong-type tektites differ from other forms by having irregular shapes and a layered structure. This has led to speculation that the Muong Nong-type tektites have been formed under somewhat different conditions than other tektites and likely formed closer to the center of the impact site [4, 5]

Tektites are found dispersed in well defined strewn fields: moldavites in Europe, around Moldavia, Czech Republic (the only ones that are light-green and translucent); bediasites and georgiaites in North America (Texas and Georgia, respectively); ivorites in Ivory Coast; tibetanites, chinites, indochinites, rizalites (from the Philippines), and australites, in the largest strewn field, which ranges from Tibet to the Antarctic Ocean [1].

High temperature processes in silicate glasses are followed by redox reactions involving variable-valence elements, iron in first. Study of valent and structural state of iron atoms can lead to a better understanding of the glasses formation process. Mössbauer spectroscopy is one of the most powerful methods for determination of the iron atoms valence and coordination [6, 7].

One particular feature commonly observed in tektites is an anomalously low Fe^{3+}/Fe^{2+} ratio [8] and this suggests that they were solidified in a low oxygen environment. Some evidence suggests that Muong Nong-type tektites may have a higher Fe^{3+}/Fe^{2+} ratio than splash form tektites and that this may be consistent with their general morphology and solidification from the melt at a lower temperature [9].

For the present work we have used Mössbauer spectroscopy to study Fe-site properties in a number of sub-spherical samples of indochinites, with diameters between 1.5 and 2.0 cm, collected near Khon Kaen, NE Thailand.

Early in the investigations of tektites, chemical analysis has been mainly used for determination of oxidation state of iron. The results of that kind of analysis became controversial and unreliable [10]. Some different chemical and instrumental methods have been used to prove usability of chemical analysis. Fudali et al. [11] used four wet chemical techniques (two standard volumetric methods, an experimental volumetric method for direct determination of Fe^{3+} and a colorimetric method) and two instrumental techniques (Mössbauer spectroscopy and electron spin resonance). They found that the Fe^{3+}/Fe^{2+} ratios are 0.02–0.12 and are in agreement with what they believe to be the best previously published analysis for tektites (review in [12]).

Another reason for inaccurate measurements of the iron oxidation state is the sample preparation. Usually the samples are cut and grinded in acetone or alcohol to prevent the contact with atmosphere. In our study we used the unique MIMOS II instrument operating in backscattering geometry to collect Mössbauer spectra, with no need of sample preparation.

In general, the results of Mössbauer effect spectroscopy for the analysis of Fe-oxidation state ratios in tektites depend critically on the method of spectral analysis. In recent years, more sophisticated and accurate methods of spectral analysis based on distributed hyperfine parameters, have been used by Dunlap and co-workers [13, 14] and Rossano et al. [15]. Those analyses gives consistently low values of the Fe^{3+}/Fe^{2+} ratio while an analysis using discrete spectral components, can yield anomalous results. In the present work we used discrete spectral components using Voigt functions for the analysis of the tektites spectra and the results were compared with the ones obtained by more sophisticated methods and found to be consistent with them.

2 Experimental procedure

The Khon Kaen indochinites, whose results are present in this study, have been all collected in the same location and are all identical. They were labeled accordingly to the recollection and storing, and we have kept their identification numbers in this paper. The samples were studied as a whole, without being cut, as Mössbauer spectra were recorded for all the samples at room temperature with a MIMOS II spectrometer [16] detecting 14.4 keV radiation in backscattering mode, and a ^{57}Co(Rh) source with an activity of about 30 mCi. No sample preparation was done to carry out the measurements. This is a big advantage relatively to the transmission geometry method, found in literature. In this later case, the sample needs to be cut and crushed into powder, with possible changes in oxidation ratios. The fitting procedure of the spectra was carried out using a set of Voigt lines and parameters determined by least squares method. For comparison, we have also used the NORMOS program as a fitting method of distribution of quadrupole splitting with a linear correlation between quadrupole splitting and isomer shift. The isomer shifts are given relative to α-Fe at RT.

3 Results and discussion

Room temperature ^{57}Fe Mössbauer spectra of the tektites are illustrated in Fig. 1, and their Mössbauer parameters are displayed in Table 1. In Fig. 1, it can be observed that for all the spectra there is a small misfit in the total fitted spectra for the more positive spectral component. This is possibly a result of constraints placed on the fitting parameters because of using Voigt functions.

^{57}Fe Mössbauer spectra of non-magnetic glasses have a characteristic form of asymmetrical, broad doublets. Broadening effects are important in glass spectra with line-widths as large as 0.6 mm/s against 0.3 mm/s in crystals.

All the spectra show the common feature having a strong quadrupole doublet with IS around 1 mm/s and QS around 1.9 mm/s, which are typical parameters for Fe^{2+} in octahedral coordination in tektites. The doublet corresponding to Fe^{2+} in tetrahedral coordination is not always seen in spectra of tektites. However some authors have seen it already, eg [14, 15, 17]. This same structure has been seen in the Mössbauer spectra of glassy andesite [18] and glassy basalt [19]. The values

Fig. 1 Room temperature ^{57}Fe Mössbauer backscattering spectra of the tektites studied. The fitresults to the spectra are also shown (color on-line). The codes shown on the left side of each spectrum are the names given to the studied samples

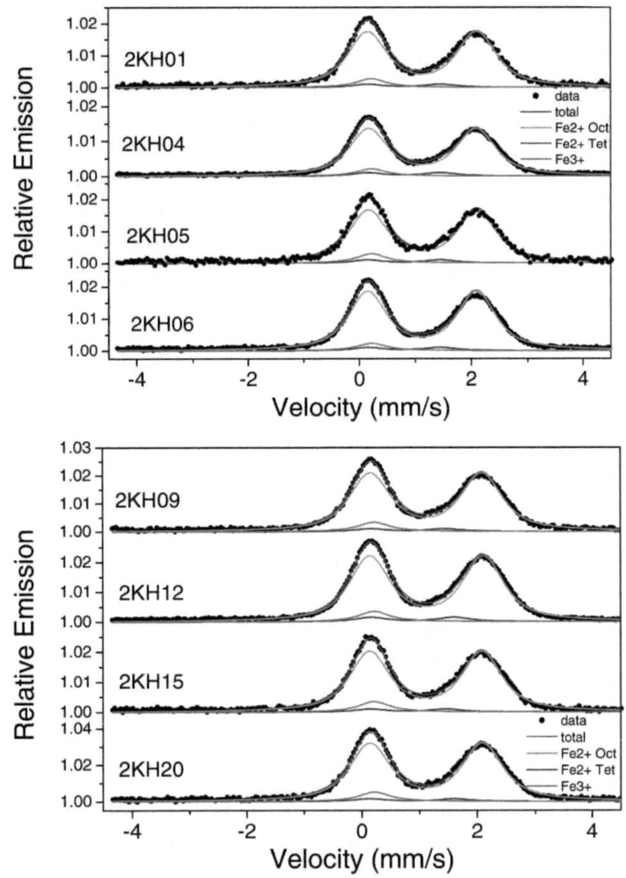

obtained by us (IS around 0.9 mm/s and QS around 1.3 mm/s) are higher than the ones obtained by Rossano and Dunlap for different tektites (IS around 0.6 mm/s and QS around 1 mm/s). Our values approach the ones obtained in other natural and synthetic glasses in which QS usually ranges between 1.4 and 1.8 mm/s [20]. The parameters obtained for the Fe^{3+} site, IS from 0.13 to 0.18 mm/s and QS of 0.02 mm/s, are in good agreement with the values referred in literature for tektites.

The site occupancy ratios agree in general with the ones obtained by Dunlap and Rossano.

The Fe^{2+}(tet)/Fe^{2+}(oct) ratio agree with the ratio for Rizalite tektites [13], Thailandites and Tibetanites [14]. The values obtained for Fe^{3+}/Fe^{2+} ratio are low and agree in general with the values obtained for tektites, in particular they are similar to the values obtained for Moldavite [13].

In the case of our tektites study, the oxygen partial pressure is presumably the most important factor in determining the Fe^{3+}/Fe^{2+} ratio, although chemical composition, maximum temperature and heating/cooling histories cannot be ruled out as possible contributing factors [13]. The larger Fe^{3+}/Fe^{2+} ratio obtained by us, relatively to the studies of Rossano and Dunlap for other tektites rather than the

Table 1 Hyperfine parameters for Fe^{2+} and Fe^{3+} sites and ratios of site occupancies in the tektites studied

	IS(mm/s)	QS(mm/s)	WID(mm/s)	%	$Fe^{2+}_{tetr}/Fe^{2+}_{oct}$	Fe^{3+}/Fe^{2+}
2KH01					0.049	0.070
Fe2+ Oct	1.14(1)	1.96(1)	0.65(1)	89.10		
Fe2+ Tetr	0.89(1)	1.30(1)	0.65(1)	4.36		
Fe3+	0.16(1)	0.02(1)	0.65(1)	6.54		
2KH04					0.063	0.066
Fe2+ Oct	1.14(1)	1.94(1)	0.65(1)	88.21		
Fe2+ Tetr	0.91(1)	1.30(1)	0.65(1)	5.57		
Fe3+	0.18(1)	0.02(1)	0.65(1)	6.22		
2KH05					0.051	0.072
Fe2+ Oct	1.16(1)	1.98(1)	0.65(1)	88.79		
Fe2+ Tetr	0.87(1)	1.30(1)	0.65(1)	4.49		
Fe3+	0.17(1)	0.02(1)	0.65(1)	6.72		
2KH06					0.048	0.073
Fe2+ Oct	1.13(1)	1.96(1)	0.65(1)	88.94		
Fe2+ Tetr	0.91(1)	1.30(1)	0.65(1)	4.29		
Fe3+	0.13(1)	0.02(1)	0.65(1)	6.77		
2KH09					0.041	0.074
Fe2+ Oct	1.14(1)	1.95(1)	0.65(1)	89.49		
Fe2+ Tetr	0.91(1)	1.30(1)	0.65(1)	3.63		
Fe3+	0.18(1)	0.02(1)	0.65(1)	6.88		
2KH12					0.058	0.066
Fe2+ Oct	1.14(1)	1.98(1)	0.65(1)	88.74		
Fe2+ Tetr	0.89(1)	1.44(1)	0.65(1)	5.11		
Fe3+	0.17(1)	0.02(1)	0.65(1)	6.15		
2KH15					0.043	0.077
Fe2+ Oct	1.13(1)	1.96(1)	0.65(1)	89.05		
Fe2+ Tetr	0.91(1)	1.30(1)	0.65(1)	3.80		
Fe3+	0.16(1)	0.02(1)	0.65(1)	7.15		
2KH20					0.040	0.071
Fe2+ Oct	1.14(1)	1.96(1)	0.65(1)	89.78		
Fe2+ Tetr	0.89(1)	1.48(1)	0.65(1)	3.60		
Fe3+	0.16(1)	0.02(1)	0.65(1)	6.62		

IS given relative to α-Fe at RT

ones mentioned above that give values closed to ours, suggests a greater oxygen partial pressure during formation of our tektites that also cause the Fe^{2+} to favor Fe^{2+} octahedral sites.

For comparison with the results obtained with the fitting carried out using Voigt functions, we fitted the spectra also using a distribution of QS with a linear correlation between QS and IS. The results obtained in this way, with our program, were not very satisfying. Although the parameters obtained for the main component of the spectra (the Fe^{2+} octahedral site) were in good agreement with the ones presented in Table 1, the Fe^{2+} tetrahedral component was not observed, and erroneous values of <QS> were obtained for the Fe^{3+} coordination site (values between 0.01 and 0.08 mm/s). The percentage obtained for this later site was found to be between 4 and 6 % of the total spectrum.

4 Conclusions

1. Our method of fitting tektites data with spectral components using Voigt functions, gives similar results as many other sophisticated methods used in recent years.
2. A distinction between tetrahedral and octahedral components of the Fe^{2+} sites is observed and this is consistent with other analysis using distribution parameters and analysis of spectra of glasses synthesized from silicate rocks with compositions that are similar to tektites.

Acknowledgements This work was supported from funds from FEDER (Programa Operacional Factores de Competitividade COMPETE) and from FCT—Fundação para a Ciência e Tecnologia under the project PEst-C/FIC/UI0036/2011. B.F.O. Costa also gratefully acknowledges the DAAD grant received for a Study Visit in Joh. Gutenberg-Universität Mainz.

References

1. O'Keefe, J.A.: Tektites and their origin. NASA Goddard Space Flight Center (1975)
2. Koeberl, C.: Geol. Soc. Amer. Spec. Pap. **239**, 133–151 (1994)
3. Faul, H.: Science **152**, 1341–1345 (1966)
4. McCall, G.J.M.: Tektites in the Geological Record: Showers of Glass from the Sky. The Geological Society, London (2001)
5. Koebert, C.: Geochim. Cosmochim. Acta. **56**, 1033 (1992)
6. Thorpe, A.N., Senftle, F.E., May, L., Barkatt, A., Adel-Hadadi, M.A., Marbury, G.S., Izett, G.A., Maurrasse, F.R.: J. Geophys. Res. **99**, 10881–10885 (1994)
7. Orkasson, N., Helgasson, O., Sigurdsson, H.: Geol. Soc. Amer. Spec. Pap. **307**, 445–452 (1996)
8. Chao, E.C.T.: The petrographic and chemical characteristics of tektites. In: O'Keefe, J.A. (ed.) Tektites, pp. 51. University of Chicago, Chicago, IL (1963)
9. Son, T.H., Koeberl, C.: Met. Planet. Sci. **46**(6), 805–815 (2005)
10. Volovetsky, M.V., Rusakov, V.S., Chistyakova, N.I., Lukanin, O.A.: Hypfine Interact. **186**, 965–970 (2008)
11. Fudali, R.F., Dyar, M.D., Griscom, D.L., Schreiber, H.D.: Geochim. Cosmochim. Acta **51**, 2749–2756 (1987)
12. Lukanin, O.A., Kadik, A.A.: Geochem. Int. **45**, 857–881 (2007)
13. Dunlap, R.A., Eelman, D.A., MacKay, G.R.: J. Non-Crystal. Solids **223**, 141–146 (1998)
14. Dunlap, R.A., Sibley, A.D.E.: J. Non-Crystal. Solids **337**, 36–41 (2004)
15. Rossano, S., Balan, E., Morin, G., Bauer, J.-P., Calas, G., Brouder, C.: Phys. Chem. Miner. **26**, 530–538 (1999)
16. Klingelhöfer, G., et al.: J. Geophys. Res. **108**, 8067 (2003)
17. Dunlap, R.A.: Hyperfine Interact. **110**, 217–225 (1997)
18. Dyar, M.D., Naney M.T., Swanson, S.E.: Am. Mineral **72**, 792 (1987)
19. Helgason, O., Steinthorsson, S., Morup, S.: Hyperfine Interact. **45**, 287 (1989)
20. Dickson, B.L., Smith, G.: Can. Miner. **14**, 206–215 (1976)

DOI 10.1007/978-94-007-6482-8_7
Reprinted from *Hyperfine Interactions*, DOI 10.1007/s10751-013-0767-z

Formation of magnetic nanoparticles studied during the initial synthesis stage

M. Kraken · I.-C. Masthoff · A. Borchers · F. J. Litterst · G. Garnweitner

Published online: 29 January 2013

Abstract The formation of iron oxide nanoparticles in course of a sol-gel preparation process was traced by UV/Vis and ^{57}Fe Mössbauer absorption spectroscopy. Samples were extracted at different stages of the reaction. While spectra measured on samples extracted at low reactor temperatures showed the starting materials $Fe(acac)_3$ diluted in benzyl alcohol undergoing slow paramagnetic relaxation, a sample extracted at a reactor temperature of 180 °C gave clear evidence for emerging iron oxide nanoparticles. A prolonged stay at 200 °C results in a complete transformation from $Fe(acac)_3$ to maghemite nanoparticles.

Keywords Non-aqueous sol-gel method · Mössbauer spectroscopy · Spin-spin relaxation

1 Introduction

The scientific interest in magnetic nanoparticles has increased continuously over the last years, resulting from their distinct physical properties leading to numerous technological applications [1]. A reproducible preparation technique, yielding

Thirteenth Latin American Conference on the Applications of the Mössbauer Effect, LACAME 2012, Medellín, Columbia, 11–16 November 2012.

M. Kraken (✉) · A. Borchers · F. J. Litterst
Institut für Physik der kondensierten Materie, Technische Universität Braunschweig, Mendelssohnstrasse 3, 38106 Braunschweig, Germany
e-mail: m.kraken@tu-bs.de

I.-C. Masthoff · G. Garnweitner
Institut für Partikeltechnik, Technische Universität Braunschweig, Volkmaroder Str. 5, 38104 Braunschweig, Germany

monodisperse, spherical particles is essential, since the physical properties of the particles strongly depend on these features. The non-aqueous sol-gel method is a bottom-up synthesis that allows preparation of particles with the desired properties. For this method, a precursor is dispersed in a high-boiling, organic solvent and following heat-up, the reaction is left to proceed for a variable span of time. Our aim is, to monitor the thermal decomposition of $Fe(acac)_3$ in benzyl alcohol (BA) and triethylene glycol (TEG) in order to gain an understanding about the formation of the magnetic nanoparticles. Based on the system described in [2], we performed further measurements to enlarge our knowledge of this synthesis route. This work focuses on the heat-up phase to 200 °C in a reactor using BA as solvent. Samples taken during the heating up process, as well as the mixed starting materials are discussed. Details about the reaction after reaching 200 °C and a comparison to TEG as a solvent will be presented elsewhere.

2 Experimental

Synthesis of magnetic nanoparticles The iron oxide nanoparticles were synthesized in a 1.5 L reactor (Polyclave, Typ 3/1 Büchi Glas Uster) using $Fe(acac)_3$ as only iron source and benzyl alcohol as solvent [2]. In a typical procedure the reactor was filled with 50 g $Fe(acac)_3$ and 1 L solvent and the reaction was left to proceed whilst stirring at 200 °C for 23 h. Samples were taken every 20 °C during the heating process and immediately cooled to room temperature (RT) in an ice bath. In the following, the reactor temperature at which the sample was taken will be called T_R. The samples were used for analysis without any further modifications.

For the ultraviolet-visible spectrophotometry (UV/Vis) the UV-3100PC Spectrophotometer from VWR is used, the samples being measured in a wavelength range from 300 to 500 nm.

^{57}Fe Mössbauer absorption measurements were performed on a standard spectrometer using a ^{57}CoRh source in a temperature range from 4 to 300 K. Absorbers were prepared from samples drawn from the reactor by rapidly cooling them to 80 K.

3 Results and discussion

The UV/Vis spectra, shown in Fig. 1, obtained for the samples taken during the heating process display the conversion of the precursor iron(III) acetylacetate in BA to Fe_2O_3 nanoparticles, being described by

$$\mathrm{Fe(acac)_3} \xrightarrow{solvolysis} \mathrm{Fe(OH)_{3-x}(OR)_x} \xrightarrow{condensation} \mathrm{Fe_2O_3}.$$

Spectra taken from samples between $T_R = \mathrm{RT}$ up to 140 °C show two distinct adsorbance maxima at the wavelengths 440 nm and at 360 nm. The absorbance is typical for $Fe(acac)_3$ [3]. For the samples taken above 180 °C the absorbance at 440 nm decreases reaching a constant value of 1.1. By comparing these curves, it is concluded that at a temperature of 180 °C the conversion from $Fe(acac)_3$ to iron oxide nanoparticles sets in.

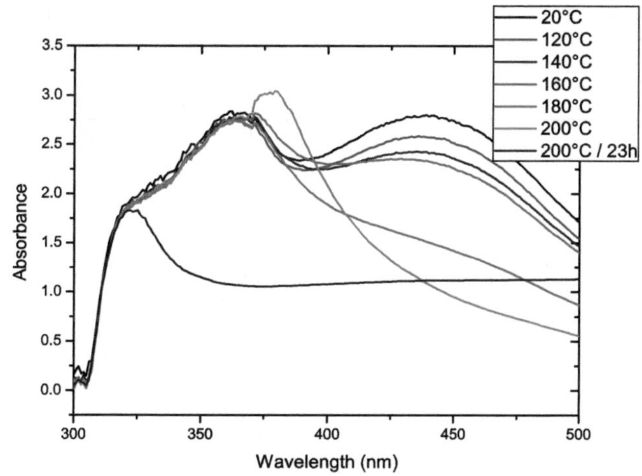

Fig. 1 UV/Vis spectra for $Fe(acac)_3$ dissolved in BA taken from the reactor at different temperatures

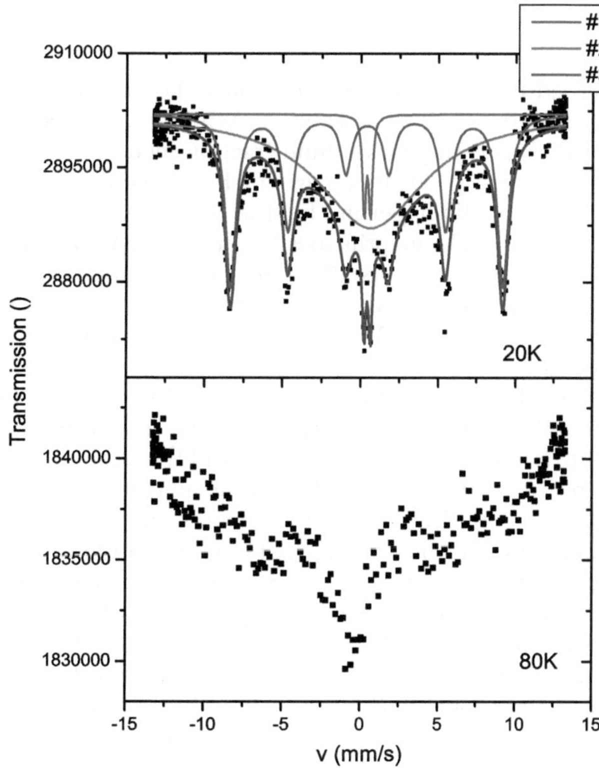

Fig. 2 Mössbauer spectra of $Fe(acac)_3$ dispersed in BA with a concentration of 50 g/L measured at 20 K and 80 K for $T_R = \mathrm{RT}$. The parameters of the fit to the measurement at 20 K can be found in Table 1

In order to gain more information about this transformation, we performed ^{57}Fe Mössbauer absorption measurements. In a first step, we examined a sample extracted from the reactor at $T_R = \mathrm{RT}$, i.e. directly after preparing the solution. The spectra are shown in Fig. 2. At low temperatures, a well-resolved sextet with

Table 1 Parameters of the spectra shown in Fig. 3: centershift CS relative to metallic Fe at RT, magnetic hyperfine field B_{hyp}, quadrupole splitting $e^2qQ/2$, linewidth W (HWHM) and the relative spectral area

T_R (°C)	#	CS (mm/s)	B_{hyp} (T)	$e^2qQ/2$ (mm/s)	W (mm/s)	Area (%)
RT	1	0.52	0.00	0.42	0.14	4
RT	2	0.74	0.00	0.00	4.15	60
RT	3	0.51	54.3	0.00	0.42	36
160	1	0.51	0.00	0.50	0.19	9
160	2	0.51	0.00	0.00	3.58	55
160	3	0.52	54.3	0.01	0.32	36
180	1	0.51	0.00	0.45	0.16	16
180	2	0.44	0.00	0.00	5.68	33
180	3	0.54	54.4	−0.03	0.27	8
180	4	0.45	46.3	0.01	0.56	43
200/30 min	4	0.43	50.4	0.00	0.41	100

The site numbers in the second column are given in the same color as the corresponding subspectra in Fig. 2 (T_R =RT), Fig. 3 (T_R = 160 °C and T_R = 180 °C) and Fig. 4 (after 30 min at 200 °C)

a broad background and a non-magnetic doublet component can be observed (fit parameters are given in Table 1). With increasing absorber temperature, the sextet broadens and the background fraction increases. The magnetic hyperfine splitting at 20 K corresponds to a magnetic hyperfine field of $B_{\text{hyp}} \sim 54.3$ T. The magnitude of the magnetic hyperfine field as well as the temperature dependent collapse of the sextet could in principle be interpreted with the typical behavior of superparamagnetic, well crystallized magnetic nanoparticles. However, since the UV/Vis measurements only indicate the formation of magnetic nanoparticles for reactor temperatures far above RT, superparamagnetic particles cannot be the origin of these spectra.

Therefore the spectra shown in Fig. 2 need to be interpreted in a different way. As has been demonstrated already in the very early days of Mössbauer spectroscopy, dilute $Fe(acac)_3$ in an organic solvent [4] reveals very similar spectra at low temperatures. Also the magnetic hyperfine splitting reported in [4] is in good agreement with our data. Reason for the apparent magnetic hyperfine splitting is the strong suppression of spin-spin relaxation between the relatively distant Fe^{III}. In this case the hyperfine patterns are determined by the coupling of the nuclear spin with the electronic ground state spin of high-spin $S = 5/2$ in a weak axial crystalline electric field. Whereas the $|S_z = \pm 5/2\rangle$ and $|S_z = \pm 3/2\rangle$ Kramers doublets lead to sextet patterns, the $|S_z = \pm 1/2\rangle$ doublet results in a more complex and broad structure since relaxational fluctuations within this electronic state are still finite even at low temperature (see, e.g., [5]). Therefore the magnetic hyperfine structure visible in Fig. 2 is not related to the formation of magnetic nanoparticles, but is characteristic of the frozen solution of $Fe(acac)_3$ and thus is also not in contradiction to our conclusions drawn from the UV/Vis spectra.

We want to note that the simplistic fit strategy used here (comprising a sextet, a broad background and a doublet) is inadequate for a proper description for the slow dynamic hyperfine patterns. It only shall be used for a rough estimate of the hyperfine parameters.

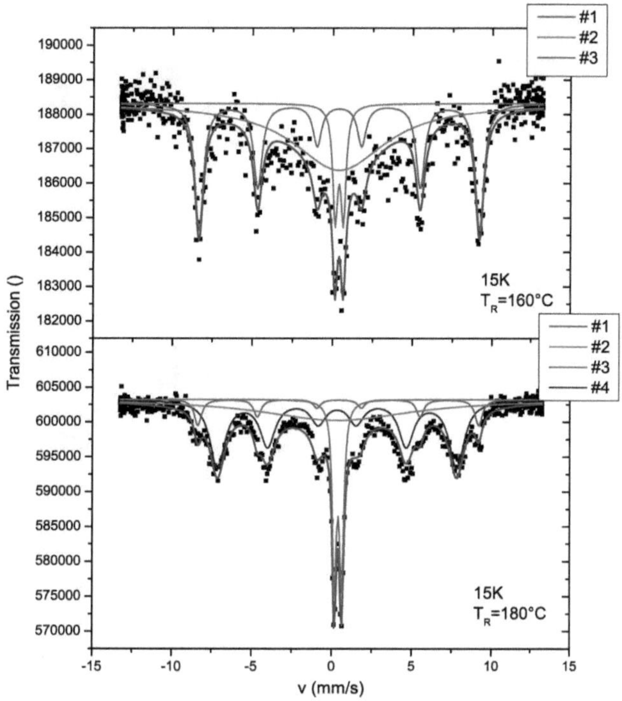

Fig. 3 Mössbauer spectra measured at 15 K for the solutions with $T_R = 160$ °C and $T_R = 180$ °C

A discussion regarding the origin of the sharp doublet patterns visible in Fig. 2 will follow below.

The Mössbauer spectra for samples extracted at $T_R = 160$ °C and for $T_R = 180$ °C and measured at 15 K are shown in Fig. 3, the corresponding parameters in Table 1. The spectrum for $T_R = 160$ °C exhibits a similar structure like the 20 K spectrum in Fig. 2 it can be fitted with the same strategy. The Mössbauer spectrum for $T_R =$ 180 °C was fitted including an additional sextet with a magnetic hyperfine field of 46.3 T. In accordance with the UV/Vis data, this sextet is interpreted as resulting from the emerging ferrite nanoparticles. The hyperfine field is smaller compared to the typical hyperfine field of nanoscaled ferrite particles (~52 T for maghemite [6]). This can be explained by an increased amount of spin canting within the particles due to the yet poor crystallinity, as may also be seen from the broad line width W.

The area of the doublet in the $T_R = 180$ °C spectrum is bigger than the area of the doublet in the spectrum for $T_R = 160$ °C. We attribute this doublet to an intermediate stage in the development of particles, where agglomerates, comprising several Fe-ions are formed. The distance between individual Fe^{III}-ions in these agglomerates becomes shorter than in the completely dilute solution. Therefore, paramagnetic relaxation is enhanced. For relaxation frequencies higher than the nuclear Larmor precession frequency, the magnetic hyperfine splitting collapses into a non-magnetic doublet pattern. This is the same doublet as the one observed in Fig. 2 and it indicates that already the start solution ($T_R =$ RT) contains a small amount of these aggregates. Figure 4 shows the temperature dependent spectra of a sample extracted after reaching the final temperature of 200 °C and staying at this

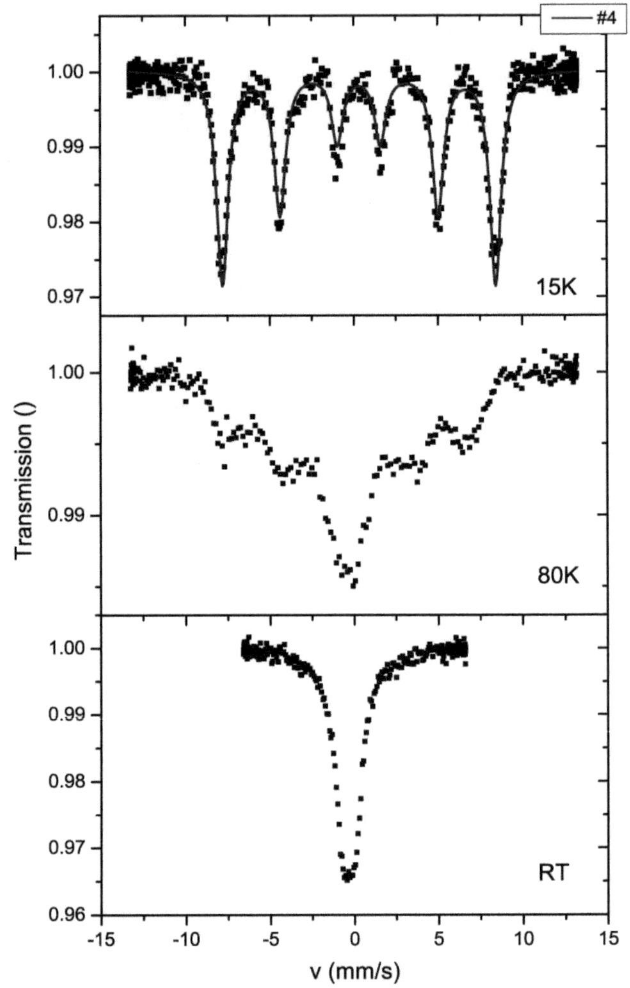

Fig. 4 Temperature dependent Mössbauer spectra of a sample extracted after staying at $T_R = 200\,^{\circ}\mathrm{C}$ for 30 min. The fit parameters for the measurement at 15 K can be found in Table 1

temperature for 30 min. The spectra are now typical for nanocrystalline maghemite. No relicts of the subspectra #1, 2 and 3 found at lower T_R can be traced. However, the crystallinity of the nanoparticles is still rather poor, as indicated by the magnetic hyperfine field with a value of ~50.4 T, which is smaller than the value for well-crystallized maghemite. The magnetic freezing temperature is about 80 K.

4 Conclusion

We investigated the formation of iron oxide nanoparticles during the initial synthesis stage by means of UV/Vis spectroscopy and Mössbauer spectroscopy. While the spectra of samples extracted at low reactor temperatures T_R up to 160 °C were typical for dilute $Fe(acac)_3$ in solution undergoing slow paramagnetic relaxation, the measurements at $T_R = 180\,^{\circ}\mathrm{C}$ gave clear evidence for emerging magnetic nanoparticles.

Prolonged reaction at 200 °C yields formation of maghemite particles, that still are not well crystallized as can be seen from the reduced value of the magnetic hyperfine field. The gradual growth and improvement of crystallinity of the nanoparticles and its dependence on solvents will be presented elsewhere.

References

1. Frey, N.A., Peng, S., Cheng, K., Sun, S.: Magnetic nanoparticles: synthesis, functionalization, and applications in bioimaging and magnetic energy storage. Chem. Soc. Rev. **38**, 2532–2542 (2009)
2. Grabs, I.-M., Bradtmöller, C., Menzel, D., Garnweitner, G.: Formation mechanism of iron oxide nanoparticles in different nonaqueous media. Cryst. Growth Des. **12**, 1469–1475 (2012)
3. Muhammad, A.: Lewis Acidity of Tris(acetylacetonate) iron (III) Interactions with some Phosphine Derivative in Methylene Chloride Solution. J. Chem. Soc. Pak. **14**, 88 (1992)
4. Wignall, J.W.G.: Mössbauer line broadening in trivalent iron compounds. J. Chem. Phys. **44**, 2462 (1966)
5. Wickman, H.H., Wertheim, G.K.: Spin relaxation in solids and aftereffects of nuclear transformations. In: Goldankii, V.I., Herber, R.H. (eds.) Chemical Applications of Mössbauer Spectroscopy, chapter 12. Academic Press, London (1968)
6. Serna, C.J., Bødker, F., Mørup, S., Morales, M.P., Sandiumenge, F., Veintemillas-Verdaguer, S.: Spin frustration in maghemite nanoparticles. Solid State Commun. **118**, 437–440 (2001)

DOI 10.1007/978-94-007-6482-8_8
Reprinted from *Hyperfine Interactions*, DOI 10.1007/s10751-013-0818-5

Implementation of a preamplifier-amplifier system for radiation detectors used in Mössbauer spectroscopy

A. A. Velásquez · M. Arroyave

Published online: 12 February 2013

Abstract We report the assembly and testing of a preamplification and amplification system for pulses produced by gaseous radiation detectors commonly used in Mössbauer spectroscopy. The system is composed by a pair of commercial integrated circuits A203 and A206, which operate as charge sensitive preamplifier-shaping amplifier and linear amplifier-low level discriminator, respectively. The integrated circuits were interconnected in the unipolar output mode and placed inside a metallic shielding, which prevents noise amplification for a suitable signal-noise ratio. The system was tested by irradiating a proportional counter LND-45431 with characteristic X rays of 6.3 keV and gamma rays of 14.4 keV emitted by a Mössbauer radioactive source of ^{57}Co (Rh). Unipolar pulses with Gaussian profile were obtained at the output of the linear amplifier, whose amplitudes were close to 0.4 V for 6.3 keV X rays and 1.4 V for 14.4 keV gamma rays. Pulse height spectra showed that the system allows a satisfactory identification of the X-rays and gamma rays emitted by the ^{57}Co source, giving the possibility to make a good selection of the 14.4 keV peak for having a suitable signal-noise ratio in the Mössbauer spectra. Absorption percentages of 14 % were found by taking the Mössbauer spectra of a natural iron absorber. The assembly and tests of the system are presented through this paper.

Keywords Mössbauer spectroscopy · Preamplifier and amplifier system · Radiation detectors

Thirteenth Latin American Conference on the Applications of the Mössbauer Effect, LACAME 2012, Medellín, Columbia, 11–16 November 2012.

A. A. Velásquez (✉) · M. Arroyave
Grupo de Electromagnetismo Aplicado, Universidad EAFIT, A. A 3300, Medellín, Colombia
e-mail: avelas26@eafit.edu.co

1 Introduction

Pulse amplifiers play an important role in the signal conditioning systems used in spectrometry experiments, such as radiation gaseous detectors and solid state detectors [1–3]. The main function of these electronic systems is to convert in electrical pulses the electrical charges collected in the electrodes of the detectors after an ionizing event. The physical phenomenon that gives rise to the generation of electric charge inside the detector depends on the type of detector, as well as the energy of the incident radiation, being the photoelectric effect the main phenomenon exploited in the gaseous detectors, the production of electron–hole pairs exploited in the semiconductor detectors and the emission of fluorescent radiation exploited in the scintillation detectors. For certain ranges of energy and operating voltage, characteristics of each detector, the total charge generated by the interaction between the radiation and the material of the detector is proportional to the energy carried by the incident radiation. In order to obtain the best advantage of this sensing property is necessary to select a convenient amplification system, with the lowest possible noise level.

In an effort to develop efficient and low cost systems to satisfy the particular needs of our Mössbauer laboratory, we made the assembly of a preamplifier-amplifier system based on a charge sensitive preamplifier A203 and a linear amplifier A206 supplied by Amptek Inc. The integrated circuit A203 was connected in the unipolar mode to the output terminal of a proportional counter LND-45431, looking for the best ratios of voltage-charge and signal-noise for the output pulses. The linear amplifier A206 was connected in cascade with the A203 in order to obtain additional amplification. The whole system allowed us to obtain pulses with Gaussian profile, whose amplitudes are in the scale of 1 V, which were suitable for the pulse discriminator and single channel analyzer modules of our Mössbauer spectrometer. The system has a modular structure, allowing easy monitoring of the signals. The details of the circuit and the testing of the system are presented through this paper.

2 Experiment

2.1 The detection system

The system used for the detection of ionizing radiation is a gaseous proportional counter LND-45431 developed by LND, INC. [4]. The counting gas of this detector is 97 % Kr and the quenching gas is 3 % CO_2, the internal pressure of the mixture is 800 Torr. The detector has an aluminum case and a beryllium window for the entry of the radiation. The high voltage applied to the electrodes of the detector was of 1.5 kV, which was supplied by a high voltage source Wissel, model HVS2 with range of 3 kV. The incidence of ionizing radiation over the detector generates, via photoelectric effect, primary electron-ion pairs, which are accelerated by the high voltage in opposite directions towards the electrodes of the detector. During their travel towards the anode, the primary electrons produce secondary ionizations in the filling gas, which amplify by a factor close to 10^5 the charge generated in the primary ionization.

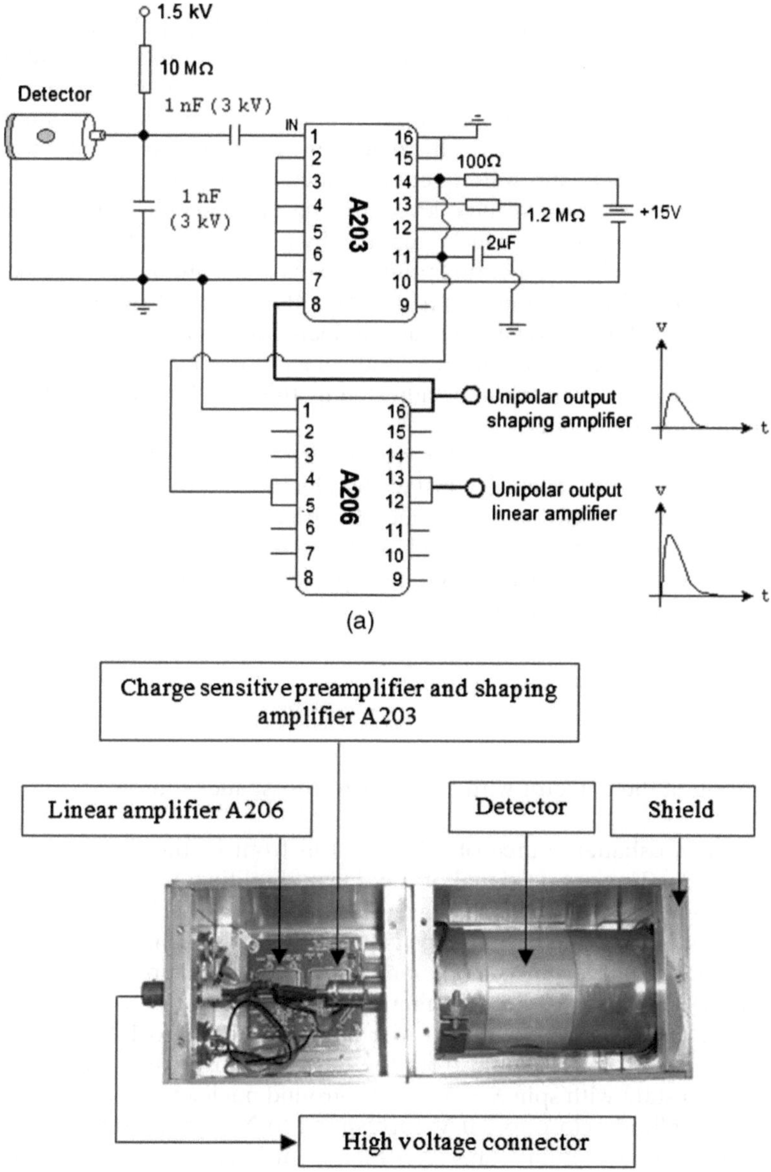

Fig. 1 **a** Scheme of connections between the detection system and the amplification system, **b** view of the real system with the aluminum case

2.2 The preamplifier-amplifier pulses system

The charge sensitive preamplifier present in the integrated circuit A203 [5] collects the electrons accumulated in the anode of the detector after an ionizing event, and converts each discharge in a voltage pulse with cusp shape through a capacitive circuit. The shaping stage of the A203 converts the cusp pulse in a pulse with Gaussian profile in order to reduce the harmonic content of the signal, allowing for an efficient propagation of the pulse through coaxial cables and prevent interference between consecutive pulses. The preamplifier also has other functions, among them to provide high input impedance for the detection system and low output impedance for driving coaxial cables and subsequent amplification stages.

In order to have suitable voltage levels for the pulse height analysis system, we use the linear amplifier A206, which amplifies by a factor of 10 the pulses obtained with the shaping amplifier stage. Integrated circuits A203 and A206 were connected in cascade in the printed circuit board PC-236, also supplied by Amptek [5]. The integrated A206 includes a lower level discriminator stage (LLD), which generates a logical pulse of 5 μs width when the amplitude of a pulse in its input terminal exceeds a threshold voltage V_T, which can be adjusted by the user. This last module was not used in our implementation.

To prevent electromagnetic interference coming from external sources, the whole system A203-A206 was enclosed inside an aluminum box with walls of 2 mm thickness. The aluminum is one of the metals with best screening properties against electromagnetic interference of far field [6]. The scheme of the connections between the detector and the amplification system is presented in Fig. 1a, and a view of the real system is presented in Fig. 1b.

3 Results and discussion

3.1 Irradiation of the detector with a ^{57}Co (Rh) Mössbauer source

We placed a Mössbauer source of ^{57}Co (Rh) in front of the detection system, as shown in Fig. 2a. The output signal of the linear amplifier A206 was observed with a digital storage oscilloscope Tektronix TDS 3052B-500 MHz. A series of pulses captured and stored by the oscilloscope is presented in Fig. 2b. Four kinds of pulses were clearly identified in this signal, namely: pulses with amplitude close to 0.4 V correspond to X-rays of 6.3 keV coming from the deexcitation of ^{57}Fe atoms after the electronic capture of ^{57}Co; pulses with amplitude close to 1.4 V correspond to gamma rays of 14.4 keV, coming from the deexcitation of the ^{57}Fe from the first nuclear excited state with spin $S = 3/2$ to the ground nuclear state with spin $S = 1/2$; pulses with amplitude close to 2.0 V correspond to X-rays of 20.5 keV due to the deexcitation of the Rhodium atoms after their excitation by gamma rays of 122 keV and 136 keV, coming from decaying of the ^{57}Fe from the nuclear excited state with spin $S = 5/2$ to the nuclear state with spin $S = 3/2$ and $S = 1/2$, respectively. The pulses with amplitude close to 0.2 V correspond to escape pulses of 1.8 keV, coming from electronic transitions in the Kr atoms after excitation by 14.4 keV gamma rays. Escape peaks have been reported in the pulses height spectra of Kr filled detectors irradiated by ^{57}Co sources [7].

The ratio between the energy of the photons and the amplitude of the pulses is not constant because the detection efficiency of the proportional counter LND-45431 is a nonlinear function of the energy of the incident radiation [4], being its value close to 90 % for photons of 6 keV and 14.5 keV, lower than 20 % for photons with energies below 2 keV and lower than 65 % for photons with energies above 20 keV.

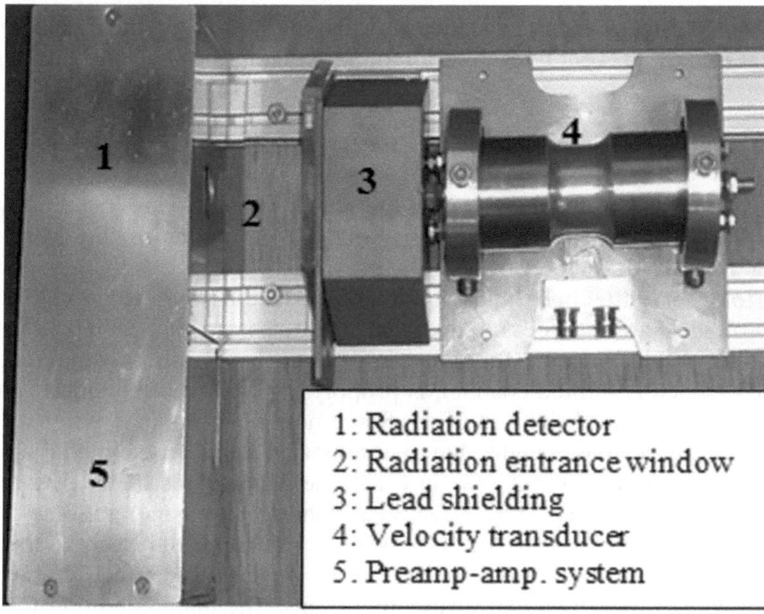

(a)

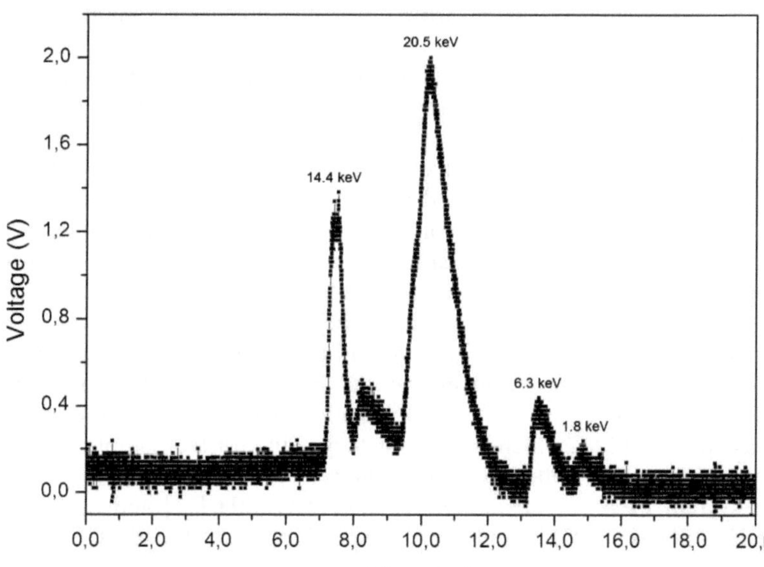

(b)

Fig. 2 **a** Experimental component mounting, **b** voltage signal at the output of the linear amplifier, registered by a digital storage oscilloscope

The ^{57}Co (Rh) source is threaded to the oscillation axis of a Doppler velocity modulator for Mössbauer effect, and can move within a lead shield to prevent user exposure to ionizing radiation it emits. The Doppler modulator was kept at rest during the experiment and the distance between the source and the detector window was put at 6 cm to prevent the saturation of the preamplifier by high counting rates.

3.2 Pulse height analysis and Mössbauer spectra acquisition

A pulse height spectrum of the pulses obtained with the amplification system is presented in Fig. 3a. The pulse analysis system is integrated to the Multichannel Analyzer (MCA) developed by us in a previous work [8, 9]. In the spectrum is possible to observe the three kinds of pulses described before, with exception of the escape peak, because this does not overcome the noise gap voltage of 0.25 V selected in the circuit of our pulse height analyzer. The memory position where each pulse is counted, which is termed channel, is proportional to the amplitude of the pulse measured by the pulse height analyzer. Some nonlinearities may occur during the measurement of the pulse height, due to nonlinear elements used in the pulse height analysis circuit, among them diodes, in which the dynamic resistance decreases with the amplitude of the signal. These deviations from linearity generate some distortion of the positions of the peaks in the pulse height spectrum with respect to the original amplitudes of the pulses observed in Fig. 2b. However, these effects do not play an important role in Mössbauer spectroscopy because the main condition for making a satisfactory selection of the energy window is to have a well resolved Mössbauer peak in the pulse height spectrum, from which the nuclear resonant events can be determined and counted univocally.

From a graphic interface developed in the software LabVIEW [10], the user can select the 14.4 keV peak by dragging a system of bars labeled LL (Lower Level) and UL (Upper Level), located in the plot where the pulse spectrum is presented. Before acquiring a Mössbauer spectrum, the user sends from the computer to the MCA the values in channels indicated by the bars, which are placed at the ends of the 14.4 keV peak approximately, both values are converted to respective voltage levels by the Single Channel Analyzer (SCA) of the spectrometer in order to fix the limits of pulse discriminator, the details of the SCA operation are described in [8].

After selecting the Mössbauer peak in the pulse height spectrum, a standard absorber of α-Fe with 28 μm thickness was placed between the source and the detector and the velocity transducer was put in motion, the Mössbauer spectrum obtained is presented in Fig. 3b. The absorption percentage obtained for 9×10^5 counts was close to 14 %, which is a satisfactory value for this sample.

Similar results can be obtained with commercial systems, such as the preamplifier-amplifier system PEA-6A developed by Wissel GmbH [11] and the preamplifier and amplifier systems 142PC and 575A, respectively, developed by ORTEC® [12]. However the system A203-A206 offers some advantages, such as no need to use bipolar voltage sources to supply power to the integrated circuits, only one voltage source with output between 10 V and 18 V is required; possibility of unipolar and bipolar outputs; the power consumption of the A203-A206 system is lower than 3 mA when operated at 15 V, being this current lower than several tens of mA required by commercial systems; the fall time of the pulses at the output of the preamplifier is 30 μs, being this time lower than the time of 100 μs characteristic of the commercial systems. This dynamic characteristic allows for an efficient counting of events, reducing the dead time of the counting system. Another important advantage of the linear amplifier A206 is the inclusion of a lower level discriminator (LLD) with

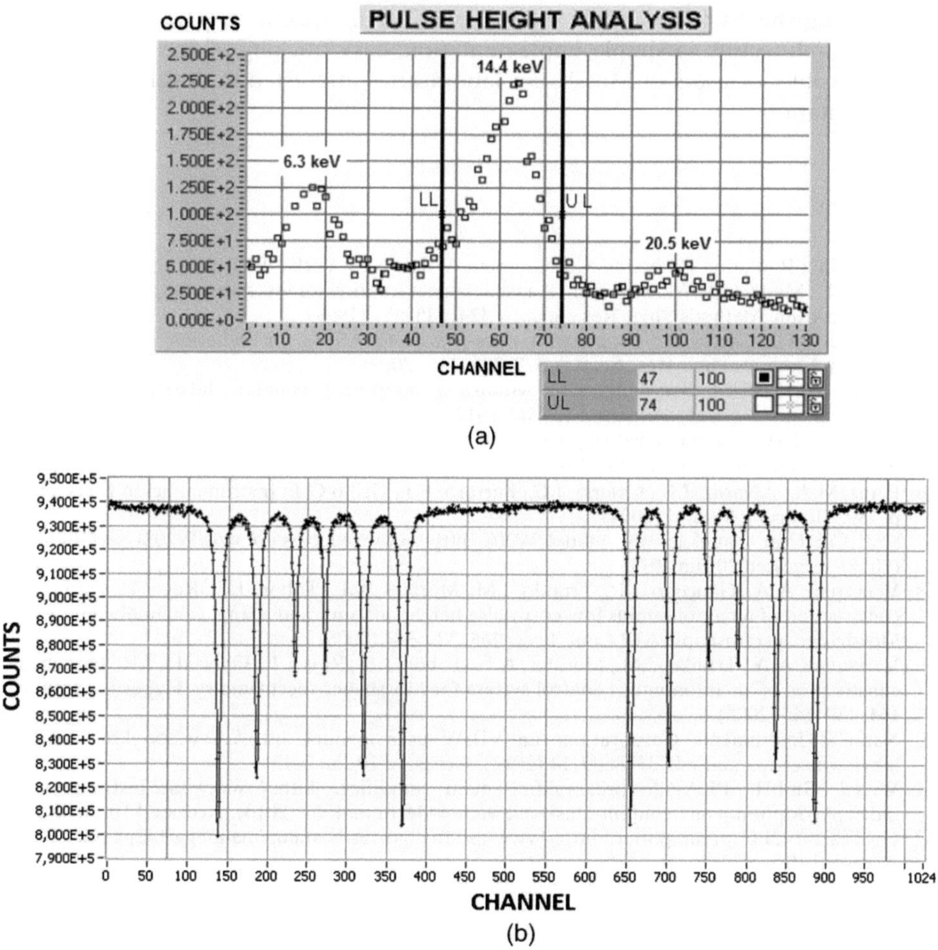

Fig. 3 **a** Pulse height spectrum of a ^{57}Co (Rh) source, **b** Mössbauer spectrum of a α-Fe absorber obtained with the implemented system

logical pulses at the output, which can be used as a high pass energy filter in the Single Channel Analyzer of the spectrometer.

4 Conclusions

From commercial integrated circuits, we have implemented a system for conditioning the electrical pulses produced by gaseous radiation detectors used in Mössbauer spectroscopy. The system allows to obtain a satisfactory signal-noise ratio, giving Gaussian profile pulses with amplitudes of the order of 1.4 V for the 14.4 keV Mössbauer gamma rays, which are suitable for the Pulse Height Analysis and Single Channel Analyzer modules of the Mössbauer spectrometer. A satisfactory absorption percentage, close to 14 %, was obtained for a natural iron absorber by selecting the 14.4 keV peak in the pulse height spectrum. The system has a modular architecture, a simple implementation, easy tracking of the signals and a cost accessible for users of Mössbauer laboratories interested to develop their own instrumentation.

References

1. Knoll, J.F.: Radiation Detection Measurement. Wiley, New York (1979)
2. Fabris, L., Madden, N.W., Yaver, H.: A fast, compact solution for low noise charge preamplifiers. Nucl. Instrum. Methods Phys. Res., Sect. A **424**, 545–551 (1999)
3. Bertuccio, G., Caccia, S.: Progress in ultra-low-noise ASICs for radiation detectors. Nucl. Instrum. Methods Phys. Res., Sect. A **579**, 243–246 (2007)
4. LND, INC.: 45431 beryllium side window proportional counter. http://www.lndinc.com /products/pdf/41/ (2012). Accessed 12 May 2012
5. AMPTEK INC.: Charge sensitive preamplifier/shaping amplifier A203. http://www.amptek.com/ products.html (2012). Accessed 12 Jul 2012
6. Pérez, M.A., Álvarez, J.C., Campo, J.C., Ferrero, F.J., Grillo G.J.: Instrumentación Electrónica, p. 789. Thomson, Madrid (2004)
7. SEE Co: User maual for the model W202. http://www.seeco.us/uploads/W202UserGuide.pdf (2011). Accessed 20 Jun 2011
8. Velásquez, A.A., Gancedo, J.R., Trujillo, J.M., Morales, A.L., Tobón, J.E., Reyes L.: Design and construction of an autonomous low-cost pulse height analyzer and a single channel analyzer for Mössbauer spectroscopy. AIP Conf. Proc. **765**, 389–384 (2005)
9. Velásquez, A.A., Trujillo, J.M., Morales, A.L., Tobón, J.E., Reyes, L., Gancedo, J.R.: Design and construction of an autonomous control system for Mössbauer spectrometry. Hyperfine Interact. **161**, 139–145 (2005)
10. National Instruments Corporation: LabVIEW user manual. http://www.ebookbrowse.com/ labview-user-manual-pdf-d318609414 (2013). Accessed 10 Jan 2013
11. Wissel GmbH: PEA-6A preamplifier+main amplifier. http://www.wissel-instruments.de/ index.php?option=com_content&task=view&id=44&Itemid=67 (2013). Accessed 10 Jan 2013
12. ORTEC: 142PC preamplifier. http://www-esd.fnal.gov/esd/catalog/main/egg/142pc.pdf (2013). Accessed 10 Jan 2013

DOI 10.1007/978-94-007-6482-8_9
Reprinted from *Hyperfine Interactions*, DOI 10.1007/s10751-013-0771-3

Self-tuning digital Mössbauer detection system

**A. Veiga · C. M. Grunfeld · G. A. Pasquevich ·
P. Mendoza Zélis · N. Martínez · F. H. Sánchez**

Published online: 29 January 2013

Abstract Long term gamma spectroscopy experiments involving single-channel analyzer equipment depend upon thermal stability of the detector and its associated high-voltage supply. Assuming constant discrimination levels, a drift in the detector gain impacts the output rate, producing an effect on the output spectrum. In some cases (e.g. single-energy resonant absorption experiments) data of interest can be completely lost. We present a digital self-adapting discrimination strategy that tracks emission line shifts using statistical measurements on a predefined region-of-interest of the spectrum. It is developed in the form of a synthesizable module that can be intercalated in the digital processing chain. It requires a moderate to small amount of digital resources and can be easily activated and deactivated.

Keywords Mössbauer · Instrumentation · Spectrometer · SCA · Constant-velocity

---

Thirteenth Latin American Conference on the Applications of the Mössbauer Effect, LACAME 2012, Medellín, Colombia, 11 16 November 2012.

A. Veiga (✉) · C. M. Grunfeld · G. A. Pasquevich · P. Mendoza Zélis
Facultad de Ingenieria, Universidad Nacional de La Plata, La Plata, Argentina
e-mail: veiga@fisica.unlp.edu.ar

A. Veiga · C. M. Grunfeld · G. A. Pasquevich · P. Mendoza Zélis · N. Martínez · F. H. Sánchez
Instituto de Física La Plata, CONICET, La Plata, Argentina

C. M. Grunfeld · G. A. Pasquevich · P. Mendoza Zélis · N. Martínez · F. H. Sánchez
Facultad de Cs. Exactas, Universidad Nacional de La Plata, La Plata, Argentina

N. Martínez
Comisión de Investigaciones Científicas de la Provincia de Buenos Aires,
Buenos Aires, Argentina

1 Introduction

A Mössbauer detection system is a classical single energy selector operating on low energy gamma events [1–3]. It is composed of four main elements: a proportional counter with high-voltage (HV) supply, a charge preamplifier, a shaping amplifier and a single channel analyzer (SCA), as shown in Fig. 1a. Long term stability of the system is an important factor, especially in high resolution experiments, being critical in constant-velocity applications [4–9]. In these cases transmission is recorded at a single energy for a long period of time. Under unstable conditions, an output rate variation cannot be distinguished to be the effect of the resonant process under study or a consequence of a shift in spectrometer operating point.

The primary source of long term instability is temperature dependence of both the HV-supply and the gaseous detector gain [10]. A temperature change displaces the emission line out of the SCA window center, changing the proportion between resonant and non-resonant photons. In precision single energy experiments either temperature control must be implemented or a reference background rate must be periodically recorded, with a consequent loss in efficiency.

Digitization of the detection system provides a powerful environment where a compensation algorithm for temperature drifts can be devised. In a regular digital arrangement, as shown in Fig. 1b, output of the shaping amplifier is sampled by an analog-to-digital converter (ADC) and sent to a digital processing module. Every detected event is digitized and translated into energy by a shaping algorithm. A digital output is generated if SCA selection criteria on energy is met (e.g. amplitude window). The selection criterion is provided by user after initial full spectrum inspection (MCA) and remains unaltered during the experiment. This setup is well addressed in bibliography [11–15] and several FPGA implementations have proved to be successful.

We will show how, with a moderate amount of additional digital resources, the processing module can be extended to automatically monitor the status of the emission spectrum, detect potential temperature drifts and adjust the selection criteria, in order to keep a constant background rate. The proposed tracking module, shown in Fig. 1c, can be enabled or disabled as required. When disabled, the fixed window criterion initially selected by the user can be applied.

In the next sections a window-based tracking strategy design is presented and the main implementation aspects of the proposed techniques are discussed. Experimental results illustrate the performance of the tracking module when perturbations are induced in the detection system.

2 Proposed tracking technique

As a first approximation, it can be assumed that the thermal drift in the detector will only displace the emission line center (14.4 keV in ^{57}Fe Mössbauer experiments), keeping its variance unaltered. In consequence, a fixed width, moving energy window can provide a simple way to implement an adaptive selection criterion. Then, the tracking problem is reduced to finding the proper displacement of the window, whose center will be coincident with the actual emission line center. Figure 2 shows the

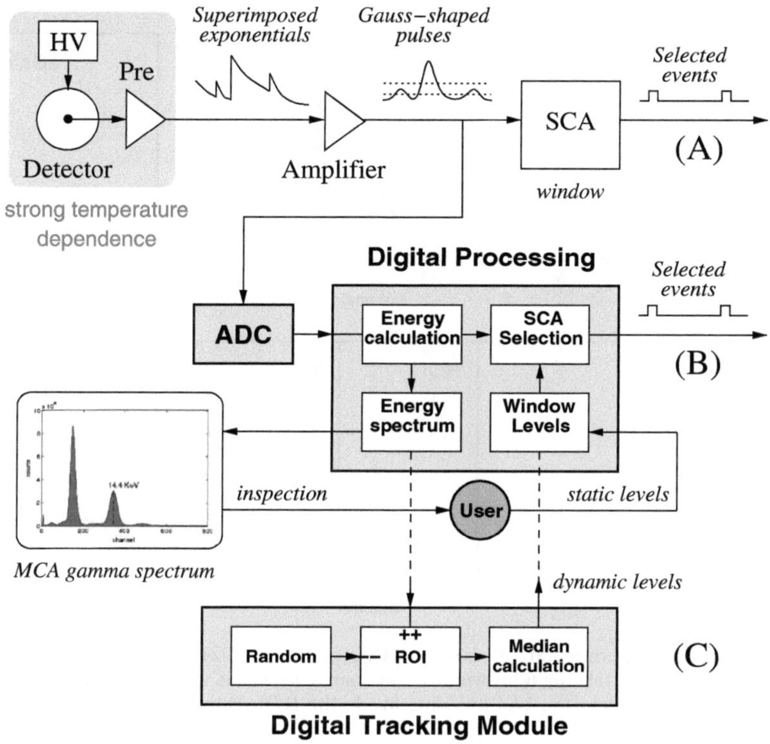

Fig. 1 Mössbauer detection system composed of proportional counter and preamplifier, followed by: **a** shaping amplifier and single channel analyzer (SCA) in analog classical NIM layout, or **b** analog-to-digital converter (ADC) and digital processing module in digitized version. User interaction for window selection is schematized. The proposed tracking module **c** can be designed to coexist with previous configuration

behavior of a section of ^{57}Fe source spectrum when a change in amplifier gain is forced.

We define the tracking window as the range of spectrum channels comprised between C–W and C+W, being C the window central channel and W the window half-width. C is designed to be initially set by user inspection of the whole spectrum and afterward continuously relocated by the tracking algorithm. W remains unaltered during the tracking process, being advisable to be set to the optimal value for the emission line [16, 17].

The region inside which the window can be moved is defined as the region of interest (ROI) of the digital processing module. This region (a fixed subset of channels larger than the window) can be selected upon experimental evidence. An empirical value of twice the window can be used, from channel C−2W to C+2W. The ROI spectrum is continuously updated by the digital processing module. An algorithm is applied to the window section in order to locate the emission line properties. Using this information, window levels are dynamically adjusted inside the ROI. The resulting center is used to compensate SCA discrimination levels.

For example, given a 12-bit digitizer (4096 levels) a 10-bit spectrum (1,024 channels) can be built and a 7-bit ROI (128 channels) can be selected. Inside it, a

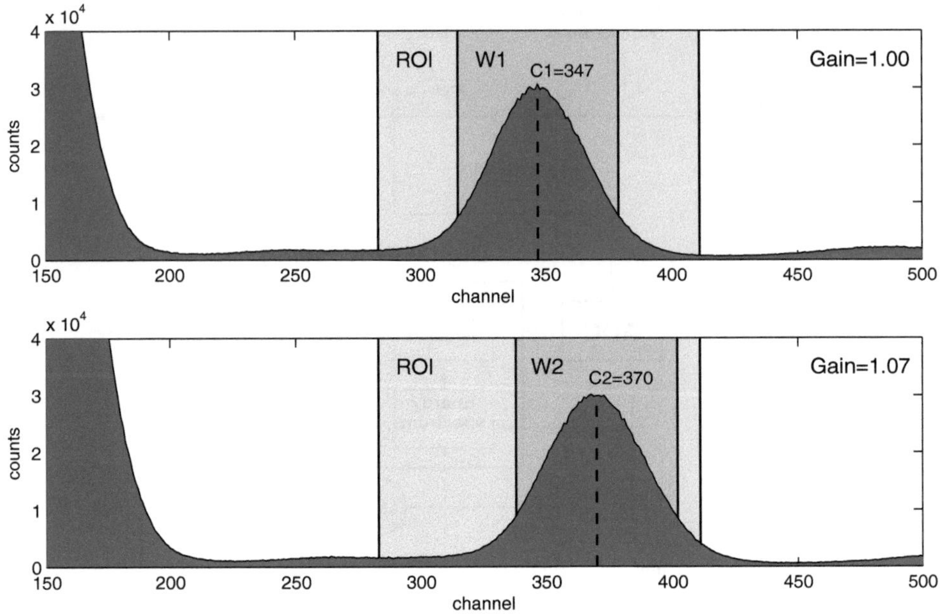

Fig. 2 Section of an experimental ^{57}Fe source MCA spectrum (1,024 channels), showing fixed 7-bit ROI (128 channels) and tracking 6-bit window (64 channels), when a 7 % increase in amplifier gain is forced. The window tracks the 14.4 keV line inside the ROI, from channel C1 = 347 to channel C2 = 370

6-bit window (64 channels) can shift. In Fig. 2 this window tracks the 14.4 keV line inside the static ROI from channel C1 to channel C2.

Under these circumstances, digital implementation of the tracking strategy presents two main challenges:

i) In order to keep updated information that can be aware of changes, ROI must be stored and continuously populated with the newly detected events, while the older ones are discarded.
ii) A new window center must be recalculated every time ROI is updated, using stored information and an algorithm capable of being implemented with medium complexity digital resources.

These two subjects are addressed in detail in the next two sections.

3 ROI storage and update

Given the characteristics of the low level trigger algorithm (Energy Calculation module in Fig. 1), the tracking strategy is designed on an event-based operation. Its input is the processed energy of the newly detected gamma event, provided by the low level trigger. This event is used to update ROI and to initiate window recalculation. In order to keep stored ROI information updated, two different strategies can be considered.

A simplistic approach would be to record the gamma spectrum for a fixed period of time (enough to reduce statistical uncertainty, collecting N events), use it to calculate the new window center, and reset the spectrum to start a new acquisition period. However, that may result in a slow tracking response because it takes N/rate seconds to get a new spectrum from scratch to calculate the next center.

A more appropriate solution would be to keep ROI updated with the last N detected events, discarding the oldest one every time a new event arrives. A FIFO architecture could be used. Implementation of this technique would result too 'expensive', as it would require the order number to be stored for each of the N events (e.g., for a 256 channel ROI that stores one million events, 1.5 % statistical error/channel, 1 MB of memory is required). Additionally, the algorithm to find the center would involve the processing of the full FIFO.

A more efficient stochastic method is proposed: every time a new event arrives, it is stored in the ROI and, immediately, a random event is discarded, keeping ROI area equal to N. In order to preserve the spectrum properties, the event to discard must be selected using the probability density function (PDF) of the current spectrum. That can be accomplished by removing a uniformly-distributed random event from the ordered spectrum. Digital implementation of pseudo-random number generators can be addressed using linear feedback shift registers (LFSR) of appropriate length [18].

It is important to note that, to preserve PDF spectrum, it would not be valid to decrease a random channel of the ROI. More exactly, a random event must be discarded, so most populated channels are more likely to be decremented, keeping the spectrum shape unaltered. That can be accomplished by generating an N-bit uniform random number (shifting N times a maximal-length sequence LFSR) [19] and using this number to select one of the N stored events. Is the channel that contains that event the one that must be decremented.

In order to provide storage space for an M-channel ROI, capable of collecting N events with increment/decrement functionality, an M array of $log2(N)$ up/down counters can be used (256 20-bit counters are required for the previous 1 MB memory example).

4 Tracking algorithm

Once the gamma spectrum is stored in ROI and continuously updated, the current emission center can be found by calculating the mean inside the current window limits (these limits were calculated using the previous center). Newly calculated center is used to reposition the window.

If a window of W channels is currently centered in channel C, next center channel C' can be calculated as

$$\text{index}_{C'} = \frac{\sum_{i=C-W}^{i=C+W} \text{count}_i \cdot i}{\sum_{i=C-W}^{i=C+W} \text{count}_i}$$

Being i the channel number and count_i the number of events stored in it.

The mean is calculated inside the window limits, not in the ROI. That is so because window converges to a narrow (mostly symmetrical) selection of the emission peak. Being much wider, ROI can include spectrum asymmetries that can distort mean calculation (RX lateral peaks, exponential background, etc.). If such asymmetries are not present, widening the window at both sides of the center can improve precision.

Implementing this technique involves an important amount of calculations, and therefore cost, power consumption and dead time. Instead of using the first order moment to find the new center, using the median in the active window can be more efficient since its calculation is simpler and results less dependent of spectrum skewness. Several algorithms are available to calculate the median of a dataset, most of them involving sorting [20]. That is very convenient, as in our case data is already sorted in the available spectrum.

The proposed tracking technique is not sensitive to the widening (second-order moment) of the dataset, as long as the algorithm is able to find a maximum inside the window. A variance estimator can be implemented on the window channels in order to detect a change, if necessary.

5 Dynamics and simulations

The responsiveness of the algorithm will be limited by the number of events stored in the ROI (N). Such amount is imposed by the required precision in window positioning. Rising it improves window median calculation (statistical uncertainty is reduced), improving the determination of the window center. However, given an event rate m, it will take N/m seconds to the algorithm to be fully aware of an abrupt change in rate. Limitations at this point will arise from detector thermal dynamics and HV stability.

Operating parameters can be tuned using simulations in a numerical computing environment (Octave/Matlab® scripts are available). Optimal operating point results from the compromise between responsiveness and precision. For example, considering the case of Fig. 2 (128-channel ROI with 64-channel window and 1 M stored events), operating at a rate of 10^4 events/s, it will take at least 100 sec to fully refresh the ROI.

6 Algorithm implementation

Tracking was tested using a development platform comprised of a 16-bit 250 MSPS analog-to-digital converter (AD9467 from Analog Devices) and a Virtex4-FX20 FPGA from Xilinx.

Both digital processing and tracking module (all blocks in Fig. 1) were implemented as an event-triggered finite state machine (FSM). USB and UART communication modules were included to download ADC traces, monitor tracking evolution, inspect spectrum and configure parameters at run time. There was enough room in the device to accommodate a 64-channel ROI of 16-bit counters, with a 32-channel window ($N = 64$ k, $M = 64$, $W = 16$). The whole system consumed

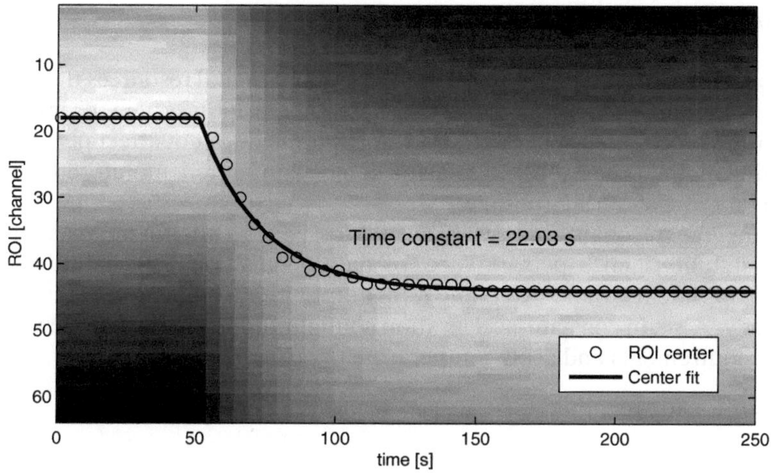

Fig. 3 Step response of the tracking algorithm, given an induced perturbation of the emission center of 26 channels. The ROI channels (background image) and center (o) were recorded using the communications module of the digital system. The fitted exponential (–) is an approximation to exponential response with a single time constant determined by the ROI area

approximately 90 % of available resources (7,756 of 8,544 slices). It was programmed using a behavioral high-level description language (source code available).

Considering implementation in a more recent Spartan6-LX45 (low-power, low-cost, more advanced FPGA device), the tracking module alone requires 1,707 of 6,822 available logic cells, representing only 25 % of device resources.

7 Experimental results

In order to characterize the system, its behavior was tested under different perturbations. Instruments used for the experiment were a proportional counter, a static ^{57}Fe source, an Ortec 142PC preamplifier and an Ortec 572A amplifier (Ortec 420A single channel analyzer was used as a reference). The digital selection system was set to operate at a rate of 3,000 events/s.

As an example a fast 14.4 keV emission line tracking is presented in Fig. 3, where amplifier gain is instantly changed from 55.94 to 56.10 (0.3 %). Information of the ROI and the calculated center time evolution were obtained from the digital module through its serial interface. Figure shows an exponential response when line center jumps from ROI channel 18 to 44 (26 channels, outside initial W) in $t = 50$ s.

Although the algorithm involves several nonlinearities, its step response can be approximated as a single exponential, evidencing that the ROI area is the dominant factor in the dynamics of the system. The time constant can be fitted to 22.03 sec, matching the value predicted from N/m(21.84 sec). This time constant can be used as an approximate limit of the operating point changes that can be compensated by the system.

8 Conclusions

The presented design involves a moderate amount of digital resources that can be appended as a module to an existing FPGA based acquisition system, or implemented as a separate block. Inclusion of the tracking module does not affect performance of the existing digital processing module and can be configured, monitored, enabled or disabled on request during runtime.

Tracking the resonant energy level in Mössbauer experiments can drastically improve stability, hence performance, of data acquisition, especially in constant-velocity experiments where SCA long term output rate is a function of the parameter under study. Also, in constant-acceleration configurations, this technique relaxes HV stability requirements and room temperature control.

References

1. Knoll, G.: Single-channel methods. In: Radiation Detection and Measurement, 3rd edn, pp. 685–690. Willey, New York (2000)
2. Ahmed, S.: Signal processing. In: Physics and Engineering of Radiation Detection, pp. 463–520. Elsevier, London (2007)
3. Leo, W.R.: Electronics for pulse signal processing. In: Techniques for Nuclear and Particle Physics Experiments, 2nd edn, pp. 277–301. Springer, Berlin (1994)
4. Veiga, A., Martínez, N., Mendoza Zélis, P., Pasquevich, G.A., Sánchez, F.H.: Advances in constant-velocity Mössbauer instrumentation. Hyperfine Interact. **167**, 905–909 (2006)
5. Mendoza Zélis, P., Pasquevich, G.A., Sánchez, F.H., Veiga, A., Martínez, N.: A new application of Mössbauer effect thermal scans: determination of the magnetic hyperfine field temperature dependence. Phys. Lett. **A298**, 55–59 (2002)
6. Pasquevich, G.A., Mendoza Zélis, P., Fernández van Raap, M.B., Sánchez, F.H.: Hyperfine field temperature dependence of Fe3Si from Mössbauer thermal scans. Physica **B354**, 369–372 (2004)
7. Mendoza Zélis, P., Pasquevich, G.A., Veiga, A., Fernández van Raap, M.B., Sánchez, F.H.: A quasi-continuous observation of the α-transition of Fe1+xS by Mössbauer line tracking. Hyperfine Interact. **195**, 161–165 (2010)
8. Pasquevich, G.A., Mendoza Zélis, P.P., Sánchez, F.H., Fernández van Raap, M.B., Veiga, A., Martínez, N.: Magnetic and thermal scans: a new Mössbauer effect approach. Hyperfine Interact. **167**, 839–844 (2006)
9. Veiga, A., Pasquevich, G.A., Mendoza Zélis, P., Sánchez, F.H., Fernández van Raap, M.B., Martínez, N.: Experimental design and methodology for a new Mössbauer scan experiment: absorption line tracking. Hyperfine Interact. **188**, 137–142 (2009)
10. Vanha-Honko, V.: The temperature dependence of the gas gain in sealed proportional counters. Nucl. Instrum. Methods **176**, 213–219 (1980)
11. Sampietro, M.: A digital system for "optimum" resolution in x-ray spectroscopy. Rev. Sci. Instrum. **66**, 975–982 (1995)
12. Angelo Geraci, A.: Adaptive digital spectroscopy in programmable logic. IEEE Trans. Nucl. Sci. **47**, 2765–2772 (2000)
13. Abbiati, R.: Self-configuring digital processor for on-line pulse analysis. IEEE Trans. Nucl. Sci. **51**, 826–830 (2004)
14. Pasquali, G.: A DSP equipped digitizer for online analysis of nuclear detector signals. Nucl. Instrum. Methods Phys. Res. A **570**, 126–132 (2007)
15. Pechousek, J.: Virtual instrumentation technique used in the nuclear digital signal processing system design: energy and time measurement tests. Nucl. Instrum. Methods Phys. Res. A **637**, 200–205 (2011)
16. Nagy, S.: SCA window optimization on Mössbauer spectroscopy. J. Radioanal. Nucl. Chem. **137**, 389–395 (1989)
17. Bravo, J.A.: Optimization criteria in Mössbauer spectroscopy. Hyperfine Interact. **148**, 253–261 (2003)

18. Scholefield, P.H.R.: Shift registers generating maximum-length sequences. Electron. Technol. **37**, 389–394 (1960)
19. Alfke, P.: Efficient shift registers, LFSR counters, and long pseudo-random sequence generators. Xilinx Application Note 052. http://www.xilinx.com/support/documentation/application_notes/xapp052.pdf (1996). Accessed 5 November 2012
20. Schönhage, A.: Finding the median. J. Comput. Syst. Sci. **13**, 184–199 (1976)

DOI 10.1007/978-94-007-6482-8_10
Reprinted from *Hyperfine Interactions*, DOI 10.1007/s10751-013-0868-8

Formation of nanostructured ω-Al_7Cu_2Fe crystalline phase by the ball milling technique

M. Z. Pinto · M. Pillaca · C. V. Landauro · J. Quispe-Marcatoma · C. Rojas-Ayala · V. A. Peña Rodriguez · E. Baggio-Saitovitch

Published online: 9 August 2013

Abstract Crystalline ω-Al_7Cu_2Fe bulk samples were prepared by arc furnace and then by means of milling, the average grain size of these samples is reduced to the nanometer scale. The structural and magnetic properties of the nanostructured ω-Al_7Cu_2Fe phase have been studied by X-ray diffraction employing Rietveld method, Mössbauer spectroscopy and vibrating sample magnetometry. The results indicate that the average grain size of the synthesized sample (ω-phase) rapidly decreases from 79 to 12 nm after 5 h of milling. Furthermore, the hyperfine parameters of the nanostructured samples are higher than the values for the bulk ω-phase. Magnetic measurements show a weak ferromagnetic behavior with $M_s = 0.46$ emu.g^{-1} for the bulk ω-phase. After the milling process this value increases to $M_s = 1.50$ emu.g^{-1} due to the formation and growth of a magnetic interstitial region after reducing the average grain size of the sample.

Keywords ω-Al_7Cu_2Fe phase · Nanostructured materials · X-ray diffraction · Mössbauer spectroscopy

1 Introduction

The discovery of quasicrystals in 1984 opened a new field of research for the study of aperiodic solids: systems with long range order but without periodicity [1]. The

Proceedings of the Thirteenth Latin American Conference on the Applications of the Mössbauer Effect, (LACAME 2012), Medellín, Columbia, 11–16 November 2012.

M. Z. Pinto (✉) · M. Pillaca · C. V. Landauro · J. Quispe-Marcatoma · C. Rojas-Ayala · V. A. Peña Rodriguez
Facultad de Ciencias Físicas, Universidad Nacional Mayor de San Marcos (UNMSM), Av. Venezuela Cdra. 34, P.O. Box 14-0149, Lima 14, Peru
e-mail: mpintov@unmsm.edu.pe

J. Quispe-Marcatoma · C. Rojas-Ayala · E. Baggio-Saitovitch
Centro Brasileiro de Pesquisas Físicas (CBPF)
Rua Dr. Xavier Sigaud 150, Urca, Rio de Janeiro, 22290-180, Brazil

unusual physical properties (e.g., high electrical resistance, low coefficient of friction, low surface energy [2], high hardness [3], good wear resistance [4]) of these systems is normally related to this new kind of order. Moreover, it is pointed out that also the chemical order is of crucial importance to define the properties of quasicrystals. Moreover, it is remarkable that in the Al-Cu-Fe phase diagram [5], quasicrystals occupy a narrow region normally surrounded by crystalline and amorphous phases. This is the case for the well studied icosahedral $Al_{64}Cu_{23}Fe_{13}$ quasicrystal (*i*-phase) [6–8] where a small variation of the composition means a new crystalline structure like the ω-Al_7Cu_2Fe phase (tetragonal unit cell, space group P4/mnc and lattice parameters a = 6.33 Å and c = 14.81 Å) [9]. The structure and some physical properties of the ω-phase have been studied [10]. However, the magnetic behavior of this system still been an open question. Furthermore, it is interesting to note that even though the ω-phase have a chemical composition close to the *i*-phase its properties are significantly different [11, 12]. Additionally, the nanostructuration of the samples gives us the possibility to improve their physical properties as function of their average grain size [13–15].

In this sense, the main aim of the present work is to investigate the structural and microstructural order of the nanostructured ω-Al_7Cu_2Fe phase as well as its magnetic properties. As mentioned above, the study of the crystalline ω-phase is important to understand more the effect of aperiodicity on the physical properties of the *i*-phase. The paper is organized as follow. The experimental details are described in the next section. Section 3 is devoted to the results and discussion. In Section 4 we summarize the main results of the present work.

2 Experimental details

Alloys in the composition Al_7Cu_2Fe (ω-phase) were prepared from a mixture of high-purity chemical elemental powders of Al (99.999+%), Cu (99.99+%) and Fe (99.9+%) in an arc furnace in argon atmosphere. Then, the samples were heat treated at 600° C in evacuated quartz tubes for up to 48 h. In order to obtain a nanostructured ω-phase these samples were milled for five hours in a high energy ball milling SPEX 8000 employing a 7:1 charge ratio. The milling was made under argon atmosphere. Ethanol was added during the milling process as a surfactant in order to reduce the adherence of the powder to the vial walls and grinding balls.

The phase purity of the samples was confirmed by powder x-ray diffraction (XRD) using Cu-Kα radiation ($\lambda = 1.5406$ Å) and, additionally, in order to determine the percentage of the phases present in the samples as well as the accurate microstructural parameters (such as cell parameters, grain size and lattice strain) of the ω-phase nanostructured, the XRD patterns were analyzed by Rietveld method using the Topas software V10.0 (Bruker AXS) [16]. The local structure around the Fe sites was analyzed by transmission Mössbauer spectroscopy (TMS) employing a 25 mCi $^{57}Co/Rh$ radioactive source. Mössbauer spectra, recorded at room temperature, were fitted using the NORMOS fitting program [17]. The isomer shift values are given relative to the α-Fe value, obtained at room temperature. The study of the magnetic properties of the samples was done using vibrating sample magnetometer (VSM) of Quantum Design, at room temperature, with a magnetic field in the range of ± 20 kOe.

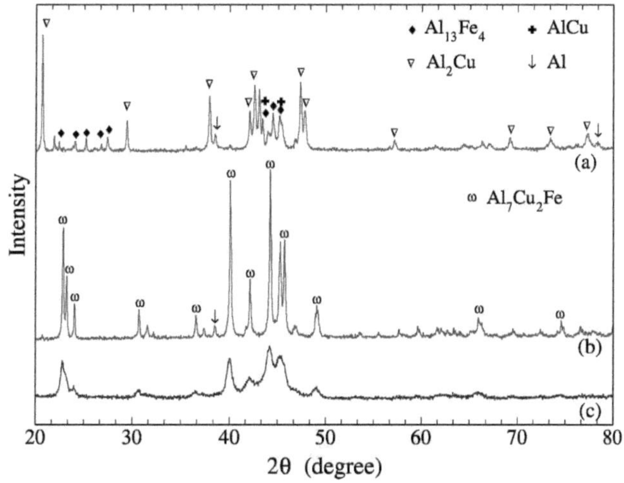

Fig. 1 (Color online) X-ray diffraction patterns of the ω-Al_7Cu_2Fe sample **a** as prepared by arc furnace, **b** after heat treatment, and **c** after 5 h milling.

3 Results and discussion

Figure 1 shows the powder XRD patterns of the Al_7Cu_2Fe sample (a) as prepared by a arc furnace, (b) after heat treatment, and subsequently (c) after 5 h milling. The X-ray diffraction peaks for the as-prepared sample show the presence of θ-Al_2Cu, λ-$Al_{13}Fe_4$, AlCu and fcc-Al phases with a weight percentage of 44.89 %, 42.51 %, 7.74 % and 4.86 %, respectively. The presence of these phases, mainly Al_2Cu and $Al_{13}Fe_4$, are usually associated to the formation process of the quasicrystalline phase [18]. In Fig. 1b, can be seen that after heat treatment the peaks corresponding to the θ, λ and AlCu phases disappear completely to form the ω-Al_7Cu_2Fe phase. This is result of the interdiffusion of the atoms (Al, Cu, Fe) of these phases which leads to the formation of a new type of structure, such as the intermetallic ω-phase [19]. In situ XRD measurements of Al-Cu-Fe thin films reported by Haidara et. al. are in agreement with this results [20]. The cell parameters corresponding to the ω-phase, obtained by Rietveld method, are a = 0.633(2) Å and c = 1.480(8) Å which is in good agreement with the literature [9]. Furthermore, the average grain size is ~ 79.2(17) nm while the lattice strain is approximately zero, 0.00(5). Additionally, in Fig. 1b can be identified traces of fcc-Al (3.80 %) which indicates that part of Al atoms still unreacted due to the larger concentration of the Al component [10]. In fact, measurements of energy dispersive X-ray spectroscopy (EDS) show the composition of $Al_{71}Cu_{19}Fe_{10}$, which is close to the ideal one.

After ball-milling a nanostructured ω-phase is formed (see Fig. 1c), which is indicated by the a reduction of the peak intensity and the broadening of the diffraction peaks. Therefore, from Rietveld refinement of the X ray diffraction pattern of the nanostructured sample, it is observed that the average grain-size reduces to 12.79(3) nm while the lattice strain increases reaching the value of 0.19(2). In the same way, a slightly increase of the cell parameters, a = 0.634(2) Å and c = 1.485(6) Å, it is observed as a result of the milling process. Moreover, in Fig. 1c can be observed that there is no evidence of the formation of another phases. It is worth mentioning that due to the lattice strain which is a result of the milling process, the

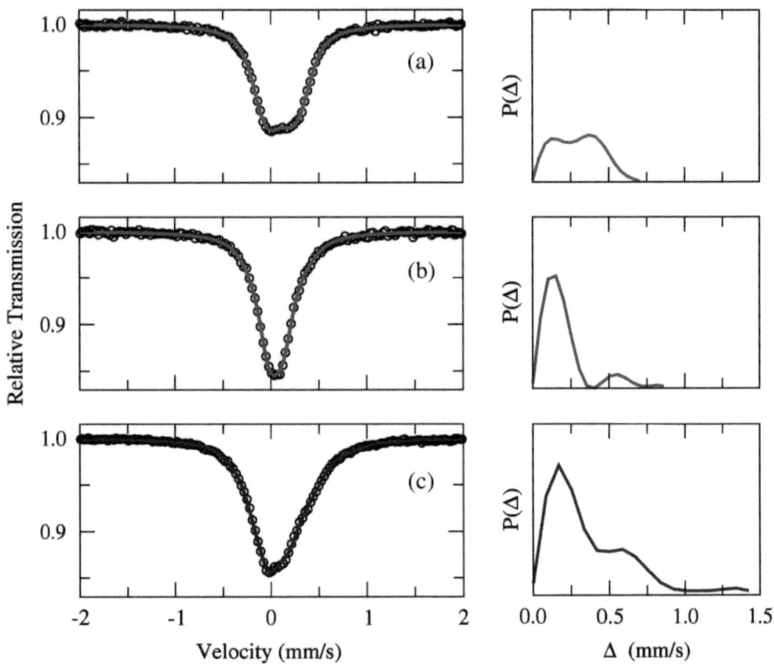

Fig. 2 (Color online) Room temperature Mössbauer spectra of the ω-Al_7Cu_2Fe sample **a** as prepared by arc furnace, **b** after heat treatment, and **c** after 5 h milling. The right panels correspond to the distributions $P(\Delta)$ for each case of the left.

Scherrer equation is not the best way to calculate the average grain size of the studied samples.

Figure 2 shows the Mössbauer spectra of the samples and the corresponding distribution of the quadrupole splitting, P(Δ). For the present fit, the full width at half maximum has been fixed at the value of $\Gamma = 0.248$ mm/s for all sites, similar as the case for the *i*-phase [8]. As can be seen, the Mössbauer spectrum in Fig. 2a corresponds to such expected for the crystalline $Al_{13}Fe_4$ phase with values of $\langle\delta\rangle = 0.222(1)$ mm/s, $\langle\Delta\rangle = 0.296(1)$ mm/s, and $\chi^2 = 0.98$, as reported in the literature [18]. Moreover, considering the crystallographic data, this system is characterized by two local arrangements for the Fe-sites: the first one with only Al neighbors and the second one with Al and Fe neighbors [18]. The local order around Fe in $Al_{13}Fe_4$ could be explained by two distribution curves between 0.13 and 0.37 mm/s, respectively.

On the other hand, Mössbauer spectra in Fig. 2b and c correspond to the bulk ω-phase and its nanostructured phase after five hours milling. The hyperfine parameters are $\langle\delta\rangle = 0.170(1)$ mm/s, $\langle\Delta\rangle = 0.198(1)$ mm/s, $\chi^2 = 1.02$ and $\langle\delta\rangle = 0.208(1)$ mm/s, $\langle\Delta\rangle = 0.370(1)$ mm/s, $\chi^2 = 1.66$, respectively. These values are very close to those reported in the literature [11, 21].

Furthermore, the shape of the quadrupole distribution of bulk ω-phase can be observed in the right panel of Fig. 2b. This distribution has a maximum value around $\Delta = 0.13$ mm/s which confirms that this phase has only one crystallographic site for the Fe sites [9]. However, due to the employed technique for the preparation of the samples, we considered a small distribution close to $\Delta = 0.55$ mm/s as a product

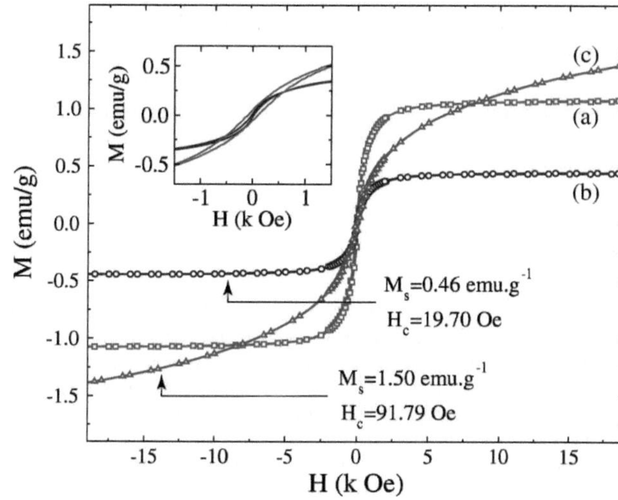

Fig. 3 (Color online) Variation of magnetization with applied field, M(H), for the ω-Al_7Cu_2Fe sample **a** as prepared by arc furnace, **b** after heat treatment, and **c** after 5 h milling. The inset shows the zoom view of the hysteresis loop near the origin.

of small disorder in the grain boundary. Thus, after five hours milling these two distributions are noticeably broadened which indicates the formation of new Fe-sites in the interstitial region due to the reduction of the grain size during the milling process.

Figure 3 shows the magnetization as a function of the magnetic field for the studied samples. The obtained saturation magnetization (M_s) and the coercive field (H_c), at room temperature, are $M_s = 1.10$ emu.g^{-1} and $H_c = 39.17$ Oe for the as-cast sample; $M_s = 0.46$ emu.g^{-1}, and $H_c = 19.70$ Oe for the heat-treated sample (bulk ω-phase); and $M_s = 1.50$ emu.g^{-1} and $H_c = 91.79$ Oe for the five hours milled sample (nanostructured phase). It is interesting to note that the magnetization values (M_s and H_c) of the nanostructured sample are higher than those obtained for the bulk ω-phase. These differences between the nanostructured and bulk samples are attributed to the formation and growth of a magnetic interstitial region after reducing the grain size of the sample. It is worth mentioning that this small magnetic contribution cannot be detected by Mössbauer spectroscopy (see Fig. 2). Similar results have been observed in the i-$Al_{64}Cu_{23}Fe_{13}$ quasicrystal [22].

4 Summary

In summary, we have shown that a nanostructured ω-Al_7Cu_2Fe phase can be obtained after milling its solid counterpart. In fact, after 5 h milling the ω-Al_7Cu_2Fe reduces its average grain size up to $\sim$ 12 nm. There are also evidences of a weak ferromagnetic behavior due to the formation and growth of a magnetic interstitial region after reducing the grain size of the sample. The next step is to find the dependence of the milling time with the grain-size in order to study the physical properties of these nanostructured ω-phase as a function of the grain-size. A systematic comparative study of the ω-Al_7Cu_2Fe and the corresponding i-$Al_{64}Cu_{23}Fe_{13}$ quasicrystal is fundamental to determine if the unusual properties of these new materials are mainly due to the aperiodic or chemical order, or both. This study is under way.

Acknowledgements M. Z. Pinto and M. Pillaca thank the "Zuñiga and Rivero Family Association" for financial support under its fellowship program for master studies in the National University of San Marcos.

References

1. Shechtman, D., Blech, I., Gratias, D., Cahn, J.W.: Phys. Rev. Lett. **53**, 1951–1953 (1984)
2. Dubois, J.M.: Mater. Sci. Eng. A **294–296**, 4–9 (2000)
3. Wolf, B., Bambauer, K.O., Paufler, P.: Mater. Sci. Eng. A **298**, 284–295 (2001)
4. Sordelet, D.J., Besser, M.F., Logsdon, J.L.: Mater. Sci. Eng. A **255**, 54–65 (1998)
5. Faudot, F., Quivy, A., Calvayrac, Y., Gratias, D., Harmelin, M.: Mater. Sci. Eng. A **133**, 383–387 (1991)
6. Tsai, A.P., Inoue, A., Masumoto, T.: Jpn. J. Appl. Phys. **26**, L1505–L1507 (1987)
7. Dolinšek, J., Vrtnik, S., Klanjšek, M., Jagličić, Z., Smontara, A., Smiljanić, I., Bilušić, A., Yokoyama, Y., Inoue, A., Landauro, C.V.: Phys. Rev. B **76**, 54201 (2007)
8. Quispe Marcatoma, J., Landauro, C.V., Taquire, M., Rojas Ayala, Ch., Yaro, M., Peña Rodriguez, V.A.: Hyperfine Interact. **195**, 105–109 (2010)
9. Bown, M.G., Brown, P.J.: Acta Crystallogr. **9**, 911–915 (1956)
10. Laplanche, G., Joulain, A., Bonneville, J., Gauthier-Brunet, V., Dubois, S.: Mater. Sci. Eng. A **527**, 4515–4518 (2010)
11. Srinivas, V., Dunlap, R.A.: Philos. Mag. B **64**, 475–484 (1991)
12. Chien, C.L., Lu, M.: Phys. Rev. B **45**, 12793–12796 (1992)
13. Koch, C.C.: Annu. Rev. Mater. Sci. **19**, 121–143 (1989)
14. Basset, D., Matteazzi, P., Mani F.: Mater. Sci. Eng. A **168**, 149–152 (1993)
15. Barua, P., Murty, B.S, Mathur, B.K., Srinivas, V.: J. Appl. Phys. **91**, 5353–5359 (2002)
16. BrukerAXS, DIFFRACPlus TOPAS: TOPAS 4.2 User Manual, Bruker-AXS GmbH,. Karlsruhe, Germany (2008)
17. Brand, R.A., Normos Mössbauer Fitting Program, User's Guide (Wissenschaftlich Elektronik GmbH, Starnberg) (1995)
18. Chittaranjan, C.M. Kumar, V., Viswanathan, B., Gopinathan, K.P.: Solid State Commun. **79**, 69–73 (1991)
19. Yong, X., Chang, I.T., Jones, I.P.: J. Alloy. Compd. **387**, 128–133 (2005)
20. Haidara, F., Duployer, B., Mangelinck, D., Record, M.-C.: J. Alloy. Compd. **534**, 47–51 (2012)
21. Kataoka, N., Tsai, A.P., Inoue, A., Masumoto, T., Nakamura, Y.: Jpn. J. Appl. Phys. **27**, L1125–L1127 (1988)
22. Pillaca, M.: Estudio de los efectos de la nanostructuración sobre las propiedades estructurales y magnéticas del cuasicristal icosaedrico *i*-$Al_{64}Cu_{23}Fe_{13}$. Master-Thesis, National University of San Marcos, Lima, Peru (2013)

DOI 10.1007/978-94-007-6482-8_11
Reprinted from *Hyperfine Interactions*, DOI 10.1007/s10751-013-0864-z

Mössbauer study of intermediate superparamagnetic relaxation of maghemite (γ-Fe_2O_3) nanoparticles

J. A. Ramos Guivar · A. Bustamante · J. Flores · M. Mejía Santillan · A. M. Osorio · A. I. Martínez · L. De Los Santos Valladares · C. H. W. Barnes

Published online: 23 May 2013

Abstract In the present work, we have synthesized and characterized magnetic nanoparticles of maghemite γ-Fe_2O_3 to study their structural and magnetic properties. For the preparation, magnetite precursor, were oxidized by adjusting the pH = 3.5 at about 80 °C in an acid medium, The mean size of the maghemite particles calculated from the X-ray diffractogram was around 5.7 nm. Mössbauer spectroscopy measurements at room temperature show their superparamagnetic behavior. Furhermore, Mössbauer measurements were carried out at 77 K and 4.2 K in order to find the typical hyperfine fields of the maghemite. Magnetite phase was not found. FC and ZFC magnetization curves measured at 500 Oe indicate a blocking temperature of 105.3 K. The magnetization measurements also show almost zero coercivity at RT.

Proceedings of the 13th Latin American Conference on the Applications of the Mössbauer Effect, (LACAME 2012), Medellín, Colombia, 11–16 November 2012.

J. A. Ramos Guivar (✉) · A. Bustamante · J. Flores
Laboratorio de Cerámicos y Nanomateriales, Facultad de Ciencias Físicas, Universidad Nacional Mayor de San Marcos, Ap. Postal 14-0149, Lima 14, Perú
e-mail: aguivar@gmail.com

M. Mejía Santillan
Laboratorio de Suelos, Universidad Nacional Mayor de San Marcos, Ap. Postal 14-0149, Lima 14, Perú

A. M. Osorio
Departamento de Química-Inorgánica, Universidad Nacional Mayor de San Marcos, Ap. Postal 14-0149, Lima 14, Perú

A. I. Martínez
Centro de Investigacón y Estudios Avanzados del Instituto Politécnico Nacional, Cinvestav, Unidad Saltillo, México

L. De Los Santos Valladares · C. H. W. Barnes
Cavendish Laboratory, Department of Physics, University of Cambridge, J. J. Thomson Av., Cambridge CB3 0HE, UK

TEM images show nanoparticles with diameter smaller than 10 nm, which are in good agreement with the X-ray pattern and the fitting of the magnetization data.

Keywords Nanomaghemite · Superparamagnetism · Blocking temperature · Mössbauer spectroscopy · ZFC · FC

1 Introduction

Currently magnetic materials at the nanoscale, such as γ-Fe_2O_3, have captured the attention of many researchers due to its important usefulness in different areas of the modern nanotechnology. Some applications are for instance, in ferrofluids, high density magnetic storage, high-frequency electronics, high performance permanent magnets, magnetic refrigerants, magnetic resonance imaging (MRI) contrast enhancement [1], biomagnetic separations, and hyperthermia treatment or magnetic drug targeting [2]. Moreover, maghemite nanoparticles are considered a significant field of investigation in medicine, since their surface can be modified (funcionalization), and they are easily controlled by an external magnetic field, so they can be driven inside blood vessels for cancer therapy [3]. In this work, we report the synthesis of γ-Fe_2O_3 nanoparticles less than 10 nm in size. Morphology and microstructure are discussed on the basis of TEM measurements, X-ray diffraction, and vibrational analysis. Besides, Magnetic studies are discussed with help of magnetic measurements and Mössbauer spectroscopy.

2 Experimental

Homogeneous maghemite nanoparticles were obtained by following the next reactions:

$$2\,Fe^{3+}_{(ac)} + Fe^{2+}_{(ac)} + 8\,OH^{-}_{(ac)} \rightarrow Fe_3O_{4(\downarrow)} + 4\,H_2O_{(\ell)}$$
$$2\,Fe_3O_{4(pp)} + {}^{1}\!/_{2}O_{2(g)} \xrightarrow[H_2O]{[H^+]} 3\,\gamma - Fe_2O_{3(\downarrow)}$$

The synthesis of the magnetite (Fe_3O_4) nanoparticles was performed in aqueous solution without any surfactants. A stoichiometric mixture of Fe^{3+}/Fe^{2+} in relation 2:1 from ferric chloride and ferrous chloride in a basic medium of pH = 11–12 was prepared. In addition, 0.85 mL of 12.1 N HCl and 25 mL of purified, deoxygenated water was added. Besides, 5.2 g of $FeCl_3$ and 2.0 g of $FeCl_2$ were dissolved in the solution with stirring. The resulting solution was dropped wise in 250 mL of 1.5 M NaOH solution under vigorous stirring. The last step generated a black precipitate (pH = 12). The paramagneticity was checked in situ by placing a magnet near the black precipitate of the Fe_3O_4. The precipitate was isolated in the magnetic field, and the supernatant was removed from the precipitate by decantation.

Purified deoxygenated water was added to the precipitate and the solution decanted after centrifugation at 4,000 rpm. After repeating the last procedures three times, 500 mL of 0.01 M HCl solution was added to the precipitate with stirring in order to neutralize the anionic charges on the nanoparticles. Immediately, the oxidation of Fe_3O_4 into γ-Fe_2O_3 was carried out by adjusting the pH of Fe_3O_4 to a pH of about 3.5 with hydrochloric acid for the moderate oxidation at about 80 °C for 30 min and with magnetic stirring with the objective to get the phase transition from magnetite to maghemite.

The sample was dried at about 40 °C with the aim of getting a reddish-brown powder [4].

The characterization was done by X-Ray Diffraction (XRD) in a Bruker D8 equipment and with the radiation $K_{\alpha 1}$ of copper ($\lambda = 1.54056$ Å) in the range from 8 to 80°, with a step of 0.02°, at room temperature (RT). The Mössbauer spectroscopy (MS) was performed with a conventional transmission Mössbauer spectrometer, operating with 1,024 channels and a Wissel INC. velocity module with a sinusoidal signal. The measurements were taken at room temperature (RT) and the obtained data were adjusted with the help of the program NORMOS [5]. In this program, the good fitting is controlled by the value of the chi^2. The source employed was a ^{57}Co in rhodium matrix with strength 25 mCi. The Isomer shifts and the velocity scale were calibrated with respect to a α-Fe film at RT. The sample holder used has a diameter 1 cm and 26 mg mass which permitted to ascertain and quantify the small systematic effects of cosine smearing which usually occur in the folded Mössbauer spectra when relatively large collection solid angles are used. The data was collected at 77 K and 4.2 K the Mössbauer measurements at 4.2 K were taken in a bath cryostat in which both, the source and the absorber, are at 4.2 K. So there is virtually no SOD compared to measurements at room temperature.

Measurements of the magnetic moment vs. temperature were carried out in a DC MPMS-SQUID magnetometer in ZFC and FC modes, in a range of temperatures of 10 K to 290 K, with a magnetic field of 500 Oe. The magnetic hysteresis measurements were taken in a range of −60, 000 Oe to 60, 000 Oe at the temperatures of 10 K and 290 K, respectively. For TEM measurements, we have used the Titan 80–300 kV microscope equipped with a special high-brightness Schottky-field emission electron source. The Raman spectrum was obtained using the Horiba LABRAM-HR equipment. The laser 532 nm Nd-Yag was used for excitation; it was calibrated with a line of Si in 521 cm^{-1}, and with a power on the sample of 0.5 mW.

3 Results and discussions

We have used Crystallographic Software to observe the mean peaks of diffraction, and we have indexed them with the PDF file 39–1346 with cubic symmetry (crystal space group $P4_132$), the average size of the nanocrystallites of $\gamma-Fe_2O_3$ was estimated from the wide of the diffraction peaks (Fig. 1) using Debye-Sherrer approximation: $d = \frac{k\lambda}{\beta \cos\theta}$, where λ is the wavelength ($\lambda = 1.54056$ Å) and θ is the Bragg angle. Assuming a spherical particle shape the particle diameter d is obtained by setting $K = 0.916$ when β is the full width at half maximum line breadth [6].

For the crystallographic plane (hkl) of the peaks in the XRD: we got grain sizes for (110)–4.2 nm, (220)–5.6 nm, (311)–5.88 nm, (400)–5.64 nm, (422)–6.41 nm, (511)–7.42 nm, (440)–6.71 nm and for (533) was 3.8 nm. The mean value of the grain size gives us a diameter of 5.7 nm.

The X-ray powder diffraction patterns show the presence of only one phase and widened peaks due to the nanometric size of the little crystallites.

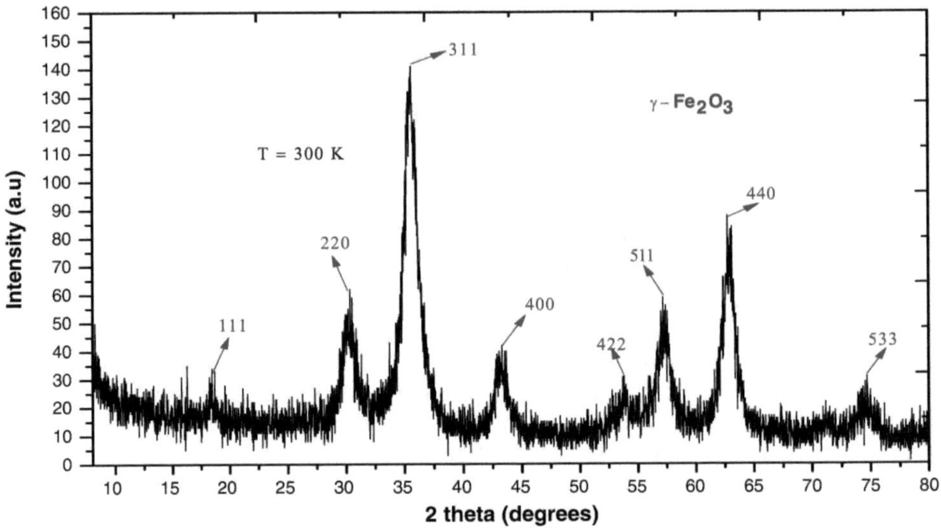

Fig. 1 X-ray powder diffraction pattern of maghemite nanoparticles

Recording that maghemite is a fully oxidized form of magnetite, has the structural formula $\left(Fe^{3+}\right)_A\left[Fe^{3+}_{5/3\,1/3}\right]_B$ O_4 where represents again the vacancies on the octahedral sites. The corresponding Mössbauer spectrum consists of a somewhat broad-lined Fe^{3+} sextet, which is in fact composed of two non-resolved sextets from Fe^{3+} in tetrahedral and octahedral sites, respectively [7].

Figure 2 shows the MS at RT and 77 K of the nanometric γ-Fe_2O (5.7 nm) sample and it reveals the intermediate superparamagnetic relaxation effect. At RT, the fitting was made using one distribution of hyperfine fields and one doublet representing the superparamagnetic particles. At 77 K, the spectrum was adjusted by using one distribution of the hyperfine fields of Fe^{3+}. It shows five peaks whose centers were calculated through five Gaussian curves obtaining 51.3, 48.6, 43.2, 23.4 and 17.1 T, respectively (see the right panel). In the Mössbauer spectra the blocking temperature (T_B) can be obtained when the area of the distribution is 50 %, of the superparamagnetic area [8, 9]. Thus, at 77 K the total area of the two principal peaks give 71.4 % assumed to maghemite without relaxation, and the remaining peaks give 28.6 %; then the T_B is nearly 100 K. This result is confirmed by the ZFC and FC measurement of the magnetization below. Since the initial sample consisted on magnetite and it was oxidized to obtain the maghemite, measurement at liquid helium were necessary to check if the entire sample was full oxidized. The Mössbauer spectrum of the sample obtained at 4.2 K is shown in Fig. 3. The spectrum is fitted with two magnetic fields of the characteristic maghemite, ie there was no magnetite component. Table 1 shows the Mössbauer hyperfine parameters obtained at RT, 77 K and 4.2 K.

Magnetic hysteresis loops with applied magnetic fields up to 60 kOe and at the temperatures of 10 K and 290 K (Fig. 4) reveal that the sample has superparamagnetic behavior, which is typical from nanosized magnetic particles. At nearly room temperature (290 K) which is well above the blocking temperature (T_B) the loop presents almost zero coercivity, which is caused by single magnetic domains. This

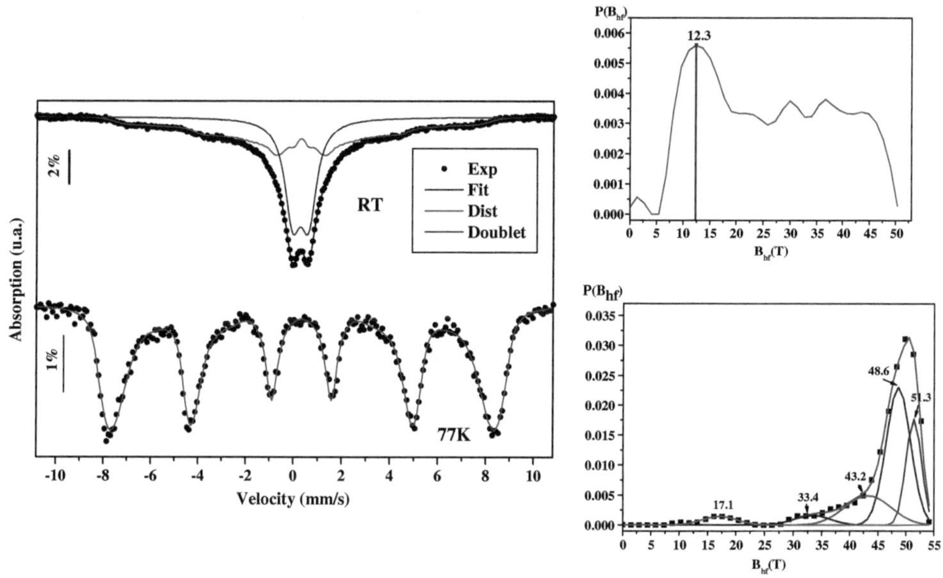

Fig. 2 Mössbauer spectra of maghemite nanoparticles at RT and 77 K. The hyperfine magnetic field distribution are on the right panel

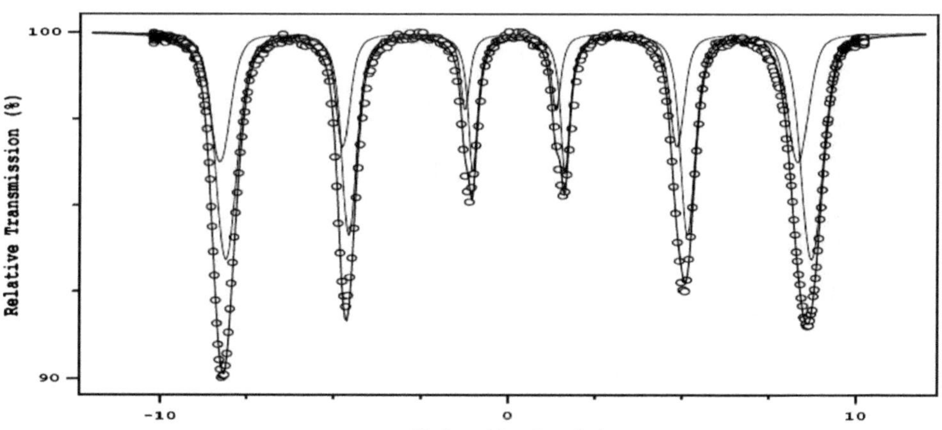

Fig. 3 MS at 4.2 K showing typical two magnetic fields of maghemite

kind of magnetic behavior means the loss of coercitivity and remanence. The coercive field found at 10 K was 229 Oe indicating that the blocking temperature is much higher than this temperature.

The data obtained was fitted using the Langevin equation, $M(H) = M_S\left[\coth\left(\frac{M(R)\mu_0 H}{K_B T}\right) - K_B T/M(R)\,\mu_0 H\right] + \lambda\mu_0 H$, where $k_B = 1.38 \times 10^{-23}$ J/K, the measurement temperature $T = 290$ K. The integral nanoparticle moment is $M(R) = M_s V_{mag}$ where V_{mag} corresponds to the volume of the nanoparticle and M_s is the saturation magnetization, λ the excess susceptibility including a diamagnetic

Table 1 Mössbauer hyperfine parameters obtained at RT, 77 K and 4.2 K

Sites	Temp. (K)	δ (mm/s) ± 0.03	ε, Δ (mm/s) ± 0.02	B_{hf}(Tesla)(+)	Area (%) ± 3
Fe^{3+} distribution	RT	0.403	0.00	12.3	54.8
SP Doublet		0.397	0.71(*)		45.2
Fe^{3+} Tetra	77	0.447	−0.002	48.6	48.1
(peak 1)					
Fe^{3+}-Fe^{2+} Oct		0.447	−0.002	51.3	23.3
(peak 2)		0.477	−0.002	43.2	19.3
Peak 3		0.477	−0.002	33.4	5.0
Peak 4		0.447	−0.002	17.1	4.3
Peak 5					
Fe^{3+} (Tetra)	4.2	0.136	−0.014	51.4	35.4
Fe^{3+}-Fe^{2+} (Oct)		0.406	0.006	52.1	64.6

(*) Representing the superparamagnetic doublet
δ is the isomer shift relative to α-iron, ε and Δ are the quadrupole splitting and A is the spectral fraction as obtained from the fit
(+) B_{hf} is the most probable hyperfine field. *SP* Superparamagnetic

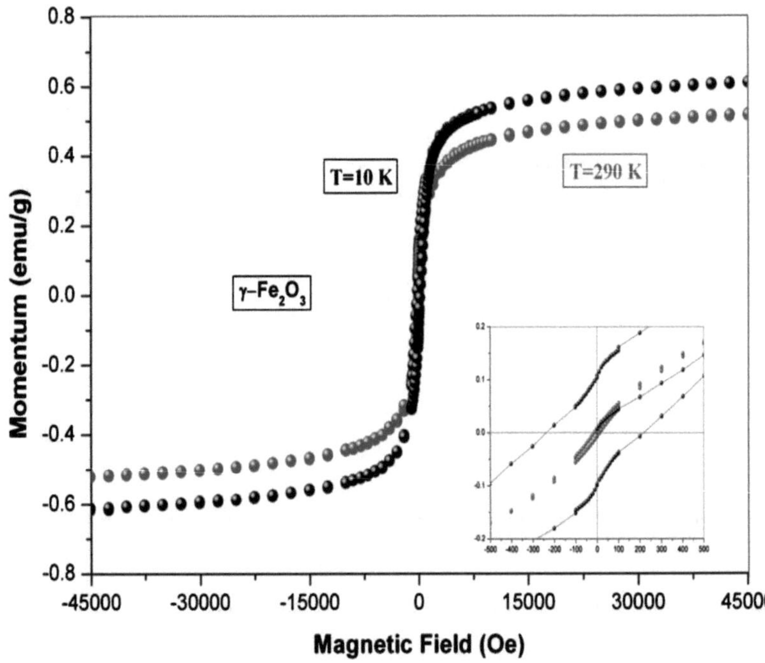

Fig. 4 Hysteresis loops of maghemite nanocrystallites with at 10 K and 290 K

contribution and μ_0H the applied field with the permeability of the free space $\mu_0 = 4\pi \times 10^{-7}$ T-m/A [10]. From Fig. 5, the saturation magnetization is M_s=0.035 A-m^{-1}. Three parameters with initial values of: P1 = 0.035, P2 = 22.87, P3 = 0.0015 were used reducing the Langevin equation into $y = P1^* \left(1/\tanh(P2^*x) - 1/(P2^*x)\right) + P3^*x$, where P1 = M_s, P2 = M_sV_{msg}/K_BT, λ = P3 and $\chi = \mu_0$H. Thus, the final

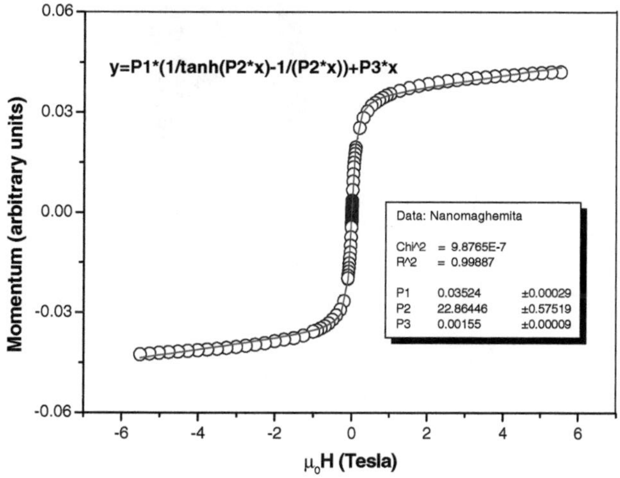

Fig. 5 Fit of the hysteresis loop at RT of maghemite nanoparticles with and applied field of 6 T at 290 K

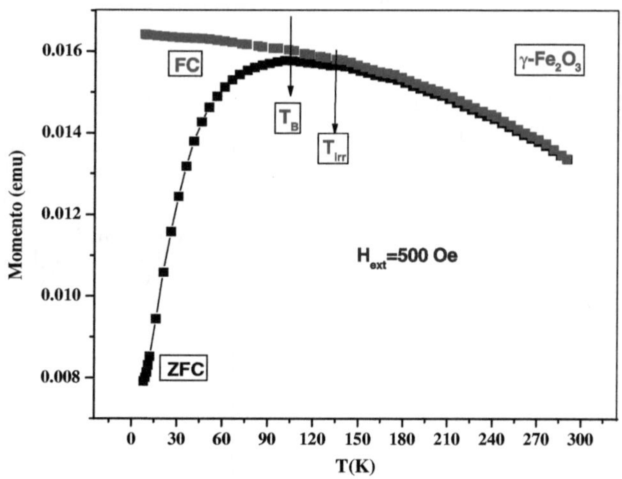

Fig. 6 Temperature dependence of the magnetic moment in the ZFC and FC cases with applied field $H = 500$ Oe for particles of 5.7 nm

values of the fitting parameters are: P1 = 0.03524 ± 0.00029, P2 = 22.86446 ± 0.57519, P3 = 0.00155 ± 0.00009. The obtained V_{mag} is 261 nm^3. Following the procedure of reference [11], assuming that all the particles are spherical without mutual interactions and taking into account that $V_{mag} = \pi D^3/6$, then the magnetic diameter is $D = 7.93$ nm which is closer to the size of the particle.

Figure 6 shows the zero-field-cooled (ZFC) and field cooled (FC) curves obtained by applying a magnetic field of 500 Oe showing a typical broad maximum in the ZFC curve of superparamagnetic particles. The blocking temperature (T_B) is defined as the temperature at which the ZFC curve exhibits a cusp, in our case T_B is 105.3 K [12]. Similar calculations of the blocking temperature are given for magnetic nanoparticles of maghemite in [13]. The FC curve shows irreversible behavior with respect to the ZFC curve until the temperature $T_{irr} = 135$ K. When the T_{irr} is larger than T_m suggests the occurrence of moderate particle-particle interaction [14].

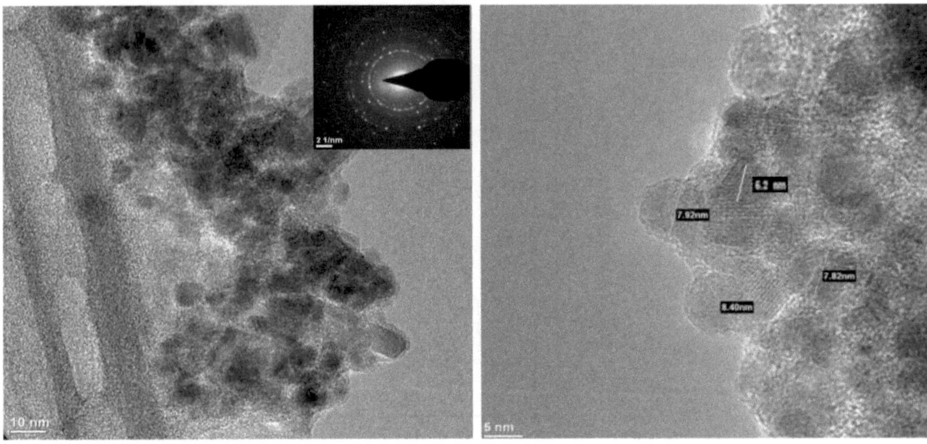

(a) (b)

Fig. 7 **a** TEM image of γ-Fe_2O_3 nanoparticles with its respective electron diffraction pattern. Bar length of 10 nm. **b** TEM image of γ-Fe_2O nanoparticles showing, different sizes of grain. Bar length of 5 nm

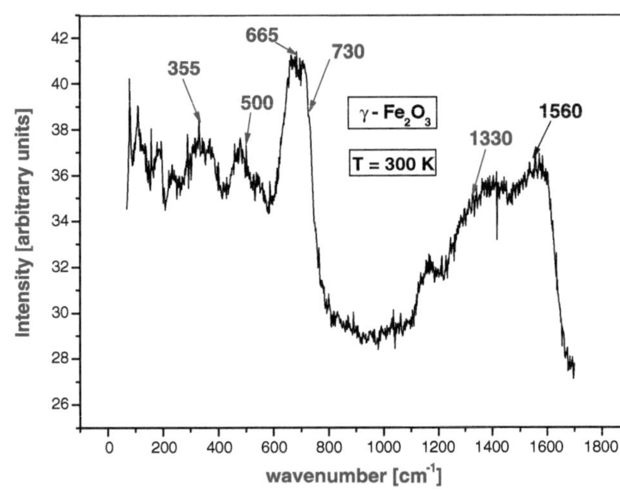

Fig. 8 Raman spectrum of maghemite nanoparticles

Figure 7a shows the distribution of γ-Fe_2O_3 nanoparticles. They are nearly spherical in shape, but without homogenous distribution. Our γ-Fe_2O_3 nanoparticles agglomerate. However, agglomeration could be preventing by using oleic acid [15, 16]. Figure 7b shows different diameters of γ-Fe_2O_3 nanoparticles, some highlighted are: 5.2 nm, 7.82, 7.92 and 8.4 nm. The mean grain size is 7.3 nm, which is closer to that obtained by the X-ray measurements above (5.7 nm)

The Raman spectrum (Fig. 8) of our synthetic maghemite nanoparticles was indexed taking the mean bands from the references [17, 18]. The phases are indexed by red arrows: 355, 500, 665, 730 and 1330 cm^{-1}, the last additional peak at 1,560 cm^{-1} belongs to ferrihydrite [19–21].

4 Conclusions

Nanomagnetite with stoichiometric mixture Fe^{3+}/Fe^{2+} in the relation 2:1, Fe_3O_4 was oxidized to form maghemite nanoparticles of around 5.7 nm. The Mössbauer spectrum and the magnetic measurements indicate that the sample is superparamagnetic. Measurements of magnetic moment vs. temperature indicates blocking temperature of $T_B = 105.3$ K. At room temperature, it is difficult to identify whether the samples are magnetite or maghemite. However, the Mössbauer measurements at 77 K was fitted using hyperfine fields distribution and confirming the presence of two magnetic fields of maghemite (48.6 T for A site and 51.3 T for B site) and finally a 4.2 K the MS is well resolved with two magnetic fields sites. Despite the fact that TEM images show that the nanoparticles of γ-Fe_2O_3 are not well-dispersed, it reveals good crystallinity in agreement with the XRD.

Acknowledgements We are indebted to Dr. F. E. Wagner from the Physics Department of the Technical University of Munich, 85747 Garching, Germany for carrying out the Mössbauer measurements at 4.2 K and Dr. J. Coaquira and L. León from the Universidade de Brasilia for carrying out the Mössbauer measurements at 77 K. The work in Cambridge was supported by a grant from the Engineering and Physical Science Research Council (EPSRC).

References

1. Kluchova, K., Zboril, R., Tucek, J.: Biomaterials **30**, 2855–2863 (2009)
2. Nijdam, A.J., Nicholson, T.R., Shapiro, J.P.: Curr. Nanosci. **5**, 88–102 (2009)
3. Goya, G.F., Grazú, V., Ibarra, M.R.: Curr. Nanosci. **4**, 1–16 (2008)
4. Kang, Y.S., Banerjee, D., Risbud, S., Rabolt, J., Stroeve, P.: Chem. Mater. **8**, 2209–2211 (1986)
5. Brand, R.A.: Normos Programs. Duisburg University (1989)
6. Klug, H.P., Alexander, L.E.: X-ray Diffraction Procedures, p. 514. Wiley, New York (1970)
7. Vandenberghe, R.E., De Grave, E.: In: Yoshida, Y., Langouche, G. (eds.) Mössbauer Spectroscopy – Tutorial Book, chap. 3. Springer-Verlag, Berlin Heidelberg (2013)
8. Gütlich, P., et al. (eds.): Mössbauer Spectroscopy and Transitions Metal Chemistry: Fundamentals and Applications. Springer (2011)
9. Morup, S.: Proceedings of the Indian National Science Academy, pp. 91–107. ICAME (1981)
10. Morup, S., Oxborrow, C.A., Hendriksen, P.V., Pedersen, M.S., Hanson, M., Johanson, C.: J. Magn. Magn. Mater. **140**, 409–410 (1995)
11. Disch, S., et al.: New J. Phys. **14**, 013025, supplementary information (2012)
12. Pascal, C., et al.: Chem. Mater. **11**, 141–147 (1999)
13. Coaquira, J.A.H., et al.: IEEE Trans. Magn. **45**, 10 (2009)
14. Georgia, C., Papaefthymiou, et al.: Phys. Rev. B **80**, 024406 (2009)
15. Chen, D.-X., et al.: J. Appl. Phys. **105**, 083924 (2009)
16. Hyeon, T., Lee, S.S., Park, J., Chung, Y., Bin Na, H.J.: Am. Chem. Soc. **123**, 12798–12801 (2001)
17. Rebbouh, L., Hermann, R.P., Grandjean, F.: Phys. Rev. B **76**, 174422 (2007)
18. Hanesch, M.: Geophys. J. Int. **177**, 941–948 (2009)
19. De Faria, D.L.A., Silva, V., et al.: J. Raman Spectrosc. **28**, 873–878 (1997)
20. Mazzeti, L., Thislethwaite,: J. Raman Spectrosc. **33**, 104–111 (2002)
21. Wei, L., Hervé, M., Edouard, P.: J. Cryst. Growth **342**, 21–27 (2012)

DOI 10.1007/978-94-007-6482-8_12
Reprinted from *Hyperfine Interactions*, DOI 10.1007/s10751-013-0829-2

Mössbauer study of a Fe_3O_4/PMMA nanocomposite synthesized by sonochemistry

H. Martínez · L. D'Onofrio · G. González

Published online: 20 March 2013

Abstract Magnetite nanoparticles of 10 nm average size were synthesized by ultrasonic waves from the chemical reaction and precipitation of ferrous and ferric iron chloride ($FeCl_3\cdot 6H_2O$ y $FeCl_2\cdot 4H_2O$) in a basic medium. The formation and the incorporation of the magnetite in PMMA were followed by XRD and Mössbauer Spectroscopy. These magnetite nanoparticles were subsequently incorporated into the polymer by ultrasonic waves in order to obtain the final sample of 5 % weight Fe_3O_4 into the polymethylmethacrylate (PMMA). Both samples Fe_3O_4 nanoparticles and 5 % Fe_3O_4/PMMA nanocomposite, were studied by Mössbauer spectroscopy in the temperature range of 300 K–77 K. In the case of room temperature, the Mössbauer spectrum of the Fe_3O_4 nanoparticles sample was fitted with two magnetic histograms, one corresponding to the tetrahedral sites (Fe^{3+}) and the other to the octahedral sites (Fe^{3+} and Fe^{2+}), while the 5 % Fe_3O_4/PMMA sample was fitted with two histograms as before and a singlet subspectrum related to a superparamagnetic behavior, caused by the dispersion of the nanoparticles into the polymer. The 77 K Mössabuer spectra for both samples were fitted with five magnetic subspectra similar to the bulk magnetite and for the 5 % Fe_3O_4/PMMA sample it was needed to add also a superparamagnetic singlet. Additionally, a study of the Verwey transition has been done and it was observed a different behavior compared with that of bulk magnetite.

Proceedings of the Thirteenth Latin American Conference on the Applications of the Mössbauer Effect, (LACAME 2012), Medellín, Colombia, 11–16 November 2012.

H. Martínez (✉) · L. D'Onofrio
Centro de Física Experimental del Sólido, Escuela de Física, Facultad de Ciencias, Universidad Central de Venezuela, Caracas, Venezuela
e-mail: hmartinez@fisica.ciens.ucv.ve, heidimg1@gmail.com

G. González
Centro de Materiales y Nanotecnología, Instituto Venezolano de Investigaciones Científicas, Caracas, Venezuela

Keywords Nanocomposites · Magnetite · Mössbauer spectroscopy

Abbreviations

PMMA	Polymethylmethacrylate
XRD	X-ray diffraction
Fe_3O_4	Magnetite
IS	Isomer shift
QS	Electrical quadrupole splitting
HF	The hyperfine magnetic field
SP	Superparamagnetic

1 Introduction

In recent years, the synthesis and analysis of materials in the nanometer scale has been an area of intense study. Nanostructured materials often exhibit chemical, electrical, magnetic and optical properties different from those of bulk materials. There is a great interest in iron oxides because of its importance in the technology of magnetic information storage, magnetic resonance imaging and ferrofluids, Skomski [1], Bahadur and Giri [2] and Neuberger et al. [3]. At room temperature bulk magnetite (Fe_3O_4) has an inverse spinel structure with eight Fe^{3+} ions in tetrahedral sites (A) and sixteen $Fe^{2.5+}$ ions in octahedral sites (B). Above the Verwey temperature, $T_V \sim$ 120 K, magnetite has a cubic spinel structure, with lattice parameter $a_0 \sim 8.397$ Å and the O atoms arranged in a face centered—cubic (fcc) lattice. The lattice can accommodate Fe^{3+} on the tetrahedral site (A) and Fe^{3+} and Fe^{2+} on the octahedral site (B) in antiparallel arrangement, yielding ferrimagnetic order below T_C (Curie's temperature, ≈860 K). Below the temperature T_V, 120 K, there is a change from cubic to orthorhombic structure, a lower symmetry structure, and a more complex Mössbauer spectrum is observed, and can be fitted by at least five subspectra due to 3d electron localization, leading to discrete Fe^{2+} and Fe^{3+} spectral contributions of the B sites, [4, 5]. In this work, we study the Mössbauer Spectroscopy of magnetite nanoparticles embedded in a polymeric matrix of PMMA because it allows a good dispersion of nanoparticles in the polymeric matrix, and a superparamagnetic behavior can be observed.

2 Experimental

Magnetite nanoparticles (Fe_3O_4) included in PMMA were synthesized by the reaction of iron chloride salts ($FeCl_3{\cdot}6H_2O$ and $FeCl_2{\cdot}4H_2O$) with pH = 10. The reaction process was done in an ultrasound equipment (Fisher, Model 150 W), at a fixed frequency of 20 KHz for 15 min. In this period of time a solution NH_3 (10 ml) was added to keep the a pH = 10 and thus ensure the formation of nanoparticles Fe_3O_4.

The formation of Fe_3O_4 nanoparticles was according to the following reaction:

$$FeCl_{2(aq)} + 2FeCl_{3(aq)} + 8NH_{3(aq)} + 4H_2O + \text{Ultrasound} \rightarrow Fe_3O_{4(sol)} + 8NH_4Cl_{(aq)}$$

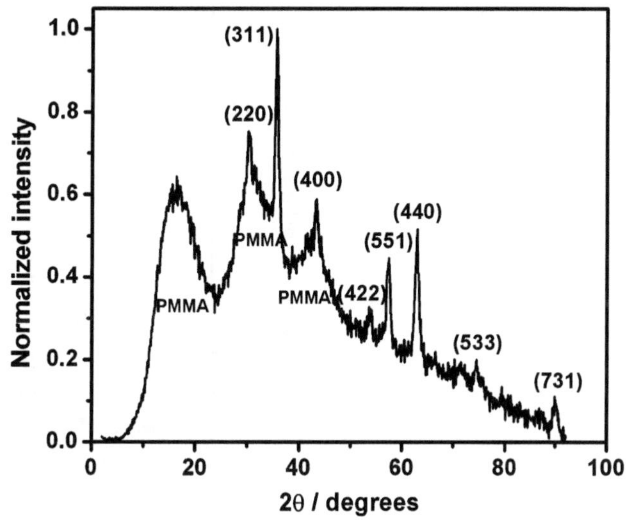

Fig. 1 XRD of synthesized 5 % Fe_3O_4/PMMA

The product was centrifuged and washed with distilled water so that all the NH_3 disappeared from the obtained nanoparticles. To incorporate these nanoparticles into the polymer, the PMMA (120.000 MW), was dissolved in acetone, and the magnetite powder was dispersed in this solution using the same conditions of the sonochemical synthesis. The relative amount of iron ions and PMMA was calculated in order to obtain 5 % weight Fe_3O_4 in PMMA. The polymer solution and the nanoparticles were subjected to ultrasonic irradiation for 15 min at a frequency of 20 KHz.

XRD measurements were performed using a Bruker D8 powder diffractometer using CuK_α ($\lambda = 0.1548$ nm). The XRD allowed us to identify the precipitate as magnetite, with an average particle size of 10 nm as deduced from the Scherrer formula for the crystallite size.

The Mössbauer spectra were obtained using a source of ^{57}Co in a Rh matrix, a driving unit running in the triangular symmetric mode for velocity and a multichannel analyzer (Wissel Instruments). The measurements were performed both at room temperature and at liquid nitrogen temperature. The fitting parameters are: the isomer shift (IS) calibrated with α-Fe, the electrical quadrupole splitting (QS) and the hyperfine magnetic field (HF).

3 Results and discussion

The XRD pattern of 5 % magnetite/PMMA is shown in Fig. 1, and the pattern of nanoparticles magnetite is shown in Fig. 2. The indexation was done by reference to the magnetite in bulk proving the presence of magnetite only. Also it is seen that the peaks are broadened indicating that the sample is composed of small crystals, where peaks broadening are inversely proportional to grain size. An average grain size of 10 nm was found using the Scherrer equation.

While magnetite and maghemite are indistinguishable by XRD, the Mossbauer spectroscopy is a useful tool to differentiate unambiguously these iron oxides. The

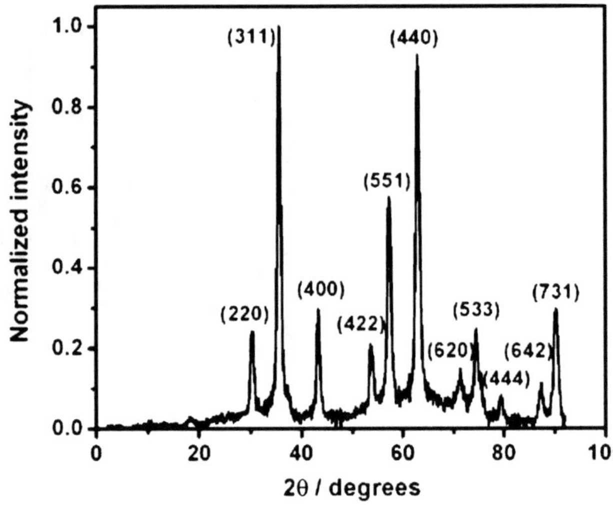

Fig. 2 XRD of bulk Fe_3O_4

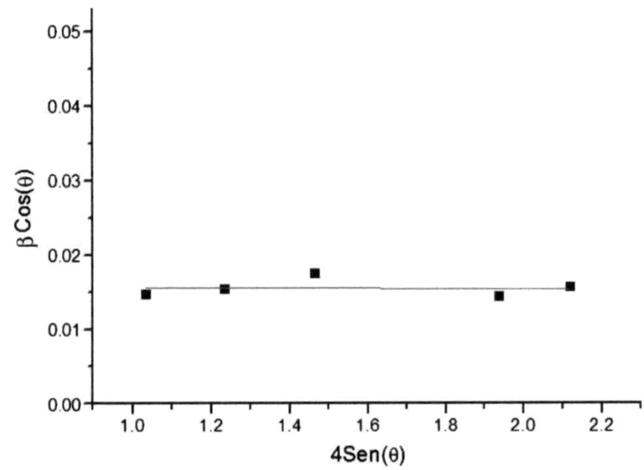

Fig. 3 Williamson-Hall plot for the 10 nm Fe_3O_4 nanoparticules

HF values at room temperature for magnetite are slightly lower than those of the maghemite. Vandenberghe et al. [6].

By Hall Williamson analysis the grain size and the microdeformation (ε) were calculated. The average particle size obtained from Fig. 3 was (10 ± 2) nm, and the microdeformation $\varepsilon = 4.2 \times 10^{-5}$ with respect to the magnetite in bulk.

The TEM micrograph of Fig. 4 corresponds to the 5 % Fe_3O_4/PMMA sample, and it shows the crystallites of magnetite with a nearly spherical morphology. The electron diffraction pattern in Fig. 4, shows intense rings corresponding to the same families of planes that characterize Fe_3O_4.

Figure 5 shows a histogram of sizes of magnetite nanoparticles obtained from microscopy analysis. The particle distribution histogram was obtained counting 250 particles from the micrograph. The solid line corresponds to the log—normal distribution fit, and give us an average particle size of 11 nm and a standard deviation $\sigma = 0.22$.

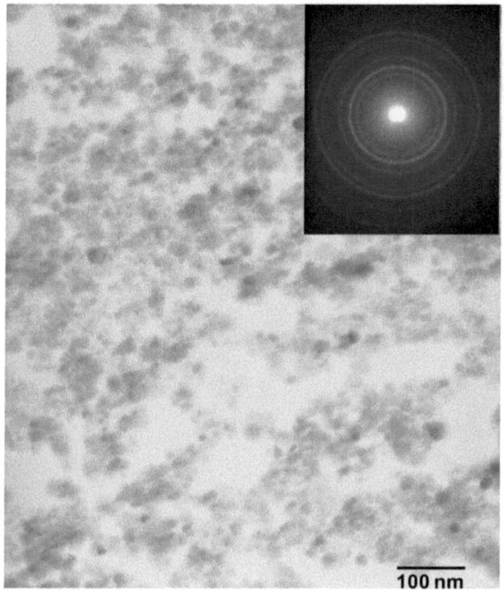

Fig. 4 TEM micrograph of 5 % de Fe_3O_4/PMMA sample

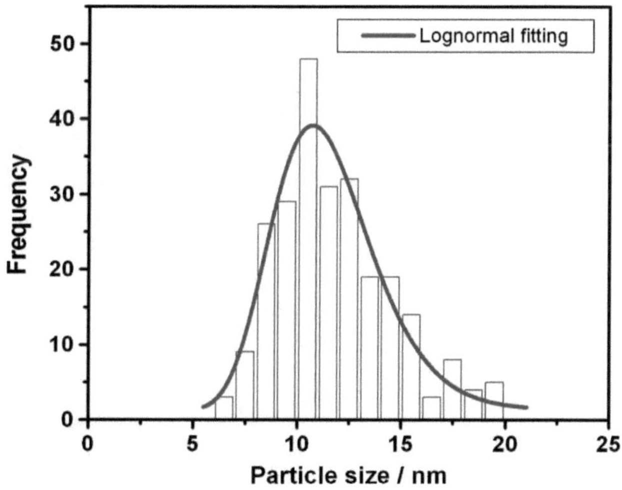

Fig. 5 Particle size distribution histogram and lognormal fitting

In Fig. 6 is shown the Mössbauer spectra recorded at 300 K of Fe_3O_4 nanoparticles and 5 % Fe_3O_4/PMMA nanocomposite. At 300 K the spectrum was fitted with two histogram for magnetite; in the case of magnetite nanoparticles, one histogram for tetrahedral sites with IS = 0.29 mm/s, QS = 0.00 and HF = 47.2 T and one histogram for octahedral sites with IS = 0.52 mm/s, QS = 0.00 and HF = 45.8 T. These value of hyperfine fields obtained represents a field magnetic mean, which is caused by the size distribution of nanoparticles in the sample. For the 5 % Fe_3O_4/PMMA was fitted, as in the case previous, with two histograms for tetrahedral and octahedral sites and a singlet; whose values were: for tetrahedral sites IS = 0.28 mm/s, QS = 0.00 and HF= 45.8 T and one for octahedral Sites with IS = 0.52 mm/s, QS = 0.00 and T in HF = 44.2 T, addition one singlet caused by the presence of nanoparticles of magnetite

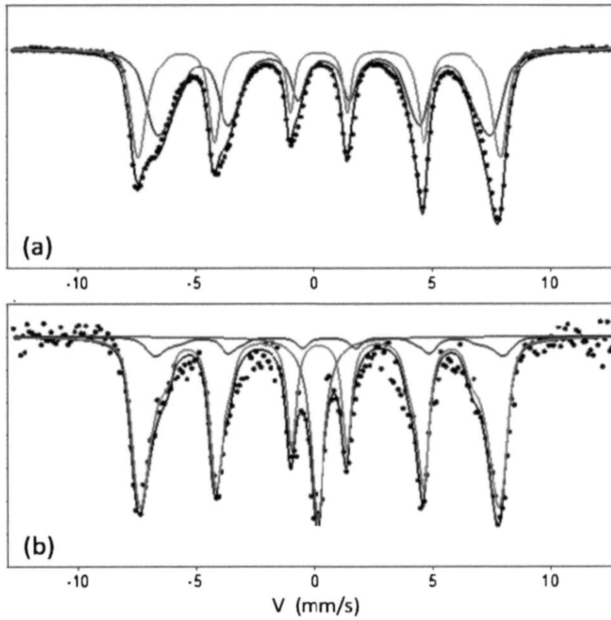

Fig. 6 Mössbauer spectrum at 300 K, (**a**) Fe_3O_4 nanoparticules, (**b**) 5 % Fe_3O_4/PMMA

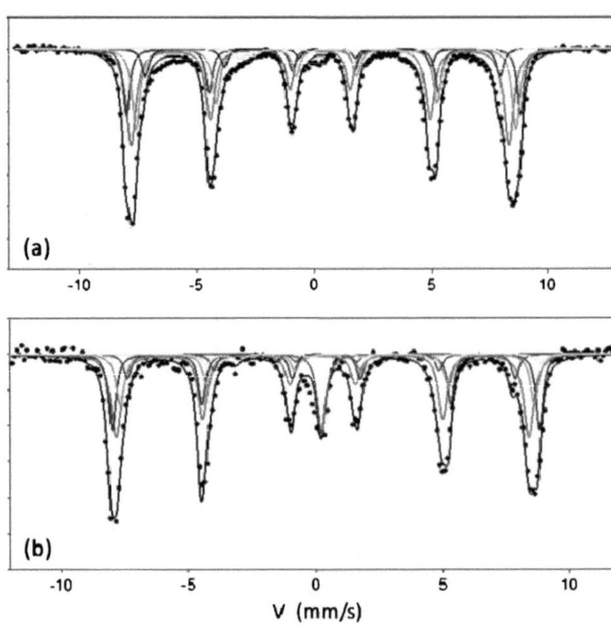

Fig. 7 Mössbauer spectrum at 77 K, (**a**) Fe_3O_4 nanoparticules, (**b**) 5 % Fe_3O_4/PMMA

present in the sample, with hyperfine parameters of IS = 0.23 mm/s, QS = 0.00 and HF = 0.00 T. The singlet that is observed in the sample is due to the superparamagnetic behavior (SP) caused by the inclusion of the magnetite obtained by sonochemistry in the PMMA. The PMMA allows for less interaction of the magnetite

Table 1 Mössbauer parameters of Fig. 7

Name	SUBESPECTRO	DI (mm/s)	QS (mm/s)	CH (Tesla)
Nanoparticules of Fe_3O_4	Tetra. (A) Fe^{3+}	0,33	0,00	50,0
	Oct. (B) Fe^{3+}—I	0,47	0,00	52,2
	Oct. (B) Fe^{3+}—II	0,52	−0,01	50,3
	Oct. (B) Fe^{2+}—I	0,48	−0,19	46,9
	Oct. (B) Fe^{2+}—II	1,62	1,80	32,5
5 % Fe_3O_4/PMMA	Tetra. (A) Fe^{3+}	0,34	0,00	50,4
	Oct. (B) Fe^{3+}—I	0,45	0,00	52,1
	Oct. (B) Fe^{3+}—II	0,50	−0,01	50,6
	Oct. (B) Fe^{2+}—I	0,33	−0,19	47,3
	Oct. (B) Fe^{2+}—I	1,51	1,50	40,8
	Singulete (SP)	0,32	0,00	0,00

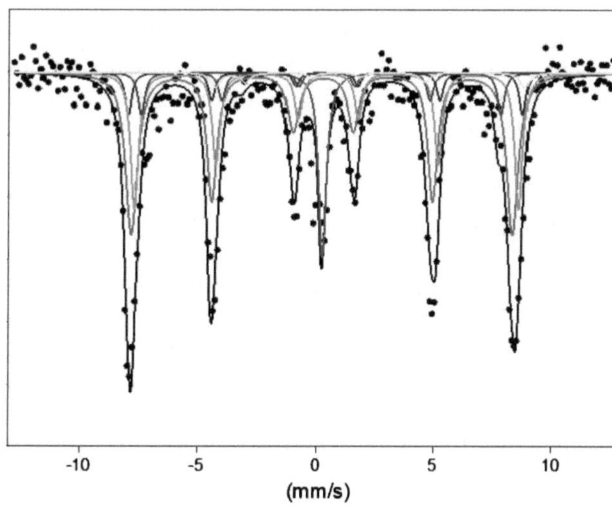

Fig. 8 Mössbauer spectrum at 110 K of the 5 % Fe_3O_4/PMMA sample

nanoparticles, in contrast to case of the nanoparticles without polymer since these are agglomerated due to the strong magnetism between these nanoparticles.

The Mössbauer spectrum at 77 K of Fig. 7a, for magnetite nanoparticle, is fitted with five sextets [3], one of them represents the tetrahedral sites of Fe^{3+} and the other four the distinction between Fe^{2+} and Fe^{3+} in the octahedral sites. The sample 5 % magnetite/PMMA, was fitted with five sextets similar to the previous case, but in addition we can observe the appearance of one singlet caused by the distribution of the magnetite nanoparticles in the polymer producing a slight separation of the nanoparticles, this happens due to the magnetism of the magnetite is very strong which makes the nanoparticles are held together. The values of parameters hyperfine shows in the Table 1. The fit of magnetite at 77 K is difficult to perform because a distinction must be made between Fe^{2+} and Fe^{3+}, although there are recent publications [4, 7], reported only in bulk magnetite at 77 K.

Figure 8 shows the 5 % magnetite/PMMA sample at 110 K, and we can observe the presence of a Verwey transition, therefore Mössbauer spectrum was fitted with six subspectra similar to the case of this sample at 77 K, and whose values of the hyperfine parameters are reported in Table 2. It has to be noted that the

Table 2 Mössbauer parameters of Fig. 8

Name	SUBESPECTRO	DI (mm/s)	QS (mm/s)	CH (Tesla)
5 % Fe_3O_4/PMMA	Tetra. (A) Fe^{3+}	0,34	0,00	50,2
	Oct. (B) Fe^{3+}—I	0,46	0,00	51,5
	Oct. (B) Fe^{3+}—II	0,50	−0,01	49,8
	Oct. (B) Fe^{2+}—I	0,31	−0,19	47,4
	Oct. (B) Fe^{2+}—I	1,50	1,50	40,8
	Singulete (SP)	0,33	0,00	0,00

Verwey transition is still observed at this temperature. Dénes Szabó et al. [7] assume that the transition disappears. However, at 77 K and 110 K the sample was fitted with five subspectra due to the suffering change of the magnetite structure, which causes a slight distortion, passing from a cubic symmetry to a monoclinic symmetry, Muxworthy and McClealland [8].

Pasternak et al. [9], Rozenberg et al. [10], reported a Tv behavior with Pressure, in which the value of T_V in absence of pressure is consistent with our results. These authors attributed the nature of this transition as due to distorted—cubic- insulater, Rozenberg et al. [10].

4 Conclusions

The sonochemical synthesis method described above provides a simple way to prepare nanoparticles of Fe_3O_4 and 5 % of Fe_3O_4/PMMA nanocomposites. For the analysis of Willamson—Hall and Scherrer equation the average grain size of the magnetite of nanoparticles is (10 ± 2)nm and the microdeformation of $4{,}21 \times 10^{-5}$ with respect to the magnetite in bulk, so we can say that the microstrain is very small. For Mössbauer spectroscopy the hyperfine magnetic field value (CH) is smaller than in the case of magnetite in bulk, this is due to the presence of magnetite nanoparticles. The Fe_3O_4 nanoparticles included in the PMMA showed a singlet superparamagnetic (SP) because the superparamagnetic polymer acted as a dispersant. At 77 K Mössbauer spectras of Fe_3O_4 nanoparticles were fitted with five sextets, of which, one represents the Fe^{3+} tetrahedral sites and the other four the distinction between (Fe^{2+} y Fe^{3+}) of octahedral sites. The 5 % Fe_3O_4/PMMA sample, additionally shows a singlet corresponding to the size distribution present in this sample. Regarding the transition of Verwey this is not suppressed, observed in both fitted as in the case 77 K and 110 K, suggesting a change in the crystal structure of the magnetite.

Acknowledgement This work was supported by the project PG 03–8224–2011 of CDCH.

References

1. Skomski, R.: Nanomagnetics, topical review. J. Phys. Condens. Matter **15**, R841–R896 (2003)
2. Bahadur, D., Giri, J.: Biomaterials and magnetism. Sadhana **28**, 639–656 (2003)
3. Neuberger, T., Schöpf, B., Hofmann, H., Hofmann, M., von Rechenberg, B.: Superparamagnetic nanoparticles for biomedical applications: possibilities and limitations of a new drug delivery system. J. Magn. Magn. Matter. **293**, 483–496 (2005)

4. Berry, F.J., Skinner, S., Thomas, M.F.: Fe57 Mössbauer spectroscopic examination of a single crystal of Fe_3O_4. J. Phys. Condens. Matter **10**, 215–220 (1998)
5. Doriguetto, A.C., Fernandes, N.G., Persiano, A.I.C., Nunes Filho, E., Greneche, J.M., Fabris, J.D.: Characterization of a natural magnetite. Phys. Chem. Miner. **30**, 249–255 (2003)
6. Vandenberghe, R.E., De Grave, E., Landuydt, C., Bowen, L.H.: Some aspects concerning the characterization of iron oxides and hydroxides in soils and clays. Hyperfine Int. **53**, 175–196 (1990)
7. Szabó, D., Czakó-Nagy, I., Zrinyi, M., Vértes, A.: Magnetic and Mössbauer studies of magnetite—loaded polyvinyl alcohol hydrogels. J. Colloid Interface Sci. **221**, 166–172 (2000)
8. Muxworthy, A.R., McClealland, E.: Review of the low-temperature magnetic properties of magnetite from a rock magnetic perspective. Geophys. J. Int. **140**, 101–114 (2000)
9. Pasternak, M.P., Xu, W.M., Rozenberg, G.K., Taylor, R.D., Jeanloz, R.: Presured—induced coordination in magnetite; the breakdown of the Verwey—Mott localization hypothesis. J. Magn. Magn. Mater. **265**, L107–L112 (2003)
10. Rozenberg, G.K., Pasternak, M.P., Xu, W.M., Amiel, Y., Hanfland, M.: Origin of the verwey transition in magnetite. Phys. Rev. Lett. **96**, 045705-1–045705-4 (2006)

DOI 10.1007/978-94-007-6482-8_13
Reprinted from *Hyperfine Interactions*, DOI 10.1007/s10751-013-0823-8

Characterization of iron in airborne particulate matter

F. V. F. Tavares · J. D. Ardisson · P. C. H. Rodrigues · W. Brito · W. A. A. Macedo · V. M. F. Jacomino

Published online: 27 February 2013

Abstract In this work soil samples, iron ore and airborne atmospheric particulate matter (PM) in the Metropolitan Region of Belo Horizonte (MRBH), State of Minas Gerais, Brazil, are investigated with the aim of identifying if the sources of the particulate matter are of natural origin, such as, resuspension of particles from soil, or due to anthropogenic origins from mining and processing of iron ore. Samples were characterized by powder X-ray diffraction, X-ray fluorescence and ^{57}Fe-Mössbauer spectroscopy. The results showed that soil samples studied are rich in quartz and have low contents of iron mainly iron oxide with low crystallinity. The samples of iron ore and PM have high concentration of iron, predominantly well crystallized hematite. ^{57}Fe-Mössbauer spectroscopy confirmed the presence of similar iron oxides in samples of PM and in the samples of iron ore, indicating the anthropogenic origin in the material present in atmosphere of the study area.

Keywords Air pollution · Particulate matter · Soil · Iron ore · ^{57}Fe-Mössbauer spectroscopy

1 Introduction

Air pollution is a severe environmental problem that has become increasingly worse in urban and/or industrial centers. Particulate matter (PM) is an important air pollutant that can cause various damages to public health (respiratory and cardiovascular

---

Thirteenth Latin American Conference on the Applications of the Mössbauer Effect, (LACAME 2012), Medellín, Colombia, 11–16 November 2012.

F. V. F. Tavares (✉) · J. D. Ardisson · P. C. H. Rodrigues ·
W. Brito · W. A. A. Macedo · V. M. F. Jacomino
Nuclear Technology Development Centre, Presidente Antônio Carlos Avenue,
6627, Campus of Federal University of Minas Gerais, 31270-901, Box 941 30123-970,
Belo Horizonte, Minas Gerais, Brasil
e-mail: ferufv@yahoo.com.br

problems), ecosystems and materials [1, 2]. Part of the Metropolitan Region of *Belo Horizonte* (MRBH) covers the region of the Iron Quadrangle—geological district of major iron production—in the State of *Minas Gerais* (MG), Brazil. The mining and beneficiation of iron ore activities can release large amounts of PM to the atmosphere, so these particles can be both an important pollutant agent and an indicator of air quality in metropolitan area of *Belo Horizonte.*

Studies in this region indicate high levels of iron (Fe) in the elemental composition of atmospheric PM [3]. Inhalation of Fe may contribute to the formation of free radicals and induce oxidative processes in the human body [4] whereas PM particles once reaching the pulmonary alveoli can cause from respiratory problems to other serious lung diseases. The main objective of this study is to investigate the chrystallochemical properties of iron based compounds present in atmospheric PM of MRBH, aiming to identify the sources of those pollutant emissions, whether from natural origin, such as the resuspension of soil particles, or from anthropogenic origin, due to mining and processing of iron ore.

2 Experiment

The region of study comprises part of the Iron Quadrangle of *Minas Gerais* that has an important large mining industry and a large population exposed to air pollution, in particular, particulate matter. The location map of the study area, the sampling points of soil, iron ore and particulate matter are shown in Fig. 1. For this work two samples of particulate matter (PM1 and PM2) were collected in the district of *Jardim Canadá* located in the *Nova Lima* municipality at approximately 16 Km linear from the city center of *Belo Horizonte. Nova Lima* is an important city of MRBH that concentrates numerous mines activities (Fig. 1).

Samples of PM (PM1, PM2) were taken with a brush from the surface of leaves of trees located in specific points of the study area and stored in properly identified polyethylene bottles. Soil samples (S1, S2, S3 and S4) were collected from the ground surface (0–10 cm) at points located in the surrounding areas of those where PM was collected. The samples taken are representative of the different types of soil present in the area of study. Samples of iron ore (M1, M2, M3 and M4) were collected in the vicinity of the collection points of PM. These areas belong to the geological *Cauê* Fomation, the most important mineralized level in the Iron Quadrangle.

Soil samples and iron ore were dried in air, when necessary, and then manually ground in an agate degree until the particle size was smaller than 200 mesh. All samples were characterized by X-ray diffraction (XRD), X-ray fluorescence (XRF) and by ^{57}Fe-Mössbauer spectroscopy (MS), an essential technique for the characterization of soil and atmospheric PM [5–10]. Measurements of XRD were performed at room temperature on an automatic Rigaku diffractometer with Cu Kα radiation from $(2\theta) = 4$ to 80°. The X-ray fluorescence analysis, were performed in a Shimadzu EDX-720, with a rhodium tube and silicon-lithium detector. The ^{57}Fe-Mössbauer spectra in transmission geometry were obtained using a conventional spectrometer, with constant acceleration, and a ^{57}Co source on a Rh matrix at room temperature. The measurements were done at room temperature (RT) and at 80 K without application of an external magnetic field. For one of the soil samples (sample S1) the measurement was done at 20 K. Isomer shift values were quoted relatively to

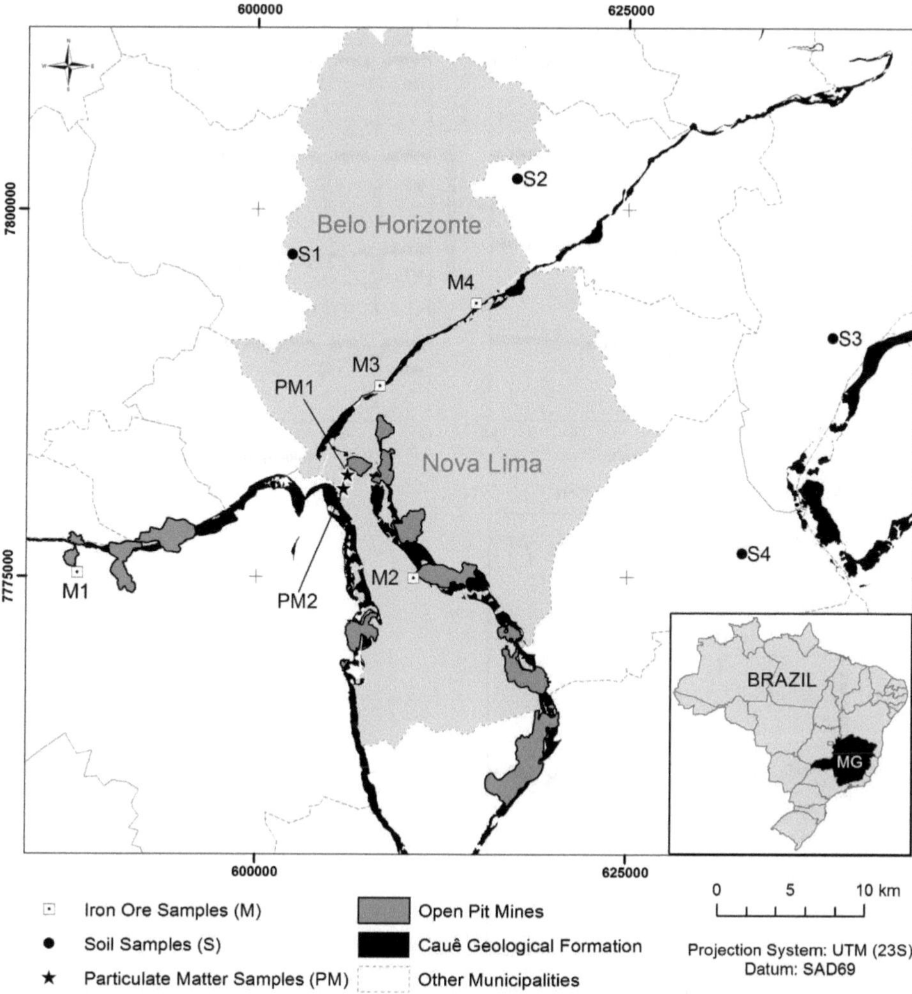

Fig. 1 Location map. Sampling of soils (S), iron ore (M) and particulate matter (PM)

the metallic α-Fe. Resonant curves were least-squares fitted to Lorentzian functions with the software NORMOS™-90 (developed by R. A. Brand, at Laboratorium für Angewandte Physik, Universität Duisburg, D-47048, Duisburg-Germany).

3 Results and discussion

Typical room temperature (RT) ^{57}Fe-Mössbauer spectra of soil samples, iron ore and PM at are shown in Fig. 2. The spectra of the soil samples, identified by S1, S2, S3 and S4 (Fig. 2a), are formed by the superposition of two contributions, one characterized by a central paramagnetic doublet, approximately symmetrical, and the other one characterized by a ferromagnetic contribution represented by a six line spectrum (sextet). The paramagnetic fraction may be assigned to the presence of Fe-silicates,

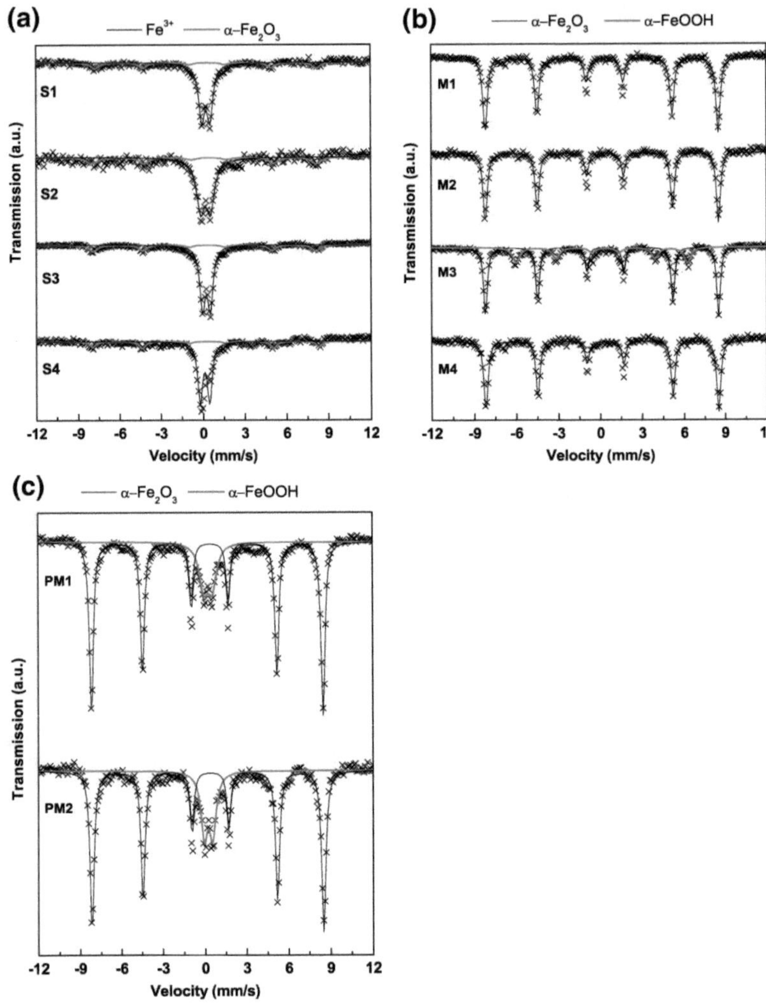

Fig. 2 Typical room temperature ^{57}Fe-Mossbauer spectra for samples of **a** soil (S), **b** Iron ore (M) and **c** particulate matter (PM)

due to the substitution of Fe^{3+} in tetrahedral sites by Al^{3+} and/or the presence of iron oxides in a superparamagnetic state [8, 9, 11–13]. The magnetic fraction has hyperfine parameters characteristics of hematite (α-Fe_2O_3) (Table 1) [11–13]. Room temperature ^{57}Fe-Mössbauer spectra of samples of iron ore, M1, M2, M3 and M4 (Fig. 2b) were fitted with a majority sextet characteristic of α-Fe_2O_3 with relative subspectral area ranging between 70 and 100 % and a minority sextet in sample M3 associated to goethite (α-FeOOH) (Table 2). The room temperature ^{57}Fe-Mössbauer spectra of PM (Fig. 2c) are were fitted with two subespectra: a sextet with ~80 % relative area identified as hematite and a doublet assigned to superparamagnetic goethite (Table 3).

^{57}Fe-Mössbauer measurements at 80 K were performed in order to better identify the mineralogical phases present in the samples. The spectra of the soil samples

Table 1 ^{57}Fe-Mössbauer parameters of soil samples at RT and 80 K

Sample	Temperature	Phase	δ (±0.05) (mm/s)	ε, Δ (±0.05) (mm/s)	BHF (±0.5) (Tesla)	RA (%)
S1	RT	$\alpha-Fe_2O_3$[a]	0.37	−0.13	49.3	25
		Fe^{3+}	0.36	0.58	–	75
	80 K	α−FeOOH[b]	0.53	−0.37	48.3	11
		$\alpha-Fe_2O_3$	0.48	−0.18	52.5	39
		$Fe_5HO_8.4H_2O$[c]	0.55	−0.30	44.2	13
		$FeTiO_3$[d]	0.54	2.38	–	12
		γ−FeOOH[e]	0.48	0.65	–	25
S2	RT	$\alpha-Fe_2O_3$	0.37	−0.38	49.9	27
		Fe^{3+}	0.31	0.68	–	73
	80 K	α−FeOOH	0.36	−0.31	48.7	14
		$\alpha-Fe_2O_3$	0.48	−0.25	52.5	24
		$Fe_5HO_8.4H_2O$	0.55	−0.24	44.2	11
		$FeTiO_3$	1.42	2.60	–	10
		γ−FeOOH	0.39	0.92	–	41
S3	RT	$\alpha-Fe_2O_3$	0.33	−0.24	49.6	26
		Fe^{3+}	0.35	0.55	–	74
	80 K	α−FeOOH	0.49	−0.28	48.3	57
		$\alpha-Fe_2O_3$	0.47	−0.14	52.5	10
		$Fe_5HO_8.4H_2O$	0.56	−0.30	43.1	9
		$FeTiO_3$	0.58	2.39	–	9
		γ−FeOOH	0.52	0.64	–	15
S4	RT	$\alpha-Fe_2O_3$	0.37	−0.18	50.5	24
		Fe^{3+}	0.23	0.68	–	76
	80 K	α−FeOOH	0.48	−0.25	49.3	28
		$\alpha-Fe_2O_3$	0.49	−0.15	52.7	10
		$Fe_5HO_8.4H_2O$	0.43	−0.28	46.4	27
		$FeTiO_3$	0.61	2.27	–	5
		γ−FeOOH	0.12	0.56	–	30

[a]Hematite, [b]Goethite, [c]Ferrihydrite, [d]Ilmenite, [e]Lepidocrocite
δ isomer shift relative to α-Fe, Δ quadrupole splitting, ε quadrupole shift, *BHF* hyperfine magnetic field, *RA* relative spectral areas

Table 2 ^{57}Fe-Mössbauer parameters of iron ore samples at RT and 80 K

Sample	Temperature	Phase	δ (±0,05) (mm/s)	ε, Δ (±0,05) (mm/s)	BHF (±0,5) (Tesla)	RA (%)
M1	RT	$\alpha-Fe_2O_3$[a]	0.37	−0.19	51.8	100
	80 K	$\alpha-Fe_2O_3$	0.49	0.35	54.0	100
M2	RT	$\alpha-Fe_2O_3$	0.38	−0.20	51.8	100
	80 K	$\alpha-Fe_2O_3$	0.49	0.40	54.2	100
M3	RT	$\alpha-Fe_2O_3$	0.37	−0.21	51.8	70
		α−FeOOH[b]	0.37	−0.26	38.0	30
	80 K	α−FeOOH	0.46	−0.21	50.0	27
		$\alpha-Fe_2O_3$	0.49	0.42	54.2	73
M4	RT	$\alpha-Fe_2O_3$	0.37	−0.20	51.8	100
	80 K	α−FeOOH	0.43	0.07	51.4	15
		$\alpha-Fe_2O_3$	0.49	0.38	54.2	85

[a]Hematite, [b]Goethite

Table 3 ^{57}Fe-Mössbauer parameters of particulate matter samples at RT and 80 K

Sample	Temperature	Phase	δ (±0,05) (mm/s)	ε, Δ (±0,05) (mm/s)	BHF (±0,5) (Tesla)	RA (%)
PM1	RT	$\alpha-FeOOH$[a]	0.35	0.61	–	20
		$\alpha-Fe_2O_3$[b]	0.38	−0.20	51.7	80
	80 K	$\alpha-FeOOH$	0.48	−0.08	51.0	29
		$\alpha-Fe_2O_3$	0.48	0.35	53.7	71
PM2	RT	$\alpha-FeOOH$	0.34	0.61	–	23
		$\alpha-Fe_2O_3$	0.38	−0.20	51.7	77
	80 K	$\alpha-FeOOH$	0.43	−0.33	49.7	29
		$\alpha-Fe_2O_3$	0.49	0.34	53.7	71

[a]Goethite, [b]Hematite

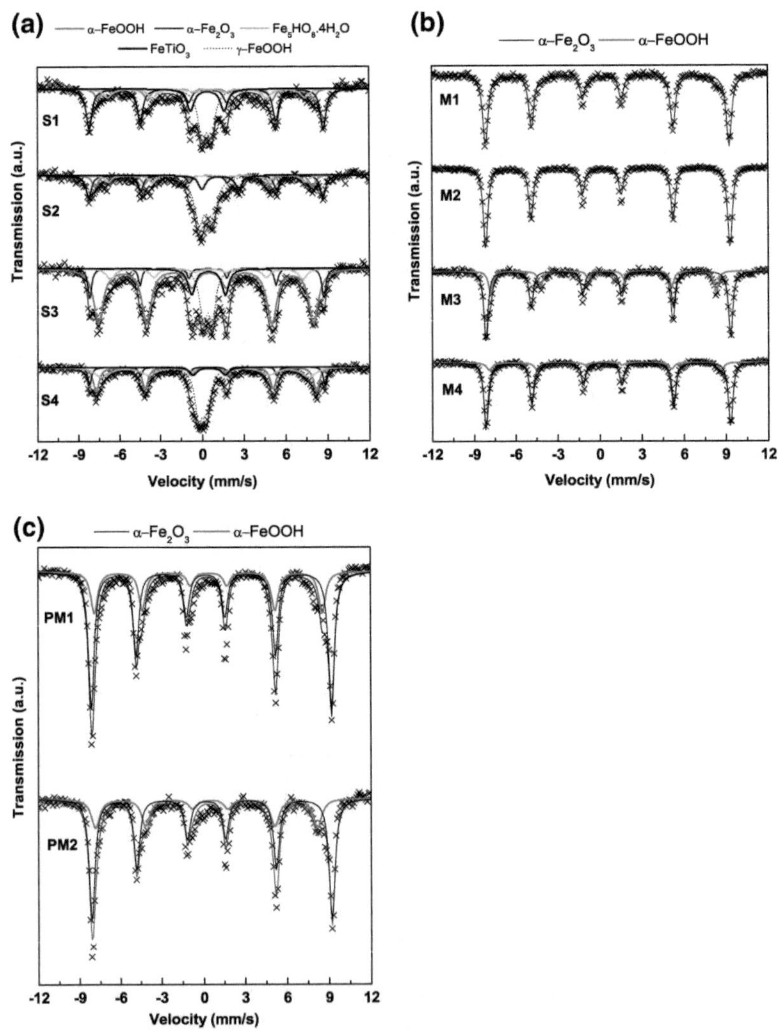

Fig. 3 Typical ^{57}Fe-Mossbauer spectra obtained at 80 K for samples of **a** soil (S), **b** iron ore (M) and **c** particulate matter (PM)

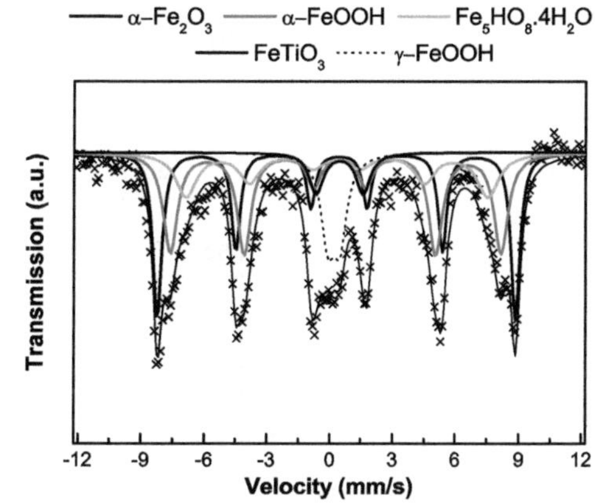

Fig. 4 Typical ^{57}Fe-Mossbauer spectra measured at 20 K on soil sample

Table 4 Chemical composition of samples of soil (S), iron ore (M) and particulate matter (PM) determined with X-ray fluorescence spectroscopy

Sample	Si (%)	Al (%)	Fe (%)	K (%)	Ca (%)	Na (%)	Mg (%)	Ti (%)	Cr (%)	Mn (%)	P (%)	S (%)	O.M.[a] (%)
S1	43.0	28.2	6.1	0.3	0.3	–	–	0.8	–	–	–	0.1	21.2
S2	53.4	25.4	2.8	3.1	0.3	–	0.3	0.3	–	–	–	0.1	14.3
S3	30.7	19.9	12.3	0.7	0.2	–	–	1.1	0.1	0.1	0.1	0.1	34.7
S4	70.2	13.8	5.7	2.1	0.2	2.2	–	0.3	–	0.3	–	0.1	5.1
M1	4.7	7.5	87.4	–	–	–	–	–	–	0.2	–	0.2	–
M2	10.0	0.9	88.6	–	–	–	–	–	–	0.3	–	0.2	–
M3	21.3	1.8	73.3	–	–	–	–	–	–	0.4	–	0.1	3.0
M4	4.6	3.3	90.5	–	0.3	–	–	–	–	0.7	0.4	0.2	–
PM1	12.8	13.0	44.6	0.6	6.5	0.0	0.0	0.8	0.1	0.3	0.3	0.5	20.6
PM2	13.8	13.0	39.9	0.9	4.3	0.0	0.0	0.7	0.1	0.3	0.3	0.6	26.1

[a] Organic matter

(Fig. 3a) were fitted with two doublets and three sextets. The two doublets were assigned to superparamagnetic ilmenite ($FeTiO_3$) [11–13] and lepidocrocite (γ-FeOOH), with very small particles size. Regarding the three sextets, two of them were assigned to hematite (α-Fe_2O_3) and goethite (α-FeOOH). The quadrupole shift (ε) of the former, with values between −0.14 and −0.25 mm/s show that such hematite in the soil has not experienced a Morin transition (T_M ~265 K), in other words, the particles are presumably small and the structure oxide has cations isomorphically substituted Fe^{3+}, most likely by Al^{3+} [14, 15]. The third sextet with hyperfine field of ~44 T was assigned to low crystallinity ferrihydrite ($Fe_5HO_8.4H_2O$). This identification is based on studies by Murad [16] in natural samples of mineral soil, which showed superparamagnetic behavior at RT and blocking temperatures ranging from 28 to 115 K. According to Carpenter et al. [17] ^{57}Fe-Mössbauer measurements at 20 K on poorly crystallized, synthetically prepared ferrihydrite presented a sextet of widened lines with negligible quadrupole splitting and average hyperfine field (BHF) of 45 T. Measurements at 20 K in one of the soil samples (sample S1) (Fig. 4) showed that one of the sextets has IS = 0.44 mm/s, ε = 0.0 mm/s and BHF = 44.7 T, confirming the presence of ferrihydrite. Although samples S1, S2, S3 and S4 are from different types of soil, the ^{57}Fe-Mössbauer spectra revealed no significant differences

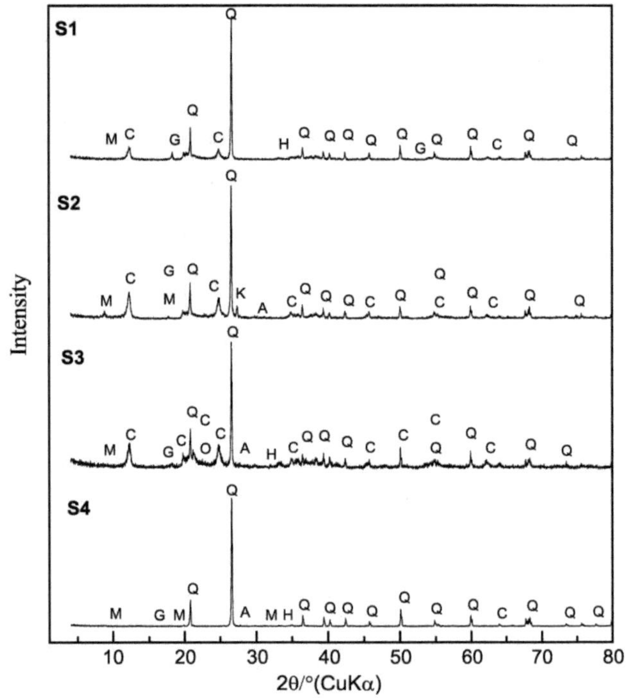

Fig. 5 Powder X-ray diffraction patterns of soil samples. *C* kaolinite, *Q* quartz, *G* gibbsite, *H* hematite, *O* goethite, *M* muscovite, *K* microcline, *A* albite

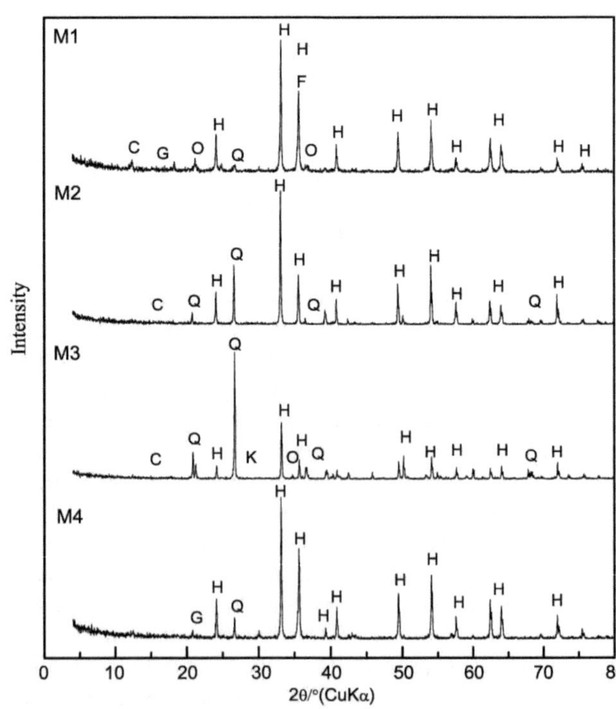

Fig. 6 Powder X-ray diffraction patterns of iron ore samples. *H* hematite, *O* goethite, *C* Kaolinite, *Q* quartz, *G* gibbsite, *K* microcline

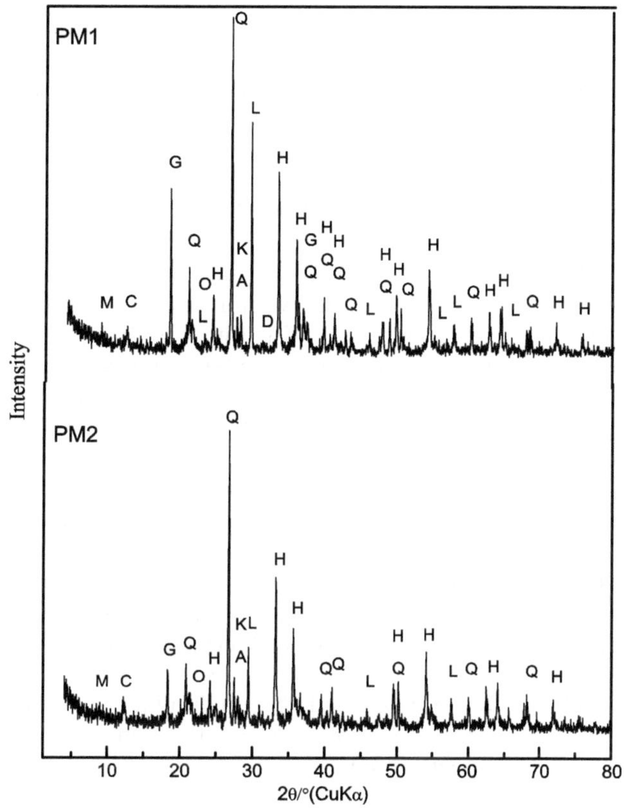

Fig. 7 Powder X-ray diffraction patterns of particulate matter. *L* calcite, *H* hematite, *Q* quartz, *G* gibbsite, *O* goethite, *A* albite, *C* kaolinite, *K* microcline, *D* dolomite, *M* muscovite

among them. The ^{57}Fe-Mössbauer spectra of samples of iron ore at 80 K (Fig. 3b) were fitted with two sextets characteristic of hematite, with subspectral area between ~70 and 100 %, and goethite. ^{57}Fe-Mössbauer spectra of PM at 80 K (Fig. 3c) were fitted with two sextets assigned to hematite and goethite with 71 % and 29 % subspectral area, respectively. Interestingly the hematite at 80 K in the iron ore and PM present positive quadrupole shift, suggesting that $T = 80$ K is below the Morin transition (T_M ~265 K), which occurs for pure and well-crystallized hematite [14].

Comparing the room temperature and 80 K results of the ^{57}Fe-Mössbauer spectroscopy for samples of soil, iron ore and PM, it can be seen that the hyperfine parameters of iron compounds in the samples of iron ore and particulate matter are similar: iron ore and PM samples showed high levels of iron, predominantly hematite with a high degree of crystallization. This similarity may be related to the common origin of the PM and iron ore, indicating that the mining and processing of iron ore in the region under study contribute to the highest percentage of PM present in the atmosphere of the district of *Jardim Canadá,* compared to the contribution of resuspension of soils particles.

Moreover, the results of the chemical analysis by XRF (Table 4) on samples S1, S2, S3 and S4 showed that the most abundant elements in the soil composition are silicon (~40 and 70 w%) and aluminum (~13 and 28 w%); iron (~3 and 12 w%) and potassium, calcium, sodium, titanium, magnesium, chromium, manganese, phosphorus and sulfur with low contributions. The content of organic matter, determined gravimetrically after heating the samples at 850 °C for 1 h, vary between 5 and 35 w%. Samples M1, M2, M3 and M4 XRF revealed high concentration of iron (~70 and 90 w%), silicon (5 and 20 w%), aluminium (1 and 8 %) and low concentration of sulfur, calcium, manganese, phosphorus and organic material. The most abundant elements in in the particulate matter samples (PM1 and PM2) were Fe (~40 w%), Al and Si (~13 w%), Ca (~5 w%), while, Ti, K, S, P, Mn, Cr were found in proportions below 1 w%.

In addition, powder XRD results for the soil samples confirmed the predominance of compounds of Si, Al and low levels of Fe, K and Na, as seen by XRF. Among the compounds identified in the XRD patterns of the soil samples (Fig. 5), quartz was the most abundant (SiO_2), followed by kaolinite ($Al_2Si_2O_5(OH)_4$), gibbsite (α-$Al_2O_3.3H_2O$), moscovite ($KAl_2(AlSi_3O_{10})(OH)_2$, hematite ($Fe_2O_3$), albite ($NaAlSi_3O_8$), microcline ($KAlSi_3O_8$) and goethite (FeOOH). For samples of iron ore (Fig. 6) it was possible to identify the predominance of hematite (Fe_2O_3), with a high degree of crystallinity and other minerals such as quartz (SiO_2), kaolinite ($Al_2Si_2O_5(OH)_4$), goethite (FeOOH), gibbsite ($Al_2O_3.3H_2O$-α) and microcline ($KAlSi_3O_8$). The XRD patterns of the PM samples (Fig. 7) revealed the predominance of hematite (Fe_2O_3) and quartz (SiO_2). Minerals such as gibbsite (α-$Al_2O_3.3H_2O$), calcite ($CaCO_3$), goethite (FeOOH), microcline ($KAlSi_3O_8$), albite ($NaAlSi_3O_8$), kaolinite ($Al_2Si_2O_5(OH)_4$), dolomite ($CaMg(CO_3)_2$) and muscovite ($KAl_2(AlSi_3O_{10})(OH)_2$) were identified.

Hence, XRD confirmed the predominance of hematite with a high degree of crystallization in the iron ore and PM samples, as observed by XRF and ^{57}Fe-Mössbauer spectroscopy.

4 Conclusion

Soil samples, iron ore and atmospheric particulate matter of MRBH were analyzed by X-ray diffraction, X-ray fluorescence and ^{57}Fe-Mössbauer spectroscopy. Our results show that the soils studied are rich in quartz and have low contents of iron mainly due to poorly crystallized iron oxide. The iron ore and PM samples showed high concentration of iron, predominantly hematite, with a high degree of crystallization. ^{57}Fe-Mössbauer spectroscopy showed that the hematite present in the soil has not experienced a Morin transition (T_M ~265 K), possibly due to the presence of Al in the structure. On the other hand, the Morin transition (T_M ~265 K) observed on the hematite present in the iron ore and atmospheric airborne particulate matter (PM) is indicative of hematite of high purity and degree of crystallinity. These differences, only observed by ^{57}Fe-Mössbauer spectroscopy, are essential to evaluate the contribution of the mining and processing of iron ore on the air quality of MRBH. These results confirm the hypothesis that mining and beneficiation of iron ore contribute for the largest percentage of PM in the atmosphere of the study area, *Jardim Canadá* neighborhood. Additional samplings of atmospheric PM are being collected in other neighborhoods in order to validate these results for entire region of study.

Acknowledgements We would like to thank Centro de Desenvolvimento da Tecnologia Nuclear (CDTN), Coordenação de Aperfeiçoamento de Pessoal de Nível Superior (CAPES), Conselho

Nacional de Desenvolvimento Científico e Tecnológico (CNPq) and Fundação de Amparo à Pesquisa do estado de Minas Gerais (FAPEMIG).

References

1. Pope, III, C.A., et al.: Lung cancer, cardiopulmonary mortality, and long-term exposure to fine particulate air pollution. J. Am. Med. Assoc. **287**, 1132–1141 (2002)
2. Donaldson, K., et al.: Ultrafine (nanometre) particle mediated lung injury. J. Aerosol Sci. **29**(5/6), 553–560 (1998)
3. Viana, C.O., et al.: Epiphytic lichens on air biomonitoring in Belo Horizonte City, Brazil: a preliminary assessment. Int. J. Environ. Health **5**, 324–337 (2011)
4. Garçon, G., et al.: Polycyclic aromatic hydrocarbon coated onto Fe2O3 particles: assessment of cellular membrane damage and antioxidant system disruption in human epithelial lung cells (L132) in culture. Toxicol. Lett. **117**, 25–35 (2000)
5. Muxworthy, A.R., et al.: Magnetic properties and Mössbauer spectra of urban atmospheric particulate matter: a case study from Munich, Germany. Geophys. J. Int. **150**, 558–570 (2002)
6. Kopcewicz, B., Kopcewicz, M.: Long-term measurements of iron-containing aerosols by Mössbauer spectroscopy in Poland. Atmos. Environ. **35**, 3739–3747 (2001)
7. Kopcewicz, B., Kopcewicz, M.: Ecological aspects of Mössbauer study of iron-containing atmospheric aerosols. Hyperfine Interact. **126**, 131–135 (2000)
8. Kopcewicz, B., Kopcewicz, M.: Iron-containing atmospheric aerosols. Hyperfine Interact. **111**, 179–187 (1998)
9. Kopcewicz, B., Kopcewicz, M.: Mössbauer study of iron-containing atmospheric aerosols. Struct. Chem. **2**, 303–312 (1991)
10. Souza, Jr., P.A., et al.: Precise indication of air pollution sources. Hyperfine Interact. **139/140,** 641–649 (2002)
11. Figueiredo, M. do A., et al.: Óxidos de ferro de solos formados sobre gnaisse do Complexo Bação, Quadrilátero Ferrífero, Minas Gerais. Pesqui. Agropecu. Bras. **41**(2), 313–321 (2006)
12. Ferreira, B.A., et al.: Óxidos de ferro das frações areia e silte de um nitossolo desenvolvido de basalto. Rev. Bras. Cienc. Solo **27**, 405–413 (2003)
13. Curi, N., et al.: Espectroscopia Mössbauer na caracterização de compostos ferrosos em solos e sua relação com retenção de fósforo. Quim. Nova **31**(6), 1467–1471 (2008)
14. Van der Woude, F.: Mössbauer effect in a-Fe_2O_3. Phys. Status Solidi. **17**, 417–432 (1966)
15. Murad, E., Schwertmann, U.: Influence of al substitution and crystal size on the room-temperature Mössbauer spectrum of hematite. Clays and Clay Miner. **34**(1), 1–6 (1986)
16. Murad, E.: The characterization of soils, clays, and clay firing products. Hyperfine Interact. **111**, 251–259 (1998)
17. Carpenter, E.E., et al.: Magnetic and Mössbauer spectroscopy studies of nanocrystalline iron oxide aerogels. J. Appl. Phys. **99**, 08N711 (2006)

DOI 10.1007/978-94-007-6482-8_14
Reprinted from *Hyperfine Interactions*, DOI 10.1007/s10751-012-0743-z

Oxidation states of iron as an indicator of the techniques used to burn clays and handcraft archaeological Tupiguarani ceramics by ancient human groups in Minas Gerais, Brazil

D. L. Floresta · J. D. Ardisson · M. Fagundes · J. D. Fabris · W. A. A. Macedo

Published online: 31 January 2013

Abstract Ceramics of the Tupiguarani Tradition typically have in common the burning characteristics, their forms and decoration motifs. Dating such ceramic pieces with the radiocarbon method indicate that these artifacts were probably handcrafted between 1,500 and 500 years before the present. Fragments ceramic utensils were collected in the archaeological site of Beltrão, in the municipality of Corinto, state of Minas Gerais, Brazil. A fragment of about 50 mm in diameter and 15 mm thick, with a color gradation across the ceramic wall ranging from red, on one side, grayish, in the middle, and orange, on the opposite side, was selected for a more detailed examination. The fragment was transversely cut and a series of subsamples were separated from different points along the piece wall, in layer segments of ~3 mm. All subsamples were analyzed with Mössbauer spectroscopy at room temperature and 80 K. Results reveal that hematite is the magnetically ordered phase. A Fe^{2+} component (relative spectral area, 50 %) appears for the grayish subsample. According to these first results, the red subsample seems to be the side that had direct contact with fire used to burn the precursor clay in air. The grayish middle layer is probably due to the clay mixed with some ashes. Mössbauer data reveal that the orange layer, corresponding to the opposite side of the ceramic relatively to the direct fire, does contain about the same Fe^{2+}:Fe^{3+} ratio but much lower proportion of α-Fe_2O_3 than the red layer.

Keywords Tupiguarani ceramics · Archaeometry · Iron oxides

Thirteenth Latin American Conference on the Applications of the Mössbauer Effect, LACAME 2012, Medellín, Columbia, 11–16 November 2012.

D. L. Floresta (✉) · J. D. Ardisson · W. A. A. Macedo
Laboratory of Applied Physics, Development Center of Nuclear Technology (CNEN/CDTN), 31270-901 Belo Horizonte, Minas Gerais, Brazil
e-mail: dlf@cdtn.br

M. Fagundes · J. D. Fabris
Laboratory of Archaeology and Landscapes Studies, Federal University of Jequitinhonha and Mucuri Valleys (UFVJM), Diamantina, 39100-000 Minas Gerais, Brazil

1 Introduction

Ceramics of the Tupiguarani Tradition are found in many parts throughout the Brazilian territory. They typically have many similarities mainly concerning the primitive technique of burning conditions, their forms and decoration motifs. Data on selected pieces obtained with the radiocarbon method indicate that these artifacts are dated from between 1,500 and 500 years before the present [1, 2]. Fragments of a Tupiguarani pottery found in the archaeological site known as Beltrão (location map, Fig. 1), in the municipality of Corinto, state of Minas Gerais, were identified and collected by researchers of the Laboratory of Archaeology and Landscape Studies of the Federal University of Jequitinhonha and Mucuri Valleys, in Diamantina, also in Minas Gerais. A selected fragment of about 15 mm thick, with a color gradation across the ceramic wall ranging from red, on one side, grayish in the middle and orange on the opposite side, was chosen for a more detailed examination.

2 Methods

The sample collected from the Beltrão archaeological site was first transversely cut so that a quarter of the ceramic piece (Fig. 2) was removed for further image analyses. The segmented quarter was divided into three subsamples of 5.0 mm thick each according to their colors from the (i) orange; (ii) grayish and (iii) red layers. These three subsamples were metalized, by coating them with a palladium-gold alloy, and imaged in a Jeol scanning electron microscope (SEM), Model JSM-5310. The remaining part of the ceramic sample was also sampled in a series of subsamples from different depths across the wall, in layer segments of ~3 mm, from the orange side up to the opposite red side of the ceramic wall. The subsamples were ground and labeled according to the depth across the fragment wall: 0 mm (orange side), 3 mm, 6 mm, 9 mm, 12 mm and 15 mm (red side). The subsample 9 mm was actually obtained from the grayish middle layer. These powdered subsamples were analyzed with X-ray fluorescence (XRF) and diffraction spectroscopy (XRD) and ^{57}Fe Mössbauer spectroscopy at room temperature (~298 K) and at 80 K.

The X-ray fluorescence analysis was made with an energy dispersive X-ray fluorescence spectrometer, Shimadzu EDX-720, with rhodium tube and silicon-lithium detector. Each of these subsamples was put in a sample holder with a diameter of 5 mm; data were collected with sample under vacuum.

The powder X-ray diffraction data were collected in a Rigaku diffractometer equipment, model D\MAX ULTIMA, with $\theta - \theta$ goniometer and copper X-ray tube ($\lambda_{K\alpha}$ = 1,5418740 Å), within a scanning range of 4° to 80° 2θ; the velocity of the goniometer was set to 4° 2θ/min, with a current of 30 mA and tension of 40 kV.

^{57}Fe-Mössbauer spectra were obtained with a conventional transmission geometry setup, using constant acceleration regime, with a ~50 mCi ^{57}Co/Rh gamma-ray source kept at room temperature. Measurements were made at room temperature (RT) and 80 K. Isomer shift values were quoted relatively to the metallic αFe. Resonant curves were fitted to Lorentzian functions through least-square fitting

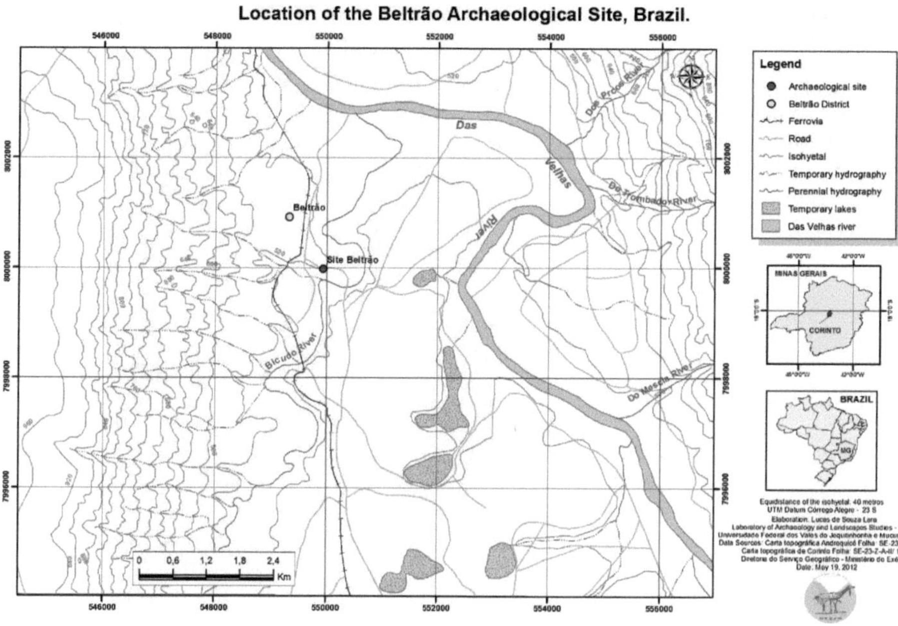

Fig. 1 Geographical location of the archaeological site of Beltrão, in Corinto, state of Minas Gerais, Brazil

Fig. 2 The ceramic fragment from the Beltrão archaeological site, with the *red* upper, the *orange* bottom and the *grayish* middle layers along the transverse cut. Colors are thought to reflect different oxidation conditions according to the local chemical environment during clay burning. Subsamples were separated from different depths all along the cut ceramic wall. The indicated scale is in cm

with software NORMOS™-90 (developed by R. A. Brand, at Laboratorium für Angewandte Physik, Universität Duisburg, D-47048, Duisburg-Germany).

3 Results and discussion

SEM images (Fig. 3) indicate that the material is uniformly similar for all depths. It was not possible to measure grain sizes as these ceramic materials consist of very finely divided and firmly compacted particles.

According to the X-ray fluorescence results, the elementary composition does not clearly vary with the depth in the sample. For the probed chemical elements Si, Al,

Fig. 3 Scanning electron microscope images for the three subsamples, according to their colors, obtained from the Tupiguarani ceramic sample of the archaeological Beltrão site: **a** *orange*, **b** *grayish* and *red* **c** layers

Fe, Cl and K there were found contents between 1 and 25 mass%. Contents for Ti, Mg, Ba, Ca and S were found to be less than 1 mass%. For all other elements contents were less than 0.1 mass% (Fig. 4).

The powder X-ray pattern obtained for each subsamples reveals essentially the same reflection peaks (Fig. 5); the pattern for the red subsample shows better resolution. An independent pattern was collected for the red subsample, with velocity of the goniometer corresponding to 0.16° 2θ/min. This pattern (Fig. 6) was used to better identify the main occurring minerals, by comparing it with the COD

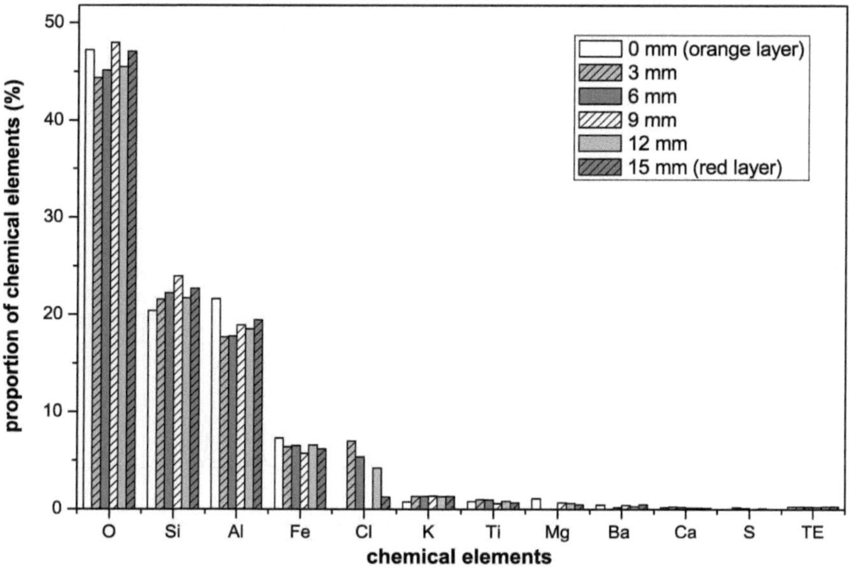

Fig. 4 Chemical composition data as determined with X-ray fluorescence spectroscopy. The labeled subsamples were collected from different depths all along the transected wall from the Tupiguarani ceramic sample of the archaeological Beltrão site. TE (meaning trace elements) = contents corresponding to less than 0.1 mass% (P, Pb, Cr, Mn, Ni, Zr, Sc, Ac, Ir, Rb, Zn, Y and Sr)

Fig. 5 Powder X-ray diffraction patterns for the subsamples of Tupigurani ceramic from the archaeological Beltrão site, collected along the transected section at different depths across the wall of the ceramic fragment

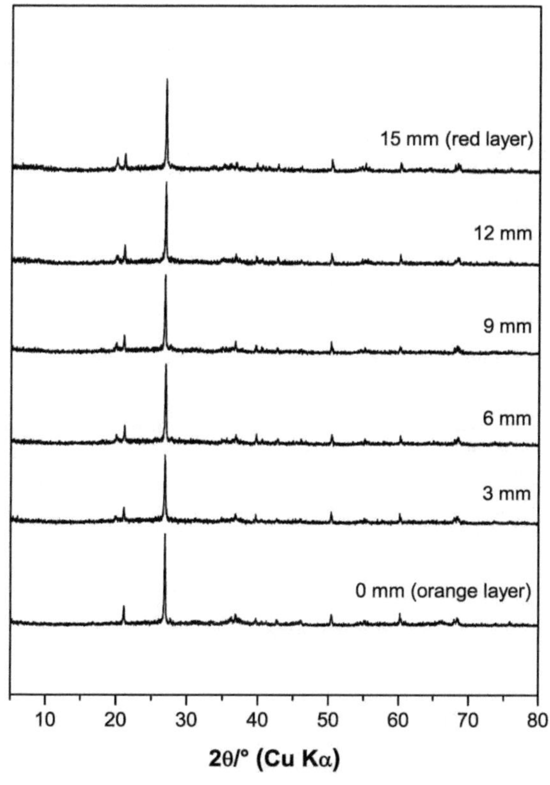

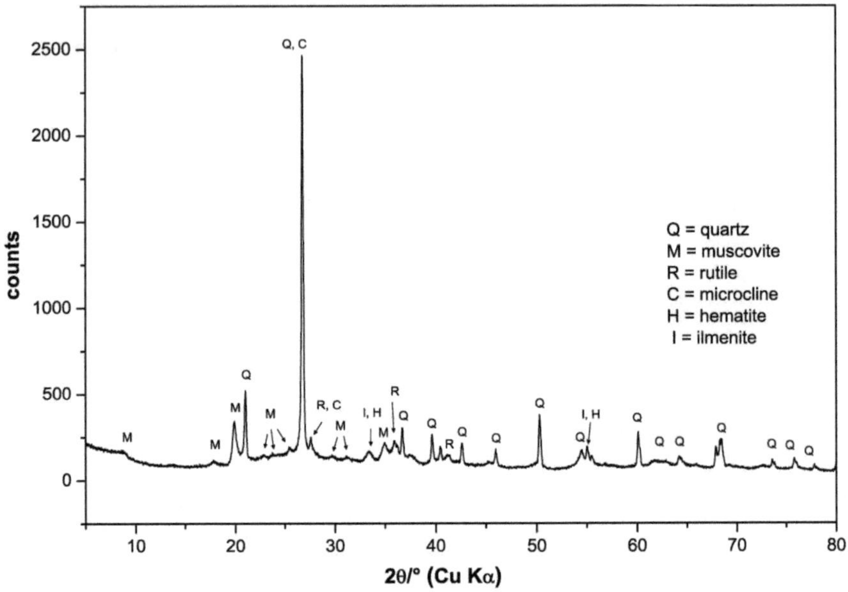

Fig. 6 Powder X-ray pattern for the subsample corresponding to the *red* layer collected from the transected wall of the Tupiguarani ceramic sample of the archaeological Beltrão site

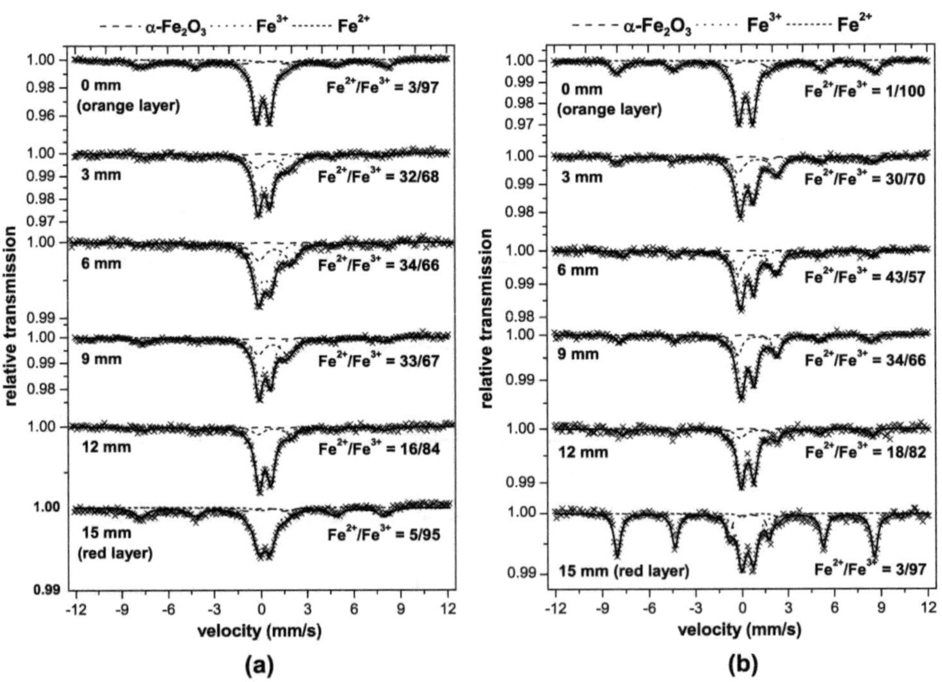

Fig. 7 ^{57}Fe-Mossbauer spectra for subsamples of materials collected from different depths across the wall of the Tupiguarani ceramic sample from the archaeological Beltrão site. Data were collected with samples at **a** room temperature (~298 K) and **b** 80 K

Table 1 ~298 K ^{57}Fe-Mössbauer parameters for subsamples from different depths along the transected wall of the Tupiguarani ceramic sample of the archaeological Beltrão site

Subsample	Site	δ (±0,0 5) mm/s	Δ, ε (±0.05) mm/s	B_{hf} (± 0,5) tesla	RA (±1) %
0 mm (orange layer)	α-Fe_2O_3	0.35	−0.14	49.1	37
	Fe^{3+}	0.27	0.83		60
	Fe^{2+}	0.98	1.80		3
3 mm	α-Fe_2O_3	0.32	−0.13	48.6	18
	Fe^{3+}	0.31	0.83		50
	Fe^{2+}	0.90	1.92		32
6 mm	α-Fe_2O_3	0.31	−0.16	48.8	15
	Fe^{3+}	0.34	0.78		51
	Fe^{2+}	1.01	1.99		35
9 mm	α-Fe_2O_3	0.30	−0.13	48.6	18
	Fe^{3+}	0.34	0.84		50
	Fe^{2+}	0.96	1.88		32
12 mm	α-Fe_2O_3	0.30	−0.16	48.9	19
	Fe^{3+}	0.34	0.76		66
	Fe^{2+}	0.92	1.99		15
15 mm (red layer)	α-Fe_2O_3	0.30	−0.16	49.0	38
	Fe^{3+}	0.29	0.73		56
	Fe^{2+}	0.96	1.83		6

δ = isomer shift relative to αFe; Δ = quadrupole splitting; ε = quadrupole shift; B_{hf} = hyperfine magnetic field; RA = relative spectral areas

Table 2 80 K ^{57}Fe-Mössbauer parameters for subsamples from different depths along the transected wall of the Tupiguarani ceramic sample of the archaeological Beltrão site

Subsample	Site	δ (±0,0 5) mm/s	Δ, ε (±0.05) mm/s	B_{hf} (±0,5) tesla	RA (±1) %
0 mm (orange layer)	α-Fe_2O_3	0.48	−0.11	52.1	40
	Fe^{3+}	0.40	0.90		59
	Fe^{2+}	1.12	2.45		1
3 mm	α-Fe_2O_3	0.47	−0.10	51.8	23
	Fe^{3+}	0.44	0.84		48
	Fe^{2+}	1.09	2.54		30
6 mm	α-Fe_2O_3	0.46	−0.08	50.9	15
	Fe^{3+}	0.47	0.85		42
	Fe^{2+}	1.08	2.53		43
9 mm	α-Fe_2O_3	0.41	−0.10	51.0	16
	Fe^{3+}	0.47	0.81		51
	Fe^{2+}	1.10	2.52		34
12 mm	α-Fe_2O_3	0.47	−0.09	51.1	18
	Fe^{3+}	0.44	0.84		64
	Fe^{2+}	1.11	2.34		18
15 mm (red layer)	α-Fe_2O_3	0.47	−0.20	52.1	60
	Fe^{3+}	0.44	0.78		37
	Fe^{2+}	1.14	2.36		3

δ = isomer shift relative to αFe; Δ = quadrupole splitting; ε = quadrupole shift; B_{hf} = hyperfine magnetic field; RA = relative spectral areas

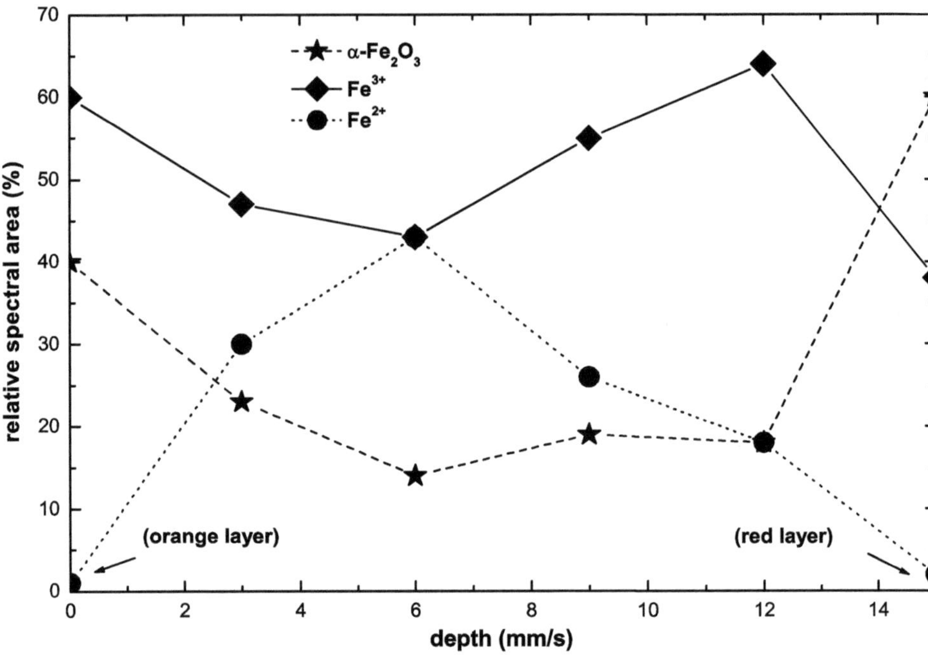

Fig. 8 Breakdown of Fe^{2+} and Fe^{3+}, from Mösbauer spectra for subsamples collected along the section across the transected wall of the ceramic sample from of the archaeological Beltrão site

(Crystallography Open Databases; updated on March 2012) database, through the software Match!® version 2.0.5, by Crystal Impact.

The powder XRD analysis revealed the occurrence mainly of quartz (60,9 %) and muscovite (28.9 %). Some reflection peaks were assigned to rutile (4.0 %), microcline (3.2 %), hematite (1.7 %) and ilmenite (1.2 %). The relative values are in agreement with the semiquantitative analysis performed by the software, based on the parameter I/I_c [3–5]. All subsamples are composed of very small particle sizes, and this tends to make a more reliable identification of other mineralogical phases more difficult.

The spectrum obtained with Mossbauer spectroscopy at room temperature and at 80 K are showed in Fig. 7 and corresponding parameters in Tables 1 and 2, respectively.

Results reveal that hematite is the magnetically ordered phase. The relative subspectral area (RA) for the Fe^{3+} distribution corresponds to between 40 % and 60 %. A ferrous component (RA = 43 %) appears in the central part of the whole spectrum for the grayish layer (Fig. 8). It is likely to be due to some reducing component occurring in this subsample [6–8].

4 Conclusion

From X-ray fluorescence spectroscopy data no significant difference was found for the elemental chemical composition along the transected section, in this archaeological ceramic sample. The main detected chemical elements, after oxygen, were aluminum, silicon and iron.

X-ray diffraction patterns reveal the main occurrence of two iron oxides: hematite (αFe_2O_3) and ilmenite ($FeTiO_3$).

The red layer is thought to be the external side of the ceramic utensil that had direct contact with fire used to burn the precursor clay in air, during its preparation.

The grayish middle layer has more probably resulted from burning clay mixed with some ashes containing residual carbon, under milder temperature than on the external parts.

Mössbauer data reveal that the orange layer, corresponding to the internal side of what is assumed to be an archaelogical ceramic pot, does contain about the same Fe^{2+}:Fe^{3+} ratio but much lower proportion of α-Fe_2O_3 than the external red layer.

The breakdown of iron chemical states as the Fe^{2+}:Fe^{3+} ratios thorough the transversal section of this pottery piece, as deduced from Mössbauer analyses, may be then taken as a primary evidence of the main conditions, meaning temperature gradient, mixture or direct contact of the precursor clay with chemical reducing agents, like carbon-rich ashes from the burnt wood, and origin and mineralogical nature of the clay itself, used to prepare the ceramic.

Acknowledgements Work financially supported by CNPq and FAPEMIG (Brazil; including grant # PPM 00419-10). CAPES (Brazil) grants the Visiting Professor PVNS fellowship to JDF at UFVJM. DLF attended the 2012 Latin American Conference on the Applications of the Mössbauer Effect, in Medellín, Colombia, with financial support by IFMG, CAPES and CDTN/CNEN.

References

1. Prous, A.: A pintura em cerâmica Tupiguarani. Ciência Hoje **36**, 22–28 (2005)
2. Cavalcante, L.C.D., Fabris, J.D., Lage, M.C.S.M.: Caracterização arqueométrica de pinturas rupestres pré-históricas, pigmentos minerais naturais e eflorescências salinas de sítios arqueológicos. (2012). 205 f. PhD Thesis - Federal University Minas Gerais, Chemistry Department. Available http://hdl.handle.net/1843/SFSA-8U6R88. Accessed 8 May 2012
3. Gražulis, S., Daškevič, A., Merkys, A., Chateigner, D., Lutterotti, L., Quirós, M., Serebryanaya, N.R., Moeck, P., Downs, R.T., LeBail, A.: Crystallography Open Database (COD): an open-access collection of crystal structures and platform for world-wide collaboration. Nucleic Acids Res. **40**, D420–D427 (2012)
4. Grazulis, S., Chateigner, D., Downs, R.T., Yokochi, A.T., Quiros, M., Lutterotti, L., Manakova, E., Butkus, J., Moeck, P., Le Bail, A.: Crystallography open database—an open-access collection of crystal structures. J. Appl. Cryst. **42**, 726–729 (2009)
5. Downs, R.T., Hall-Wallace, M.: The American mineralogist crystal structure database. Am. Mineral. **88**, 247–250 (2003)
6. Wagner F.E., Wagner, U.: Mössbauer spectra of clays and ceramics. Hyp. Interact. **154**, 35–82, (2004). doi:10.1023/B:HYPE.0000032113.42496.f2
7. Lumbreras, L.G. et al.: Mössbauer study of ceramic finds from the Galería de las Ofrendas, Chavín de Huántar. Hyp. Interact. **150**, 51–72 (2003). doi:10.1023/B:HYPE.0000007214.89534.37
8. Munayco, P., Scorzelli, R.B.: Characterization of the firing conditions of archaeological Marajoara pottery by Mössbauer spectroscopy and X-ray diffraction. Hyp. Interact. (2012). doi:10.1007/s10751-012-0656-x

DOI 10.1007/978-94-007-6482-8_15
Reprinted from *Hyperfine Interactions*, DOI 10.1007/s10751-013-0856-z

Comparison of methods to obtain ash from coal of the Southwest of Colombia

G. Medina · J. A. Tabares · G. A. Pérez Alcazar · J. M. Barraza

Published online: 27 March 2013

Abstract The method for concentration of mineral matter at low temperature (about 250 °C), called Low Temperature Ash (LTA) was applied to a sample of coal from the mine "Las Mercedes" located in Colombia southwestern. This method provides better information about the content of mineral matter in natural coal (NC), removing the organic matter more efficiently without significant transformations of mineral phases present in that coal. These results were observed through Mössbauer spectra and X-ray patterns taken from samples of NC, (LTA) and the conventional method of High Temperature Ash (HTA). The results show that the LTA process provides more representative data of the mineral phases for natural coal than that using the conventional HTA process.

Keywords Ash · Mineral · Organic · Phases

1 Introduction

Most of the minerals found in natural coal and in residuals from its combustion known as ashes, contain Fe as its more important component, which makes Mössbauer spectroscopy a suitable technique to obtain information on the main phases present in coal such as Siderite (S), Pyrite (P), Illite (I) and Jarosite (J) [1, 2].

Proceedings of the Thirteenth Latin American Conference on the Applications of the Mössbauer Effect, (LACAME 2012), Medellín, Colombia, 11–16 November 2012.

G. Medina (✉)
Departamento de Física, Grupo de nuevos sólidos (GINSAI),
Universidad Autónoma de Occidente, A. A. 2790, Cali, Colombia
e-mail: gmedina@uao.edu.co

J. A. Tabares · G. A. Pérez Alcazar
Departamento de Física, Grupo GMTF, Universidad del Valle, A. A. 25360, Cali, Colombia

J. M. Barraza
Escuela de Ingeniería, Universidad del Valle, A. A. 25360, Cali, Colombia

The most common method for determining ash content from coal is the High Temperature Ash (HTA) [3, 4]. This method permits the removal of all organic matter in a combustion process, leaving a residue associated with coal mineral matter. However, by this method, some mineral phases present in coal are mainly transformed into oxides and some other products attach to the organic matter. Therefore, the results from applying this method could be less reliable than other methods where the losses of mineral phases present in coal [5] are reduced significantly. A more reliable method is the low temperature ash (LTA) method, which consists in eliminating organic matter at a controlled temperature of 250 °C, with a constant flow of air, in a fluidized bed [6]. At this temperature phase transformations are not present [7–9], giving a more representative result of the mineral content in natural coal (NC).

The aim of this paper is to report the results of the mineral content of coal from a southwestern region of Colombia, through HTA and LTA processes. X-ray and Mössbauer spectroscopy techniques were used to probe their effectiveness.

2 Experimental procedure

A sample of natural coal (NC) from the "The Mercedes" mine (located in the southwest section of Colombia) was selected and submitted for combustion for 4 hours. This sample has a low content of mineral matter (ash 11.7 %); according to the result obtained by the HTA process (ASTM D3174). The sample used was first powdered to particles size of $\leq$200 μm. The first hour, a sample of 1.0 g was exposed to a temperature of 550 °C at a rate of 8 °C/min, the second hour, at 750 °C at a rate of 3 °C/min, and the last two hours at a constant temperature of 750 °C. The sample was cooled to room temperature in a desiccator before being weighted and examined.

Mineral matter in a sample of the same coal was also separated by the low temperature ash (LTA) process. The setup used for the LTA method consisted of a Pyrex tube reactor of 30.0 cm long × 2.5 cm diameter with a porous frit that supported a fluidized bed of pulverized coal (90–200 μm) [4]. The temperature of the reactor was monitored by a thermocouple in the exterior wall, while the coal temperature was monitored by a thermocouple located in the centre of the fluidized bed. Oxidation of the coal samples was accomplished by heating the fluidized coal at $\approx$10 °C/min up to 250 °C and maintaining this temperature for a designated period of time. Fluidization was achieved with dry air flowing at $\approx$500 cm^3/min. Mössbauer measurements at room temperature were performed using a conventional multichannel spectrometer with a ^{57}Co source. Spectra were fitted using the VARFIT program and the α-Fe as the calibration sample. The X-ray powder diffraction patterns were obtained using a conventional X-ray spectrometer, using Ni-filtered Cu K_α radiation. Typical scan speeds were 0.02° 2θ min^{-1}, and patterns were fitted using the MAUD program.

3 Results and discussion

Results from LTA and HTA processes, gave residual mineral matter. In order to establish the efficiency of the LTA process in identifying 1 the mineral matter

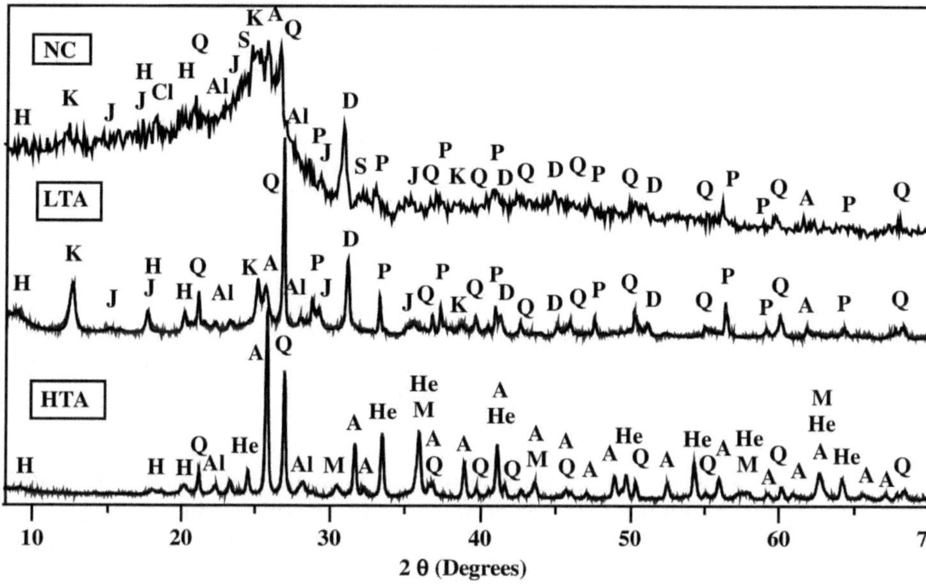

Fig. 1 X-ray diffractions results obtained on the natural coal (NC), low temperature ash (LTA) and high temperature ash (HTA), respectively. The main minerals identified by NC - LTA are: K, kaolinite; Q, quartz; D, dolomite; A, anhydrite; J jarosite; Al, albite; H, halloysite; and P, pyrite. Minerals identified by HTA are: A, anhydrite; He, hematite; Q, quartz; Al, albite; H, halloysite; and M, magnetite

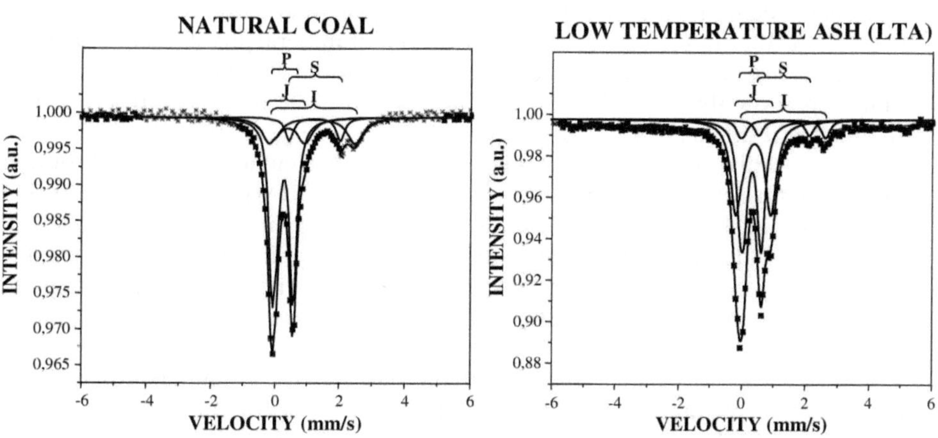

Fig. 2 Mössbauer spectroscopy results obtained on natural coal (NC) and low temperature ash (LTA) samples, respectively. Phases shown: siderite (S), pyrite (P), illite (I) and jarosite (J)

without significant phases changes, tests were carried out by x-ray diffraction and Mössbauer spectroscopy on NC, LTA and HTA samples. These results are shown in Figs. 1 and 2, respectively.

Figure 1 shows the results obtained by X-ray diffraction for the NC, LTA and HTA samples, respectively. In the diffractogram of the NC sample a background

Table 1 Mössbauer parameters and spectral areas of the phases found in the CN and CBT samples.

Mineral phases	Mössbauer parameters	Sample NC	Sample LTA
Pyrite	δ (mm/s) + 0.01	0.35	0.36
	Δ (mm/s) + 0.01	0.59	0.58
	% Spectral Area	57	56
Illite	δ (mm/s) + 0.01	1.24	1.25
	Δ (mm/s) + 0.01	2.58	3.02
	% Spectral Area	17	20
Siderite	δ (mm/s) + 0.01	1.30	1.32
	Δ (mm/s) + 0.01	1.58	1.45
	% Spectral Area	8	10
Jarosite	δ (mm/s) + 0.01	0.47	0.55
	Δ (mm/s) + 0.01	1.04	1.15
	% Spectral Area	16	13

line is observed with a widening trace in the 2theta angle region of 15–30°, which could be associated with the presence of organic and amorphous material [5].

It is also possible to identify in the LTA produced samples the major picks corresponding to the main crystalline phases, such as: kaolinite (K), quartz (Q), dolomite (D), anhydrite (A), jarosite (J), albite (Al), halloysite (H), and pyrite (P); while in the HTA sample some other mineral phases can be identified: anhydrite (A), hematite (He), quartz (Q), albite (Al), halloysite (H), and magnetite M.

In the diffraction patterns of the LTA and HTA samples it is possible to note a very flat horizontal background line, with an appreciable decrease in the widening trace respect to that observed in the X-ray diffraction pattern of the NC sample. These results demonstrate that there is a better removal of organic matter by these processes. Furthermore, it is observed that the major mineral phases identified in the natural coal, are also present in the LTA sample without significant changes. These results also show that the LTA process removes organic matter more efficiently, without significant changes in mineral phases.

However, unlike the results obtained for the LTA sample, in the HTA sample new mineral phases such as hematite (He) and magnetite (M) can be identified. The disappearance of phases for kaolinite (K), jarosite (J), pyrite (P) and dolomite (D) is notorious. A decrease of the intensity peak for halloysite (H) and an increased of the intensity peak for anhydrite (A) are observed. These results show that the HTA process removes organic matter too, but transforms the mineral phases present in coal.

Figure 2 shows Mössbauer spectra for the NC and LTA samples, respectively. These spectra show that the Fe-containing mineral phases present in the NC sample persist after a CBT process. In these spectra it is possible to observe the following phases: siderite (S), pyrite (P), illite (I) and jarosite (J). In the Table 1 the Mössbauer parameters and spectral areas of the phases found in the NC and LTA samples are shown and it is possible to observe a decrease in the spectral area of the P and J, and an increase for the spectral area of the S and I phases. These behaviors may be associated with the influence of oxygen on these phases during the LTA process.

The results found by the techniques mentioned above confirm the effectiveness of the process of LTA to remove the organic matter having a higher concentration of mineral matter, without considerable changes of their original phases.

4 Conclusions

Results obtained by X-ray diffraction and Mössbauer spectroscopy on CN, HTA and LTA samples, show that the low temperature ash (LTA) process is very reliable to eliminate organic matter and condense the mineral matter without a big loss or significant transformations of the mineral phases present in coal. This occurs because the sample is exposed to 250 °C, which is below the temperature at which transitions occur in the respective minerals phases [3].

References

1. Fajardo, M., Mojica, J., Barraza, J., Pérez Alcázar, G.A., Tabares, J.A.: Mineral identification in Colombian coals using Mössbauer spectroscopy and X-ray diffraction. Hyperfine Interact. **122**, 129–138 (1999)
2. Gracia, M., Marco, J.F., Gancedo, J.R.: Uses and perspectives of Mössbauer spectroscopic studies of iron minerals in coal. Hyperfine Interact. **122**, 97–114 (1999)
3. Vassilev, S.V., Vassileva, C.G.: Occurrence, abundance and origin of minerals in coals and coal ashes. Fuel Process. Technol. **48**, 85–106 (1996)
4. Martínez-Tarazona, M.R., Martínez-Alonso, A., Tascón, J.M.D.: Mineral matter in Spanish bituminous and brown coals. Part. 2. Mineral matter quantification. Erdol Kohle Erdgas Petrochem. **46**, 202–209 (1993)
5. Stanislav, V., Vassilev, Tascón, J.M.D.: Methods for characterization of inorganic and mineral matter in coal: a critical overview. Energy Fuels **17**, 271–281 (2003)
6. Herman, R.G., Simmons, G.W., Cole, D.A., Kuzmicz, V., Klier, K.: Catalytic action of minerals in the low temperature oxidation of coal. Fuel **63**, 673–677 (1984)
7. Martínez-Tarazona, M.R., Spears, D.A., Palacios, J.M., Martínez-Alonso, A., Tascón, J.M.D.: Mineral matter in coals of different rank from Asturian Central basin. Fuel **71**, 367–372 (1992)
8. Sentorun, C., Schobert, H.H.: Determination of the mineral matter contents of some Turkish lignites by low-temperature ashing. In: Proc. 9th Intern. Conf. Coal Science, vol. 1, pp. 401–404. Germany (1997)
9. Helle, S., Alfaro, G., Tascón, J.M.D.: Mineralogical and chemical characterisation of coals from Southern Chile. Int. J. Coal Geol. **44**, 85–94 (2000)

DOI 10.1007/978-94-007-6482-8_16
Reprinted from *Hyperfine Interactions*, DOI 10.1007/s10751-013-0859-9

Whiteness process of tile ceramics: using a synthetic flow as a modifier agent of color firing

G. R. dos Santos · M. C. Pereira · M. Olzon-Dionysio · S. D. de Souza · M. R. Morelli

Published online: 5 April 2013

Abstract Synthetic flow is proposed as a modifier agent of color firing in tile ceramic mass during the sinterization process, turning the red color firing into whiteness. Therefore, the ^{57}Fe Mössbauer spectroscopy was used to understand how the interaction of the iron element in the mechanism of color firing mass occurs in this system. The results suggest that the change of color firing can be alternatively due to two main factors: (i) diluting the hematite content in the sample because of the use of synthetic flow and (ii) part of the hematite is converted in other uncolored crystal structures, which makes the final color firing lighter.

Keywords Synthetic flow · Whiteness · Ceramic materials · Mössbauer spectroscopy

1 Introduction

Ceramic tile industries in Brazil are the greatest producers and consumers in the world competing with countries like China, Italy, Spain and others [1, 2].

The typology of higher consumption in the domestic market is the ceramic tiles manufactured by the "dry process" in Santa Gertrudes, São Paulo, Brazil, where

Proceedings of the Thirteenth Latin American Conference on the Applications of the Mössbauer Effect, (LACAME 2012), Medellín, Columbia, 11–16 Novermber 2012.

G. R. dos Santos (✉) · M. R. Morelli
Departamento de Engenharia dos Materiais, Universidade Federal de São Carlos, 13565-905, São Carlos, SP, Brazil
e-mail: geocris.rodrigues@gmail.com

M. C. Pereira
Instituto de Ciência, Engenharia e Tecnologia, Universidade Federal dos Vales do Jequitinhonha e Mucuri, 39803-371, Teófilo Otoni, MG, Brazil

M. Olzon-Dionysio · S. D. de Souza
Departamento de Física, Universidade Federal de São Carlos, 13565-905, São Carlos, SP, Brazil
e-mail: maristolzon@hotmail.com

the industries use only red color firing clay. However, the fastest growing typology globally is Brazilian porcelain tiles (or stoneware tiles), a production which reached 60 million square meters in 2010 [1]. The reason for this growth is directly related to its highly technical and aesthetic performance, and consequently, with higher added value products [1, 2].

Among the basic requirements for the masses of porcelain tile is the white color firing, which restricts the use of raw materials containing high levels of chromophore oxides (mainly Fe_2O_3 and TiO_2) so that the feldspars are used as the main fluxes of these compositions. Furthermore, clay found widely in Brazil is also not used in the porcelain tile masses due to the high content of iron and titanium oxides, thus causing a massive dependence on foreign markets where the large deposits of white clay is [3].

Knowledge in the area of developing whitening materials in ceramic is extremely limited in databases of papers, books or patents, and at the moment, there is no product on the market showing similar properties.

To the best of our knowledge, only one study about the whitening process of ceramic materials has been reported [4]. In this study, Yatsenko et al. reported that the use of 15–20 wt.% of calcite combined with clays containing large amounts of iron oxide after firing leads to the production of white ceramic. The whitening phenomenon occurs due to the crystallization of the okermanite ferrous phase, dicalcium ferrite, wollastonite and anorthite, mainly melilite, i.e., depending on the proportion of Fe^{2+} or Fe^{3+} present in the crystalline phases, the color effect performs differently. Thus, the increase in the concentration of calcium carbonate in the composition helps the growth of these crystals. On the other hand, it also increases the formation of free CaO and porosity, which damages the structure of the ceramic because it is less resistant to the compression [4].

In this short description of the state of art, it is important to reaffirm that the publications on this topic are extremely limited, and most of the knowledge that has been applied directly to traditional ceramic comes from this work [5, 6].

Thus, there is continuing research for new materials that can replace or aid the porcelain tile compositions. Therefore, this work proposes the development of a novel process to obtain white ceramic using a synthetic flow as a mediator agent of color firing. This process can partially replace feldspars and phyllites in these compositions and change the coloring of a ceramic mass during sinterization.

The synthetic flow proposed herein consists of a viscous flow (glassy phase), which has a characteristic of an amorphous solid. In general, the chemical composition of the synthetic flow is formed by SiO_2–Al_2O_3–M_2O–M'O, where M is selected from the elements Na, K or a combination of them, and M' is selected from among the elements Mg, Ca or a combination of them. This chemical composition promotes the devitrification of the flow during the firing cycle in crystals from the clinopyroxenes family [7–11].

2 Materials and methods

The ceramic sample used in this work, labeled SG, was obtained from Santa Gertrudes, São Paulo, Brazil. The iron oxide contents present in this sample are approximately 5 wt.%, which depreciates the value of these ceramics.

In order to produce an added value material, 45 wt.% of synthetic flow (SF) was added to the 55 wt.% original ceramic sample and fired at 1,100 °C for 5 min in air atmosphere. This sample was labeled 45SF/55SG. The chemical composition of the synthetic flow used in this work was 35–75 wt.% SiO_2, 0.5–5 wt.% Al_2O_3, 0.5–5 wt.% K_2O, 10–15 wt.% Na_2O, 5–12 wt.% CaO, 1–5 wt.% MgO.

X-ray diffraction (XRD) in powder form was carried out using CuKα with a Rigaku Geigerflex diffractometer equipped with a graphite diffracted-beam monochromator. Data were collected from 10 to 80° 2θ in steps of 0.02° per 5 s iteration. Silicon was used as an external standard.

Mössbauer spectra were collected in a constant acceleration transmission mode with a 10 mCi ^{57}Co/Rh source. The spectra of all samples were taken at 298 K and Doppler velocities ranged between ±11 mm s^{-1}. The data were stored in a 512-channel MCS memory unit and were fitted using the NORMOS program. Isomer shifts were quoted relative to α-Fe.

3 Results and discussion

Chemical analysis shows the SG sample to consist essentially of 68.2 wt.% SiO_2, 14.2 wt.% Al_2O_3, 5.3 wt.% Fe_2O_3, 3.6 wt.% K_2O, 2.7 wt.% Na_2O, 1.9 wt.% MgO, 0.7 wt.% CaO and 0.6 wt.% TiO_2.

45 wt.% synthetic flow were mixed to 55 wt.% SG sample and heated at 1,100 °C for 5 min to produce a white ceramic (Fig. 1). A preliminary test showed that the whitening effect is clearly observed when the synthetic flow is added in levels above 30 wt.% in the SG sample. Moreover, the fusibility of the compositions increases as the proportion of the synthetic flow increases. Consequently, this leads to a densification, reduction in porosity and mechanical strength increasing the ceramic tiles, properties which are very favorable for manufacturing ceramic tiles because of their highly technical performance, such as porcelain tiles [12–19].

To fully understand the origin of the color change in the ceramics, XRD and ^{57}Fe Mössbauer measurements were carried out. The powder XRD patterns of all samples are shown in Fig. 2. The SF sample showed an amorphous region between 15 to 40° 2θ, which is characteristic of the synthetic flow produced in this work. The SG sample basically consists of quartz. A minor contribution of illite and albite was also observed. The XRD pattern of the SG sample after heating at 1,100 °C, labeled SG1100, is very similar to the XRD pattern of the SG sample before the thermal treatment. This sample is still dominated for quartz. However, illite should be converted into cristobalite at 1,100 °C.

After heating the mixture of synthetic flow and SG ceramic at 1,100 °C, the XRD pattern of the 45SF/55SG sample (Fig. 2) was completely different from the other samples, indicating that the use of synthetic flow can promote structural changes in the original ceramic sample. The reflections of quartz decreased and new reflections due to diopside appeared in the diffraction pattern of the 45SF/55SG sample, suggesting that the synthetic flow can promote the migration of Si from quartz structure to form a new crystal, i.e. diopside. Although the XRD data were fitted assuming these dominant phases, contributions of other minor crystalline species cannot be excluded.

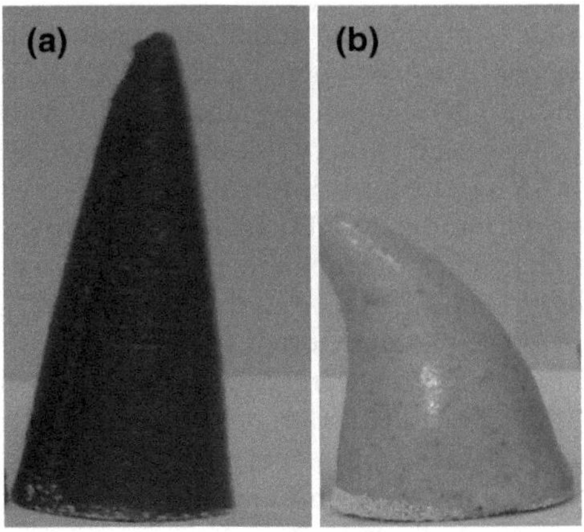

Fig. 1 Digital image of samples (**a**) SG1100 and (**b**) 45SF/55SG, after firing at 1,100 °C

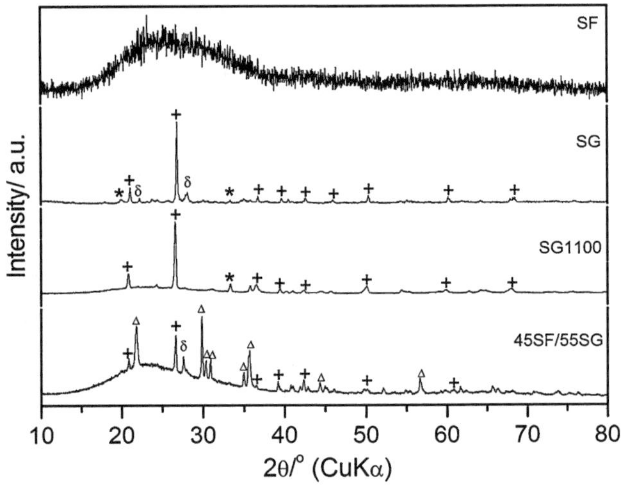

Fig. 2 X-ray diffraction patterns of samples SF (synthetic flow), SG (original ceramic, Santa Gertrudes), SG1100 (original ceramic heated at 1,100 °C) and 45SF/55SG (mixture of 45 wt.% synthetic flow +55 wt.% original ceramic heated at 1,100 °C). Legend: δ = albite, + = quartz, *= illite, Δ = diopside

Therefore the XRD data shows that the thermal treatment of the original ceramic sample (red color) in the presence of a synthetic flow induces structural changes that lead to the formation of new crystals (lighter color). In spite of this, the origin of this color change could not be completely understood based only on these data. The red color observed in the SG sample is mainly due to the presence of iron. However, it was not possible to specify iron in these samples in view of its low concentration in them.

In order to understand the interaction of iron and different matrixes during the sinterization process at 1,100 °C, ^{57}Fe Mössbauer spectroscopy measurements were carried out. The spectrum of SG1100 sample (Fig. 3, Table 1) indicated that 62 % of the total iron is in the hematite structure and this is the main reason of the red color of that sample. Approximately another 13 % is due to Fe^{3+} located in octahedral sites and about 25 % of Fe^{3+} in tetrahedral sites. However, further studies are needed to define unequivocally the Fe sites in this sample.

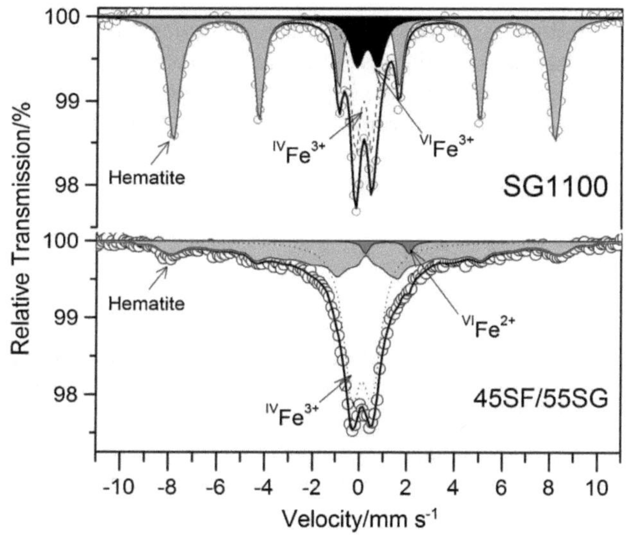

Fig. 3 298 K Mössbauer spectra of SG1100 and 45SF/55SG samples

Table 1 Mössbauer parameters of the samples SG1100 and 45SF/55SG at room temperature

Sample	δ/mm s^{-1}	Δ, ε/mm s^{-1}	B_{hf}/T	Γ/mm s^{-1}	RA/%	^{57}Fe site
SG100	0.38(2)	−0.19(2)	49.7(3)	0.37(1)	62(1)	Hematite
	0.38(1)	0.88(3)	–	0.39(1)	13(1)	$^{VI}Fe^{3+}$
	0.26(2)	0.67(3)	–	0.34(1)	25(1)	$^{IV}Fe^{3+}$
45SF/55SG	0.38(1)	−0.15(2)	50.6(4)	0.35*	45(1)	Hematite
	1.27(2)	1.85(2)	–	0.30(1)	2(1)	$^{VI}Fe^{2+}$
	0.23(1)	0.76(2)	–	0.36(1)	52(1)	$^{IV}Fe^{3+}$

*Parameter fixed during the fit

After the thermal treatment in the presence of the synthetic flow, the hematite content decreases from 75 % in the SG1100 sample to 45 % in the 45SF/55SG sample. Another 53 % of Fe was found in the tetrahedral site and only 2 % is due to Fe^{2+} in octahedral coordination.

Thus, we suggest two alternative ways that can contribute to the whitening process of ceramics:

(i) Fe can migrate from the hematite structure to other crystal structures;
(ii) The addition of synthetic flow causes a dilution effect of hematite, thus decreasing the intensity of the red color.

4 Conclusions

In this work, a novel process for ceramic whitening was developed based on thermal treatment at 1,100 °C of a mixture of red ceramic and synthetic flow. The thermal treatment in the presence of the synthetic flow suggests the reduction in the hematite content in the ceramic, which is mainly responsible for the red color in these samples. This process led to the production of a white ceramic, which has a market added value. Nevertheless, further studies are needed to completely understand the mechanism of whiteness of ceramics promoted by the synthetic flow.

Because of the huge gap of knowledge in the area of ceramic whitening, this work could bring a significant scientific-technological contribution to future studies.

Acknowledgement We would like to thank FAPEMIG, FAPESP and CNPq for the financial support of this work.

References

1. Anfacer.: Setor.www.anfacer.org.br (2012). Acessed 31 Oct 2012
2. Junior, M.C., et al.: Panorama e Perspectivas da Indústria de Revestimentos Cerâmicos no Brasil. Ceram. Ind. **15**(3), 7–18 (2010)
3. Motta, J.F.M., Zanardo, A., Junior, M.C.: As Matérias-Primas Cerâmicas. Parte I: O Perfil das Principais Indústrias Cerâmicas e Seus Produtos. Ceram. Ind. **6**(2), 28–39 (2001)
4. Yatsenko, N., et al.: Regularities of tinting ceramics based on low-melting clays. Glass Ceram. **63**(7), 265–266 (2006)
5. Santos, G.R., Salvetti, A.R., Morelli, M.R.: Estudo de um aditivo descolorante de massas para revestimentos cerâmicos. Anais 54° Congresso Brasileiro de Cerâmica: Foz do Iguaçu—PR (2010)
6. Santos, G.R., et al.: Composições de vidro soda-cal com massas cerâmicas vermelhas, processo de obtenção de artefatos cerâmicos, artefatos cerâmicos assim obtidos e uso dos mesmos. INPI. PI1102416-0 (2011)
7. Zanotto, E.D.: Crystallization of liquids and glasses. Braz. J. Phys. **22**(2), 77–84 (1992)
8. Barbieri, L., et al.: Crystallization of (Na_2O–MgO)–CaO–Al_2O_3–SiO_2 glassy systems formulated from waste products. J. Am. Ceram. Soc. **83**(10), 2515–2520 (2000)
9. Saiello, S., et al.: Crystallization of glasses in the 2 $Na_2O.CaO.3SiO_2$ -$Na_2O.2CaO.3SiO_2$ Composition Range. Thermochim. Acta **52**(1–3), 263–269 (1982)
10. Poirier, T., et al.: Formation of crystalline phases in (SiO_2-CaO-Fe_2O_3)-TiO_2 based glasses. Mater. Lett. **59**(2–3), 308–312 (2005)
11. Navarro, J.M.F.: El vidro. CSIC, España (2003)
12. Matteucci, F., Dondi, M., Guarini, G.: Effect of soda-lime glass on sintering and technological properties of porcelain stoneware tiles. Ceram. Int.**28**(8), 873–880 (2002)
13. Martín-Márquez, J., Rincón, J.M., Romero, M.: Effect of microstructure on mechanical properties of porcelain stoneware. J. Eur. Ceram. Soc. **30**(15), 3063–3069 (2010)
14. Suvaci, E., Tamsu, N.: The role of viscosity on microstructure development and stain resistance in porcelain stoneware tiles. J. Eur. Ceram. Soc. **30**(15), 3071–3077 (2010)
15. Martín-Márquez, J., Rincón, J.M., Romero, M.: Effect of firing temperature on sintering of porcelain stoneware tiles. Ceram. Int. **34**(8), 1867–1873 (2008)
16. Rambaldi, E., et al.: Using waste glass as a partial flux substitution and pyroplastic deformation of a porcelain stoneware tile body. Ceram. Int. **33**(5), 727–733 (2007)
17. Zanelli, C., et al.: Glass-ceramic frits for porcelain stoneware bodies: effects on sintering, phase composition and technological properties. Ceram. Int. **34**(3), 455–465 (2008)
18. Luz, A.P., Ribeiro, S.: Use of glass waste as a raw material in porcelain stoneware tile mixtures. Ceram. Int. **33**(5), 761–765 (2007)
19. Tucci, A., et al.: Use of soda-lime scrap-glass as a fluxing agent in a porcelain stoneware tile mix. J. Eur. Ceram. Soc. **24**(1), 83–92 (2004)

DOI 10.1007/978-94-007-6482-8_17
Reprinted from *Hyperfine Interactions*, DOI 10.1007/s10751-013-0866-x

Minerals of a soil developed in the meteoritic crater of Carancas, Peru, and evidences of phase changes on the impact event

María L. Cerón Loayza · Jorge A. Bravo Cabrejos

Published online: 23 April 2013

Abstract We report studies about the phase transformations in the soil of the Carancas meteoritic crater located in an inhabited area near the town of Carancas, in the Region of Puno, about 1,300 km southeast of Lima, Peru. The studies by energy dispersive X-ray fluorescence, X-ray diffractometry and transmission Mössbauer spectroscopy (at RT and 77 K) reveal that the sample consists mainly of quartz, albite and impactites such as coesite and stishovite (SiO_2) that have experienced phase metamorphism or alterations, related to high pressures and temperatures, forming their corresponding polymorphs. The occurrence of these phases, with high content of SiO_2, in the soil of the crater strengthens the hypothesis of its origin by metamorphism induced by impact; we observed also a magnetic sextet on the Mössbauer pattern, assigned to the Fe^{3+} in hematite, and three paramagnetic doublets, two of them associated with structural Fe^{3+} and Fe^{2+} cations, respectively, in illite and/or montmorillonite, and a third one due to an unidentified Fe^{3+} site.

Keywords Transmission Mössbauer spectroscopy · X-ray diffractometry · X-ray fluorescence · Metamorphism · Coesite · Stishovite · Impactites

1 Introduction

Continuing with our previous studies of the Carancas meteorite [1] in Puno, Peru, we now present results related to studies of three soil samples collected from different

Proceedings of the Thirteenth Latin American Conference on the Applications of the Mössbauer Effect, (LACAME 2012), Medellín, Colombia, 11–16 November 2012.

M. L. Cerón Loayza (✉) · J. A. Bravo Cabrejos
Laboratorios de Análisis de Suelos y Arqueometría, Facultad de Ciencias Físicas, Universidad Nacional Mayor de San Marcos, Apartado Postal 14-0149, Lima 14, Perú
e-mail: malucelo@hotmail.com

J. A. Bravo Cabrejos
e-mail: jbravo8@hotmail.com

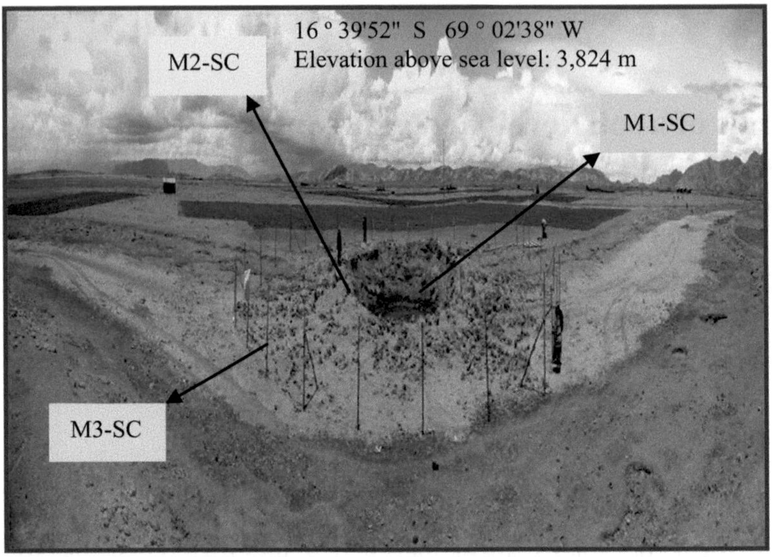

Fig. 1 The Carancas crater. It is shown the site positions from where the samples were collected (the photo was taken in November 2007)

positions within the crater as seen in Fig. 1 (located at 16° 39′52″ S, 69° 02′38″ W, 3,824 meters above sea level) which witness events of phase transformations. In general, minerals show a unique response when subjected to shock waves due to the occurrence of high pressures and temperatures. The effects caused by transforming them to other minerals are an unmistakable fingerprint of the impact event [2]. The metamorphic minerals may be regarded as a relic of the meteorite impact witnessing the age of the geological event.

A shock wave is a compression wave that propagates with supersonic velocity and abruptly affects compressing, heating and deforming plastically the solid matter. In the sense of an impact of a planetary body on the Earth surface, the shock waves are fundamentally different from seismic waves (elastic) [3]: the shock forms an internal compression wave that reacts almost instantly against an internal decompression wave that fractures the crust and projects pulverized materials into the atmosphere. At the instant of the impact, shock waves are generated from an abrupt increase of both pressure and temperature, drastically changing or virtually destroying the structure of rocks, leading to their fracture; the high temperature causes an impact metamorphism, by reforming minerals to generate new structures such as coesite y stishovite (high pressure modifications of SiO_2, with the same composition of quartz but with a more compact structure). If the meteorite is sufficiently large, the rocks may even melt and vaporize. In Fig. 2, we see the phase diagrams of coesite and stishovite, which are minerals that have been found from this study.

The purpose of this work was to study the phase changes under metamorphism of soil minerals from samples collected from the meteoritic crater of Carancas in Puno, Peru, in an attempt to characterize the main structural properties of these samples applying the analytical techniques available in our Faculty of Physical Sciences-UNMSM, such as energy dispersive X-ray fluorescence (EDXRF), X-ray diffractometry (XRD) and particularly transmission ^{57}Fe Mössbauer spectroscopy

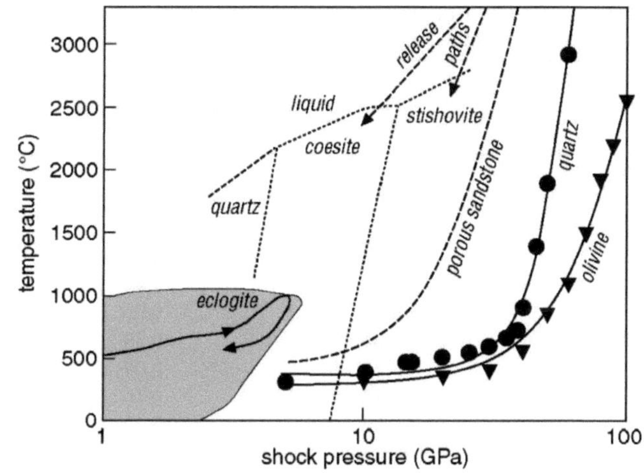

Fig. 2 Phase diagram depicting the pressure-temperature conditions reached in quartz, olivine (*solid line*), and porous sandstone (*long dashed line*) by shock compression (data after Wackerle [4], Kieffer et al. [5], Holland and Ahrens [6]). The release paths hold for porous quartz, which first melts on loading and then solidifies on cooling as coesite or stishovite. The equilibrium phase boundaries between quartz, coesite, stishovite and liquid are drawn as *dashed lines*

(TMS) for being an isotope selective technique. This meteorite impact caused a great explosion and formed a crater with approximately 20 m in diameter and 3 m in depth. At the site, the underground water table lies at a depth of about 3 m; therefore, the crater lies partially filled with water.

2 Materials and methods

The collection of the samples used in this study was undertaken two weeks after the meteorite impact that took place on September 15, 2007. These samples were obtained in collaboration with Dr. J. Itshitsuka, Instituto Geofísico del Perú, and were classified as:

M1-SC: Sample from inside the crater, reddish color,
M2-SC: Sample from the central part of the crater, grayish color, and
M3-SC: Sample from outside the crater, brown color.

These materials were ground in an agatha mortar to a fine 160 μm mesh powder for analysis by XRD and TMS at RT; pellets were prepared for EDXRF analysis. The TMS spectra at 72 K were taken in collaboration with Dr. J.M. Grenech, Laboratoire de Physique de l'État Condensé, Université du Maine, Le Mans, France, using about 150 mg of each sample powder.

2.1 Analysis by Energy Dispersive X-Ray Fluorescence (EDXRF)

The elemental composition analysis was performed with a portable EDXRF AMPTEK instrument. This instrument uses an X-ray tube with an Ag anode operating at 30 kV and 30 μA. This instrument allows the identification of elements with $Z > 12$ (magnesium). Results of these quantitative analyses are given in Table 1. This was done by fitting the experimental EDXRF spectrum with a program that simulates EDXRF spectra based on the fundamental parameters model; this program is written in FORTRAN and simulates all physical processes that affect an X-ray spectrum: energy distribution of the primary and secondary radiation,

Table 1 Results from the quantitative elemental analysis of the studied samples compared to the meteorite elemental composition

Element	Samples Concentration (% weight)			
	Meteorite	M1-SC	M2-SC	M3-SC
Al	7.50	9.0	9.8	9.7
Si	13.0	27.0	21.2	24.0
S	0.90	0.01	0.01	0.01
Cl	0.90	0.70	0.47	0.76
K	0.35	1.43	1.36	1.05
Ca	0.66	0.92	1.37	3.46
Ti	0.04	0.38	0.30	0.37
Cr	0.11	0.09	0.01	0.01
Mn	0.12	0.10	0.07	0.12
Fe	6.00	3.39	1.69	3.30
Ni	0.22	0.01	–	0.01
Cu	–	0.06	0.04	0.06
Zn	0.01	0.07	0.06	0.08
La	0.10	-	–	-
Rb	–	0.01	0.01	0.01
Sr	–	0.01	0.05	0.02

propagation, scattering, absorption and photoelectric production, which take place in an EDXRF measurement with the used experimental geometry. It is based on data of X-ray cross sections provided by NIST and atomic characteristic X ray parameters provided by IAEA. The X-ray distribution function, which includes continuous and discrete components, has been calibrated by performing scattering of these primary X-rays by known samples. The performance of the program has been checked using reference samples. The estimated uncertainty in these measurements of the elemental concentrations is about 10 %, depending on the atomic number of the element (Dr. J.A. Bravo C., private communication).

2.2 Analysis by X-Ray Diffractometry (XRD)

For the structural analysis of the minerals present in the sample, the XRD technique was applied using a BRUKER diffractometer, model D8-Focus, with CuKα (1.5406 Å) radiation (40 kV, 40 mA) and a vertical goniometer. The scanned angle interval was $4° < 2\theta < 90°$ and the 2θ advance was 0.02°/step with a time interval of 3 s per step.

2.3 Analysis by ^{57}Fe transmission Mössbauer spectroscopy (TMS)

TMS was used to obtain more detailed information about iron containing minerals. A conventional spectrometer was used with a sinusoidal velocity modulation signal and 1,024 channels. The Mössbauer spectrum at room temperature (RT:~298 K) was collected at the Laboratory of Archaeometry, Facultad de Ciencias Físicas, UNMSM. TMS measurements at liquid nitrogen temperature (LNT; ~80 K) were collected at Laboratoire de Physique de l'État Condensé, Université du Maine, Le Mans, France.

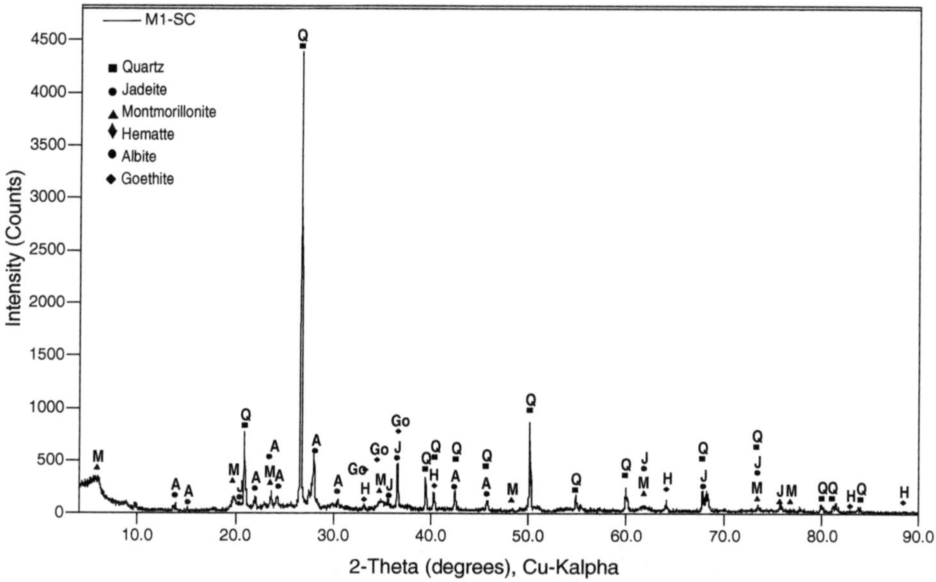

Fig. 3 X ray pattern for the sample M1-SC from the Carancas crater showing reflections of: Quartz (Q), Montmorillonite (M), Jadeite (J), Hematite (H), Albite (A) and Goethite (Go)

A ^{57}Co source in a Rh matrix was used to collect spectra, which were analyzed using the Normos program written by R. Brand (Normos Site) [7].

3 Results and discussion

Table 1 lists data from the quantitative elemental chemical analyses for the studied samples; for comparison it lists also data from the elemental composition of a sample of the Carancas meteorite whose elemental composition has already been reported [8]. The meteorite sample shows the presence of the following elements: Si, Cl, S, K, Fe, Cr, Cu, Ca, Ti, Mn, and Ni, with relatively high concentration of Fe, as well as of S, Cl, Cr and Ni, compared to soil samples from the crater. The soil samples do contain: Si, S, Cl, K, Ca, Ti, Mn, Fe, Ni, Cu, Zn, Rb, and Sr; Si, Ca, and Ti are present in higher concentrations compared to the meteorite; Ca is found in the following proportions M3>M2>M1; Fe as M1>M3>M2; and Zn and Mn as M3>M1>M2; the elements Al, S, Rb, Sr and Cu are present in about the same concentrations in the three samples. Ni is present only in samples M1-SC and M3-SC, but in low concentrations.

The diffractograms of Figs. 3, 4 and 5 are, respectively, for samples M1-SC, M2-SC and M3-SC. All of them show typical reflections of mineralogical phases that include mainly quartz (SiO_2), montmorillonite, illite, albite ($NaAlSi_3O_8$), and hematite. In samples M1-SC and M3-SC, montmorillonite has strong characteristic $(0\ 0\ \ell)$ reflection peaks; the sample M2-SC shows a very strong characteristic (0 0 1) peak of illite. The reflection peaks of hematite appear to correspond to a well crystallized mineral in samples M2-SC and M3-SC; in sample M1-SC these peaks are superposed with other wider reflection peaks. Moreover, in sample

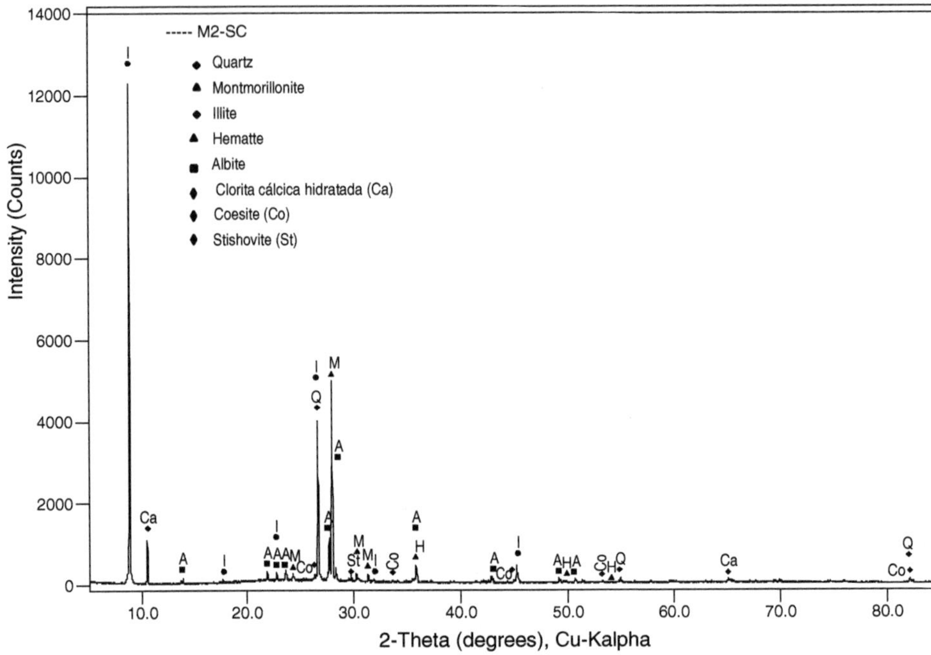

Fig. 4 X ray pattern for the sample M2-SC from the Carancas crater showing reflections of: Illite (I), Quartz (Q), Albite (A), Montmorillonite (M), Hematite (H), Hydrated Calcic Chlorite (Ca), Coesite (Co), and Stishovite (St)

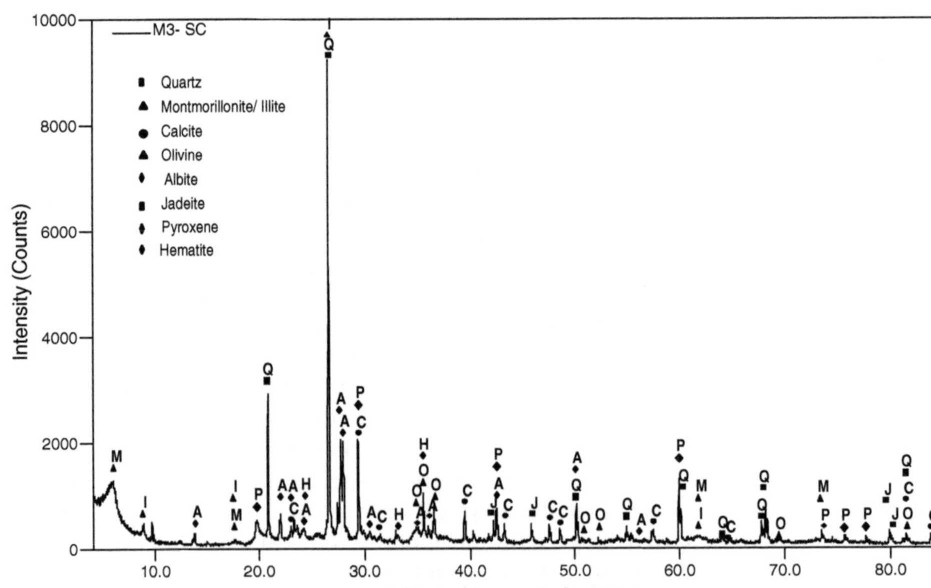

Fig. 5 X ray pattern for the sample M3-SC from the Carancas the crater showing reflections of: Quartz (Q), Montmorillonite (M), Illite (I), Calcite (C), Olivine (O), Albite (A), Jadeite (J) Hematite (H) and Pyroxene (P)

Table 2 Hyperfine parameters of the Carancas meteorite samples

Mineral phases	Spectra taken at 298 K			
	δ (mm/s) ±0.01	2ε (mm/s) ±0.01	B_{hf} (T) ±0.1	R. A. (%)
	M1-SC			
Doublet (Fe^{3+}), includes super paramagnetism	0.22	0.55		55
Hematite (Fe^{3+})	0.26	−0.22	50.8	35
Montmorillonite (Fe^{2+})	1.02	2.72		11
	M2-SC			
Illite (Fe^{+3})	0.35	0.85		35
Hematite (Fe^{+3})	0.23	−0.12	49.3	22
Illite (Fe^{2+})	1.00	2.47		11
Illite (Fe^{2+})	1.03	2.00		33
	M3-SC			
Hematite I (Fe^{+3})	0.30	−0.18	49.1	13
Hematite II (Fe^{+3})	0.27	−0.22	50.9	23
Illite (Fe^{+3}) /Montmorillonite	0.19	0.55		54
Olivine	1.22	2.30		9
	Spectra taken at 77 K			
	δ (mm/s) ±0.01	2ε (mm/s) ±0.01	B_{hf} (T) ±0.1	R. A. (%)
	M1-SC			
Goethite	0.46	−0.20	49.4	8
Hematite (Fe^{+3})	0.51	−0.17	53.1	31
Montmorillonite (Fe^{+3})	0.46	0.49		56
Montmorillonite (Fe^{2+})	1.30	2.76		5
	M2-SC			
Illite (Fe^{+3})	0.49	0.82		35
Hematite (Fe^{+3})	0.43	−0.18	51.2	17
Illite (Fe^{2+})	1.24	2.85		17
Illite (Fe^{2+})	1.19	2.26		30

δ: isomer shift, 2ε: quadrupole shift, B_{hf}: hyperfine field, RA = relative area

M1-SC we observe the mineralogical phases of goethite and jadeite; in sample M3-SC we observe the presence of the following phases; quartz, interstratified montmorillonite-illite, calcite, olivine, albite, pyroxene and jadeite; the latter, which occurs also in sample M1-SC, is a pyroxene. Under high pressure, albite may be transformed to an association of jadeite + quartz [9], which could have happened when the meteorite collided with the soils of Carancas due to the powerful shock wave. In the sample M2-SC, we observe the presence of eight primary phases associated to well defined reflection peaks of illite, albite (without the presence of jadeite as compared with the sample M1-SC), quartz, montmorillonite, hematite, hydrated calcic chlorite, coesite and stishovite (Si_2O).

Silica, in the form of trigonal alpha-quartz, is one of the most common and resistant components of the crust of the Earth. During a powerful meteoritic impact, two polymorphs denser than silica, monoclinic coesite and tetragonal stishovite, can form from quartz under pressures higher than 3 to 10 GPa [10]. Although coesite can

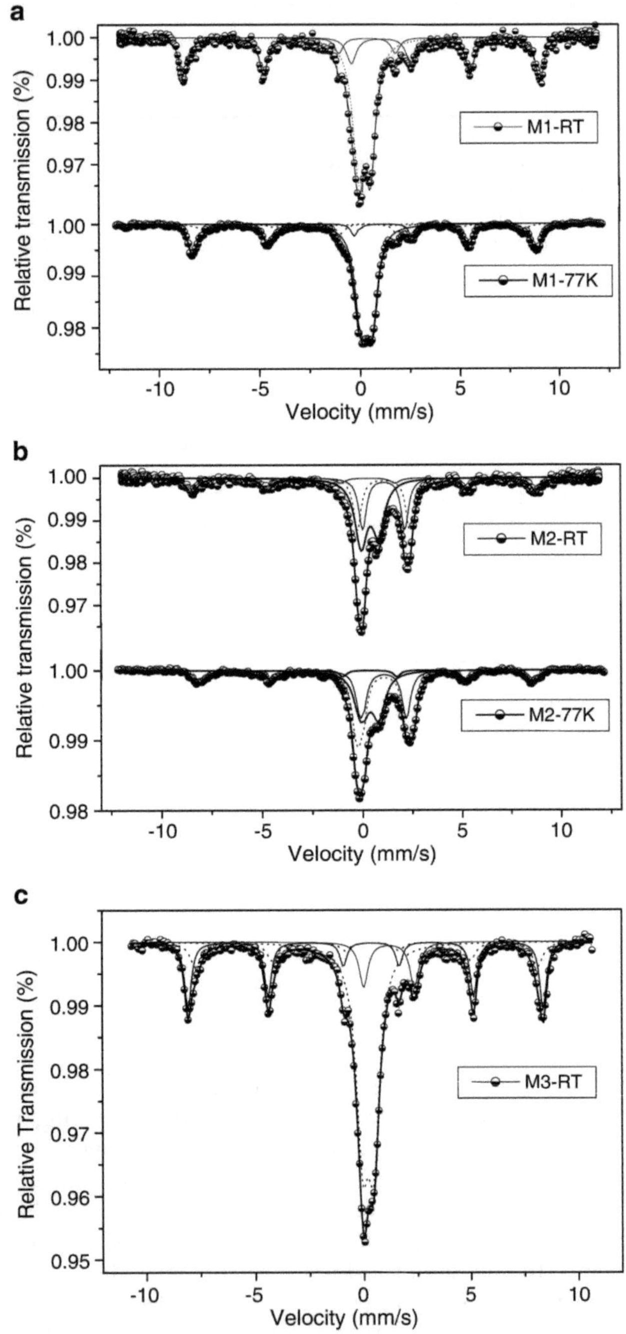

Fig. 6 **a** Mössbauer spectra of the Carancas crater sample M1-SC taken at RT and at 77 K. **b** Mössbauer spectra of the sample M2-SC taken at RT and LNT. **c** Mössbauer spectrum of the sample M3-SC taken at RT

form under conditions of extreme metamorphism deep within the Earth, these are not sufficient to produce stishovite; consequently, its formation in the Earth can only take place in sites of meteoritic impacts [11], under the required intense pressures, as can be observed for sample M2-SC (Fig. 4).

Table 2 lists the ^{57}Fe hyperfine parameters corresponding to the different subspectra for these soil samples from the crater taken at RT and LNT. As shown in Fig. 6a the spectrum taken at RT for sample M1-SC was fitted with a magnetic sextet, assigned to an Fe^{3+} site, which is characteristic of hematite, and two paramagnetic doublets. The doublet with the largest area is assigned to an Fe^{3+} site, which may include sites in a superparamagnetic state, and the second one to an Fe^{2+} site which is associated to montmorillonite. Results from the spectrum taken at LNT (see Fig. 6a) confirm the presence of a strong sextet of hematite along with a weak sextet of goethite with 8 % of relative area, which contributed to the Fe^{3+} doublet at RT; moreover, it shows the presence of two doublets associated with Fe^{3+} and Fe^{2+} sites in montmorillonite. It is interesting to observe that the quadrupole shift for hematite at 77 K is negative, typical of a hematite sample that has not undergone the Morin transition (MT); therefore, it remains in a weak ferromagnetic state.

In Fig. 6b we can observe the RT Mössbauer spectrum of the sample M2-SC; it shows one magnetic sextet assigned to hematite, with $B_{hf} = 49.31$ T and three doublets, all of them assigned to illite, in agreement with the respective difractogram, one of them associated to an Fe^{3+} site and two to Fe^{2+} sites. At LNT we observe that the above assignments are confirmed; we note that that the parameter B_{hf} of hematite increases with decreasing temperature and that the relative areas for the observed sites remain about the same at both temperatures.

In Fig. 6c we can observe the RT Mössbauer spectrum for the sample M3-SC; it shows two magnetic sextets both assigned to sites in hematite, with $B_{hf} = 50.9$ T and 49.1 T respectively; the latter corresponds to a hematite with substitution of Fe by Al, which is responsible for the decrease of the Bhf value; it shows also two doublets, one of them associated with Fe^{3+} in the montmorillonite structure, and the second one may be assigned to an Fe^{2+} site in olivine. We have no data at LNT.

4 Conclusions

The analyses of the crater soil samples by EDXRF reveal that they contain: Al, Si, S, Cl, K, Ca, Ti, Mn, Fe, Cu, Zn, Rb, and Sr; Al, S, Rb, Sr, and Cu are observed in about the same concentrations in the three samples; Ni is observed only in samples M1 y M3. The concentration of Si, Ca, and Ti in these samples are found to be higher than in the meteorite; the concentration of Si is found in the following relationship M1>M3>M2, Cl as M3>M1>M2, K as M1>M2>M3, Ca as M3>M2>M1; Fe as M1>M3>M2. The lower concentrations of Si, Cl and Fe in the sample M2-SC may be the result of the meteoritic impact.

The analyses by XRD show that the samples contain mineralogical phases consisting mainly of quartz (SiO_2), montmorillonite, illite, albite ($NaAlSi_3O_8$), and hematite, an iron oxide that is generally found in soils used for agricultural production. In the samples M1-SC and M3-SC it is observed the occurrence of jadeite; under high pressure albite decomposes to form the jadeite-quartz association, which could have happened when the meteorite collided with the soils at Carancas due to the powerful shock wave. In sample M2-SC, by XRD, we observe the presence of a high concentration of illite, and the characteristic peaks of impactites such as coesite and stishovite which are high pressure mineral polymorphs. These are generated by shock waves and up to now have only been identified in quartz containing rocks that have suffered impact induced metamorphism within craters produced by meteorites. It is known that coesite can form under extreme metamorphic conditions within the crust of the Earth; however, these conditions are not sufficient to produce stishovite. Therefore, its formation on Earth can only take place in sites that were subjected to meteorite impacts. In sample M3-SC we observe main phases assigned to olivine and pyroxene, which are common minerals in meteorites and igneous rocks; taking as reference earlier studies we may assume that these minerals were ejected outside the crater and may constitute small fragments of the meteorite.

The analysis by TMS at RT and TNL allowed the observation of superposed magnetic, paramagnetic and superparamagnetic phases. The presence of hematite is evident in all samples. In the sample M1-SC at RT we identified super paramagnetic goethite, which at LNT becomes magnetically ordered goethite. The Fe^{3+} cations present in this sample are distributed in the octahedral sites of montmorillonite and part in goethite. In samples M1-SC and M2-SC hematite does not exhibit the Morin transition (MT) when cooled down to LNT. In the case of the sample M3-SC, its TMS spectrum taken at RT shows two magnetic sextets assigned to the presence of Fe^{3+} cations: one assigned to well crystallized hematite, and the other to hematite with Al substitution in Fe sites that causes a reduction of the observed magnetic hyperfine field.

Acknowledgements The authors are grateful for the collaboration of the Laboratories of Soil Analysis and Archaeometry at San Marcos University for allowing us the use of their instruments and materials; likewise, we are grateful to Dr. J. M. Grenech for providing the Mössbauer spectra taken at LNT in his laboratory at the Institut des Molécules et Matériaux du Mans (IMMM), Université du Maine, France.

References

1. Loayza, M.L.C., Cabrejos, J.A.B.: Hyperfine Interact. **203**(1–3), 17–23 (2011)
2. Langenhorst, F.: Shock metamorphism of some minerals: basic introduction and microstructural observations. Bull. Czech. Geol. Surv. **77**(4), 265–282 (2002)
3. Langenhorst, F., Deutsch, A.: Adv. Mineral. **3**, 95–110 (1998). Springer-Verlag
4. Wackerle, J.: Shock-wave compression of quartz. J. Appl. Phys. **33**, 922 (1962)
5. Kieffer, S.W., Phakey, P.P., Christie, J.M.: Contrib. Mineral. Petrol. **1**(59), 41–93 (1976)
6. Holland, K.G., Ahrens, T.J.: Melting of $(Mg,Fe)_2SiO_4$ at the core-mantle boundary of the Earth. Science **275**, 1623–1625 (1997)
7. Brand, R.A.: Normos-90 Mössbauer Fitting Program Package. U. of Duisburg, Alemania (1994)
8. Cerón Loayza, M.L., Bravo Cabrejos, J.A.: Caracterización Mineralógica de un Impacto Meteoritico en la Localidad de Carancas–Puno. Rev. Investig. Fis. **12**(1), 5–12 (2009)
9. Vanderberghe, R.E., De Grave, E., De Geyter, G., Landuydt, C.: Clays Clay Miner. **34**(3), 275–280 (1986)
10. Hemley, R.J., Prewitt, C.T., Kingma K.J.: High pressure behavior of silica. Rev. Mineral. **29**, 41–82 (1994)
11. Chao, E.C.T., Fahey, J.J., Littler, J., Milton D.J.: Stishovite, SiO_2, a very high pressure new mineral from meteor crater, Arizona. J. Geophys. Res. **67**(1), 419–421 (1962)

DOI 10.1007/978-94-007-6482-8_18
Reprinted from *Hyperfine Interactions*, DOI 10.1007/s10751-013-0768-y

Iron-bearing minerals in ashes emanated from Osorno volcano, in Chile

Alexandre Christófaro Silva · Mauricio Escudey · Juan Enrique Förster · Carmen Pizarro · José Domingos Ardisson · Uidemar Morais Barral · Márcio César Pereira · José Domingos Fabris

Published online: 9 February 2013

Abstract A sample of volcanic ashes emanated from the Osorno volcano, southern Chile, was characterized with X-ray fluorescence, X-ray diffraction and ^{57}Fe Mössbauer spectroscopy, in an attempt to identify the iron-bearing minerals of that geologically recent magmatic deposit. X-ray patterns indicated that the sample is mainly constituted of anorthite, Fe-diopside-type and Ca-magnetite. The crystallographic structures of these dominant iron minerals are proposed on basis of their chemical composition and corresponding Mössbauer data to support models refined by fitting powder X-ray diffraction data with the Rietveld algorithm.

Keywords Iron oxides · Magnetic minerals · Iron-rich spinel · Anorthite · Diopside

Proceedings of the Thirteenth Latin American Conference on the Applications of the Mössbauer Effect, (LACAME 2012), Medellín, Columbia, 11–16 November 2012.

A. C. Silva · U. M. Barral · J. D. Fabris (✉)
Universidade Federal dos Vales do Jequitinhonha e Mucuri (UFVJM),
39100-000 Diamantina, Minas Gerais, Brazil
e-mail: jdfabris@ufmg.br

M. Escudey · J. E. Förster · C. Pizarro
Centro para el Desarrollo de la Nanociencia y Nanotecnología (CEDENNA),
Santiago 725475, Chile

M. Escudey · J. E. Förster · C. Pizarro
Facultad de Química y Biología, Universidad de Santiago de Chile, USACH,
Santiago 725475, Chile

J. D. Ardisson
Centro de Desenvolvimento da Tecnologia Nuclear,
31270-901 Belo Horizonte, Minas Gerais, Brazil

M. C. Pereira
Instituto de Ciência, Engenharia e Tecnologia, Universidade Federal dos Vales do Jequitinhonha e Mucuri, 39803-371 Teófilo Otoni, Minas Gerais, Brazil

Fig. 1 Osorno volcano in the Los Lagos Region, Puerto Montt, Southern Chile

1 Introduction

Soils derived from volcanic materials are abundant and widespread all over the southern Chile. They actually account for around 70 % of the Chilean land areas covered with agriculture crops. Blasted ashes spread out around a large perimeter and lava flows surrounding the volcanic source lead to consolidate magmatic rocks, which markedly influence the lithogenic parent materials on which such soils develop. Resulting pedons are mainly characterized by having high clay and iron (oxyhydr)oxides contents. Most of those soils are magnetic for containing important amounts of magnetite (ideal formula, Fe_3O_4) or maghemite (γFe_2O_3) [1–6].

The problem concerning the nature of precursor minerals and their chemical and mineralogical changes during pedogenesis, to render the currently found magnetic iron oxides, still remains to be better elucidated. The purpose of this work was to look somewhat closer in an attempt to identify in more detail the iron-bearing chemical species occurring in ashes forming the parent material of volcanic soils, which could be the elemental source or direct precursor of iron oxides found in the corresponding pedon.

2 Materials and methods

A sample of volcanic ashes was collected at a depth between 200 cm and 220 cm in a dug pit (geographical coordinates of the sampling site, 41° 11′ 37.47″ S 72° 28′ 31.52″ W; altitude, 104 m) on the flow way of geologically recent lavas and ashes, surrounding the volcanic source Osorno, in the Los Lagos Region, Puerto Montt, southern Chile (Fig. 1). In the laboratory, the sample was dried in air and ground in an agate mortar. This sample of volcanic ashes was characterized with powder X-ray diffraction (XRD) measurements, using a Shimadzu XRD 6000 diffractometer equipped with a Cu tube and a graphite monochromator. The scans were performed between 10 and 80° (2θ) with a scanning speed of 1°/min. Silicon was used as an external standard. The Rietveld structural refinement was performed with FULLPROF 2012 program (code available

Table 1 Crystallographic phase, lattice parameter and agreement factor values as obtained from the structural refinement with Rietveld algorithm for this volcanic ash sample

Phase	Lattice parameter/Å	Bragg R-factor	R_f-Factor	R_p	R_{wp}
Anorthite	a = 8.1681(3) b = 12.8663(4) c = 7.0980(2)	4.53	4.46	4.18	5.66
Fe-diopside-type	a = 9.7392(6) b = 8.9134(4) c = 5.2809(3)	3.69	3.83		
Ca-magnetite	a = 8.4205(5)	2.39	1.54		

at http://www.ill.eu/sites/fullprof/php/downloads.html). X-ray fluorescence (XRF) spectrometry measurement was performed on a S4 explorer X-ray spectrometer (Bruker-AXS, Germany).

Mössbauer spectra were collected in constant acceleration transmission mode with a ~30 mCi ^{57}Co/Rh source. The spectra were obtained at 298 K and 80 K using a liquid-N_2 bath cryostat. Data were stored in a 512-channel MCS memory unit. The collected experimental data were fitted with Lorentzian-shape resonance lines through a least-squares procedure of the software NORMOS™-90 (software package developed by R. A. Brand, at Laboratorium für Angewandte Physik, Universität Duisburg, D-47048, Duisburg-Germany); isomer shifts are quoted relative to α-Fe.

3 Results and discussion

XRF analysis revealed that this sample of volcanic ashes is basically constituted of 60 mass% Si; 24.5 mass% Al; 6.9 mass% Fe; 6.8 mass% Ca; 0.7 mass% Ti and 0.7 mass% K. Its mineralogical composition was first inferred from a phase search based on data of the powder XRD pattern. Further Rietveld refinement of these XRD data with pseudo-Voigt peak fitting gave the structural parameters of each crystalline phase and the refinement reliability factors summarized in Table 1. Figure 2 shows patterns from the corresponding Rietveld refinement. This analysis reveals that the volcanic ashes sample is constituted of 79 mass% anorthite (ideal formula, $CaAl_2Si_2O_8$), 17 mass% Fe-diopside-typc ($CaFeSi_2O_6$) and 4 mass% Ca-magnetite ($Fe_{3-x}Ca_xO_4$). Although the XRD data were fitted assuming three dominant phases, contributions of other minor crystalline species cannot be excluded.

Although XRD data provide information about the crystallographic structure, the Fe distribution per mineral structure may not be directly determined this way. It was therefore used ^{57}Fe Mössbauer spectroscopy data to define the Fe sites in each structure so found from the XRD analysis. Mössbauer spectra for the volcanic ashes sample collected at 298 and 80 K are shown in Fig. 3. The 298 K Mössbauer spectrum was fitted with two magnetic components and three paramagnetic components. The sextet-components are assignable to a tetrahedral Fe^{3+} site and to a mixture of octahedral $Fe^{2+/3+}$ ions, which is due to the fast electron hopping, in the magnetite

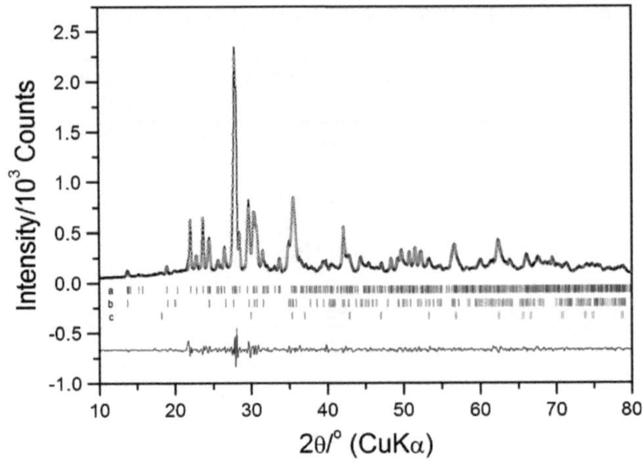

Fig. 2 Rietveld refinement on the powder XRD pattern of the volcanic ashes sample. a = Anorthite, b = Fe-diopside-type structure and c = Ca-magnetite

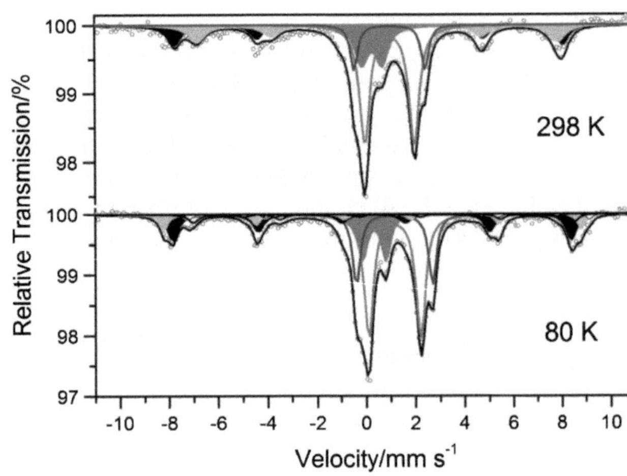

Fig. 3 298 and 80 K Mössbauer spectra for the volcanic ashes sample

structure (Table 2). Doublets are due to Fe^{2+} on octahedral (M1) and dodecahedral (M2) positions of the Fe-diopside-type structure (Table 2). The Fe^{3+} doublet could be appeared as due to the oxidation of Fe^{2+} oxidation on the M1 site in the Fe-diopside-type structure.

80 K-Mössbauer spectrum of volcanic ashes sample (Fig. 3) was fitted with three doublets corresponding to Fe in the Fe-diopside-type structure and three sextets due to iron in the magnetite. It is known that pure magnetite undergoes Verwey transition around 120 K, which is also related to the charge ordering of Fe^{2+} and Fe^{3+} ions in octahedral sites, but the mechanism is still not sufficiently known. Some reports have proposed that the Mössbauer spectrum for pure magnetite below Verwey transition should be fitted with five spectral components: one related to Fe^{3+} ions on the tetrahedral sites and four corresponding to Fe^{2+} or Fe^{3+} ions on two nonequivalent octahedral sites [7, 8]. The Mössbauer spectrum for the volcanic ashes collected with the sample at 80 K was accordingly fitted with a magnetic

Table 2 Hyperfine parameters of the fitted Mössbauer spectra recorded at 298 and 80 K

Temperature/K	δ/mm s^{-1}	ε, Δ/mm s^{-1}	B_{hf}/T	RA/%	^{57}Fe site
298	0.29(1)	−0.05	49.0(2)	16(1)	$^{IV}Fe^{3+}$
	0.59(2)	0.05(2)	45.5(2)	18(2)	$^{VI}Fe^{2.5+}$
	0.41(2)	0.83(3)	–	16(1)	$^{VI}Fe^{3+}$
	1.11(3)	2.01(3)	–	39(2)	$^{VI}Fe^{2+}$
	1.12(1)	2.86(2)	–	11(1)	$^{VIII}Fe^{2+}$
80	0.39(2)	−0.07	50.3(2)	18(1)	$^{IV}Fe^{3+}$
	0.50(2)	−0.18	52.4(1)	7(1)	$^{VI}Fe^{3+}$
	0.92(1)	0.32(2)	50.7(2)	3(1)	$^{VI}Fe^{2+}$
	0.98(2)	−0.25	48.0(1)	5(1)	$^{VI}Fe^{2+}$
	1.09(1)	0.90(2)	39.5[a]	2(1)	$^{VI}Fe^{2+}$
	0.46(3)	0.95(3)	–	13(1)	$^{VI}Fe^{3+}$
	1.27(2)	2.10(5)	–	36(2)	$^{VI}Fe^{2+}$
	1.27(3)	3.06(7)	–	17(1)	$^{VIII}Fe^{2+}$

δ = isomer shift relative to α-Fe; ε = quadrupole shift; Δ = quadrupole splitting; B_{hf} = hyperfine field and RA = relative subspectral area
[a] Parameter kept fixed during fitting convergence

Fe^{3+} component at tetrahedral coordination and four magnetic Fe components in octahedral coordination: one sextet for Fe^{3+} and three sextets for Fe^{2+} (Table 1), which correspond to the different Fe^{2+} environment in the magnetite structure. Although the proposal model involves only the occurrence of magnetite as the magnetically ordered specie, eventual contributions of maghemite and/or hematite cannot be completely discarded.

The relative area ratio (RA; Table 2) between octahedral (B) and tetrahedral sites (A), $\frac{(RA)_B}{[RA]_A} = 1.1$, at 298 K, which is far below the value expected for stoichiometric magnetite, $\frac{(RA)_B}{[RA]_A} = 1.88$. The found occupancy ratio for this magnetite is determined by taking into account that the recoilless fraction for octahedral (B) sites is 6 % lower than that for tetrahedral [A] sites [9], or $R = \frac{(Fe^{3+}_{2+})_B}{[Fe^{3+}]_A} = \frac{\frac{(RA)_B}{0.94}}{[RA]_A} = \frac{\frac{18}{0.94}}{16} = 1.2$. Actually, there are two main factors that lead to $R = \frac{(Fe^{3+}_{2+})_B}{[Fe^{3+}]_A} < 2$ in magnetite: (i) an oxidation mechanism involving $Fe^{2+} \Rightarrow Fe^{3+}$ and (ii) isomorphic replacement of Fe by other cations. In the first case, for each three Fe^{2+} that oxidizes, two Fe^{3+} are formed and one cationic vacancy is created, in order to maintain the charge balance in the chemical structure. The high spin Fe^{3+} ionic radium on octahedral coordination is $r = 65$ pm; for Fe^{2+}, $r = 78$ pm. Thus, the oxidation of Fe^{2+} to Fe^{3+} would lead to a decreasing of the cubic unit cell dimension of magnetite. Our Rietveld results evidence that the lattice parameter of magnetite (Table 1) in this sample is higher than that expected for pure and stoichiometric magnetite, suggesting an isomorphic substitution of iron by other cations with higher ionic radium than those reported for Fe^{3+} or Fe^{2+} ions. Based on the fluorescence analysis, we suggest that Fe^{2+} ions in the magnetite structure are partially replaced by Ca^{2+} (ionic radium, $r = 100$ pm) ions, which causes significant expansion of the magnetite cubic structure. Thus, the isomorphic replacement of Fe^{2+} by Ca^{2+} should increase the unit cell of magnetite, as it was observed from Rietveld analysis.

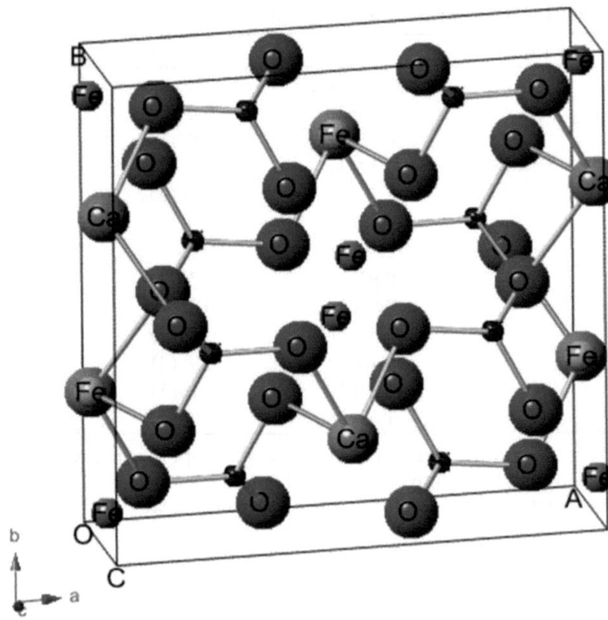

Fig. 4 Crystallographic structure of this Fe-diopside-type

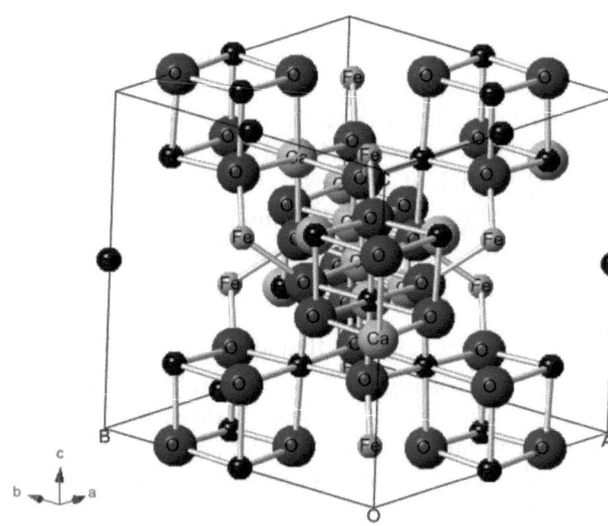

Fig. 5 Crystallographic structure of this Ca-magnetite

The Fe-diopside-type structure is monoclinic, space group $C2/c$. The Ca^{2+} cations are located in eightfold (M2 sites) oxygen-coordination sites. The Fe^{2+} cations are located in both M2 sites and in sixfold (M1 sites) oxygen-coordination sites, as confirmed by Mössbauer spectroscopy.

The crystallographic structures, as deduced from the Rietveld refinement of the powder X-ray data, for the Fe-diopside-type and Ca-magnetite of this Ca-rich sample from volcanic ashes are represented in Figs. 4 and 5, respectively.

4 Conclusions

The mineralogy of this sample of volcanic ashes sample, which is relatively rich in calcium, is dominantly constituted of anorthite, Fe-diopside-type and Ca-magnetite. Mössbauer parameters are consistent with iron in octahedral (B) and tetrahedral [A] coordination sites of the magnetite (with a ratio $R = \frac{(Fe^{3+}_{2+})_B}{[Fe^{3+}]_A} = 1.2$) and in the octahedral and at dodecahedral sites of the Fe-diopside-type structure. The crystallographic structure for these Fe-diopside-type and Ca-doped magnetite are proposed basing on fitting powder X-ray diffraction data by using the Rietveld algorithm, and chemical composition deduced from X-ray fluorescence data. Progressive weathering processes acting on this magmatic magnetite should respond for structural alterations to ultimately form the magnetic iron oxides found in Chilean soils. Further Mössbauer measurements, as with applied magnetic field, would be interesting, in order to definitively confirm this interpretation concerning the iron distribution in minerals of this sample.

Acknowledgements Work financially supported by DICYT-USACH 0221242PA, CEDENNA FB-0807 (Chile), CNPq (Brazil; grants #480295/2009-3; #302479/2010-4 and PROSUL #490096/2010-7) and FAPEMIG (Brazil; including #PPM 00419-10 and #APQ-04333-10). CAPES (Brazil) grants the Visiting Professor PVNS fellowship to JDF at UFVJM

References

1. Pizarro, C., Furet, N.R., Venegas, R., Fabris, J.D., Escudey, M.: Some cautions on the interpretation of Mossbauer spectra in mineralogical studies of volcanic soils. Bol. Soc. Chil. Quim. **45**, 243–250 (2000)
2. Pizarro, C., Escudey, M., Fabris, J.D., Almeida, A.B.: Iron-rich spinel phases from the sand fraction of three Chilean soils developing on volcanic materials. Comm. Soil Sci. Plan. **32**, 2741–2754 (2001)
3. Pizarro, C., Escudey, M., Fabris, J.D.: Influence of organic matter on the iron oxide mineralogy of volcanic soils. Hyperfine Interact. **148**, 53–59 (2003)
4. Pizarro, C., Escudey, M., Moyá, S.A., Fabris, J.D.: Iron oxides from volcanic soils as potential catalysts in the water gas shift reaction. In: Garcia, M., Marco, J.F., Plazaola, F. (eds.) Industrial Applications of the Mössbauer Effect—International Symposium on the Industrial Applications of the Mössbauer Effect, 765 ed., American Institute of Physics - AIP Conference Proceedings, New York, vol. 765, p. 56–59 (2005)
5. Pizarro, C., Fabris, J.D., Stucki, J.W., Garg, V.K., Morales, C., Aravena, S., Gautier, J.L., Galindo, G.: Distribution of Fe-bearing compounds in an Ultisol as determined with selective chemical dissolution and Mössbauer spectroscopy. Hyperfine Interact. **175**, 95–101 (2007)
6. Aravena, S., Pizarro, C., Rubio, M.A., Cavalcante, L.C.D., Garg, V.K., Pereira, M.C., Fabris, J.D.: Magnetic minerals from volcanic Ultisols as heterogeneous Fenton catalysts. Hyperfine Interact. **195**, 35–41 (2010)
7. Hargrove, R.S., Kündig, W.: Mössbauer measurements of magnetite below the Verwey transition. Solid State Commun. **8**, 303–308 (1970)
8. Berry, F.J., Skinner, S., Thomas, M.F.: ^{57}Fe Mössbauer spectroscopic examination of a single crystal of Fe_3O_4. J. Phys. Condens. Matter **10**, 215–220 (1998)
9. Sawatzky, G.A., van der Woude, F., Morrish, A.H.: Recoilless-fraction ratios for Fe^{57} in octahedral and tetrahedral sites of a Spinel and a Garnet. Phys. Rev. **183**, 383–386 (1969)

DOI 10.1007/978-94-007-6482-8_19
Reprinted from *Hyperfine Interactions*, DOI 10.1007/s10751-013-0762-4

A pre-Columbian copper alloy smelting furnace: Mössbauer and XRD study of the firing conditions

F. M. Hayashida · D. Killick · I. Shimada · W. Häusler · F. E. Wagner · U. Wagner

Published online: 23 July 2013

Abstract Samples from a pre-Columbian furnace used for copper alloy smelting on the Pampa de Chaparrí in northern Peru during the Middle to Late Sicán Period (AD 900–1375) were studied by Mössbauer spectroscopy, X-ray diffraction and X-ray fluorescence. The data thus obtained allow conclusions as to the temperatures and oxidation-reduction conditions prevaling at different positions inside the furnace during the smelting process.

Keywords Mössbauer spectroscopy · XRD · Copper smelting · Furnace

1 Introduction

Pre-Columbian copper alloy smelting on the northern coast of Peru was done in simple pit furnaces dug into the ground of the well ventilated workshops [1, 2]. Many such copper alloy smelting furnaces from the Middle to Late Sicán period (AD 900–

Proceedings of the Thirteenth Latin American Conference on the Applications of the Mössbauer Effect, (LACAME 2012), Medellín, Columbia, 11–16 November 2012

F. M. Hayashida
Department of Anthropology, University of New Mexico, Albuquerque NM 87131, USA

D. Killick
School of Anthropology, University of Arizona, Tucson, AZ 85721-0030, USA

I. Shimada
Department of Anthropology, Southern Illinois University, Carbondale IL 62901, USA

W. Häusler · F. E. Wagner · U. Wagner (✉)
Physics Department, Technical University of Munich, 85747 Garching, Germany
e-mail: urwagner@tum.de

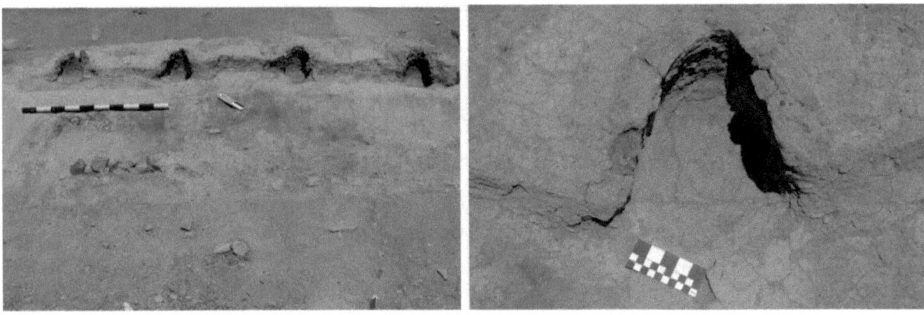

Fig. 1 View of the four furnaces on the Pampa de Chaparrí, of which the second from the left was investigated. A close-up view of this furnace is shown on the right

1375) were found in the Lambayeque region of the north coast of Peru, including several sets at the 256A01 site on the Pampa de Chaparrí, situated between the adjacent Lambayeque and La Leche valleys. Excavations and analyses of slag from this site [1] indicate that charcoal made from algarrobo (*Prosopis spp.*) hardwood, copper and possibly arsenic-bearing ores, and iron ore as a flux were combined in the furnaces and heated with the help of lung-powered blow tubes to temperatures of up to about 1150 °C, resulting in a fluid slag and possibly the separation of the metal as an ingot at the base of the furnace. Some prills, including "dirty" prills contaminated with iron and sulfides, remained in the slag. They were recovered by crushing the slag and then refined in shallow plate crucibles. ^{14}C dating of charcoal finds shows that the furnaces were used during the late Middle Sicán through Late Sicán period.

We have sampled one out of a row of four furnaces (Fig. 1) found at site 256A01 on the Pampa de Chaparrí. To this end a slice of material was cut out of this furnace, reaching from the top of the rear wall down to the bottom and from there to the mouth on the front and extending up to about 50 mm deep into the soil. The material was separated into two parts, named 612 and 613 (Fig. 2). Both parts were about 30 cm long and 15 cm wide. Part 612 is mainly the rear wall of the furnace down to the rear bottom, while part 613 reaches from the rear bottom to the front ledge at the mouth of the furnace. The two parts were separately enclosed in plaster molds for shipping to the Mössbauer laboratory at the Archeological State Collection, Munich. There a number of samples were taken from individual spots in the furnace for studies by ^{57}Fe Mössbauer spectroscopy, X-ray diffraction (XRD) and X-ray fluorescence (XRF) with the aim of obtaining information on the temperatures reached in the furnace, both on the inner surface and down to about 50 mm into the soil, and on the prevailing oxidation-reduction conditions. The probed positions range from the rear upper rim, through the bottom, to the front ledge at the mouth of the furnace.

Information on the temperature and redox conditions would mainly be drawn from typical features of the Mössbauer spectra by comparison with laboratory firing experiments on clays from the Lambayeque region [3, 4]. Additionally, re-firing experiments were conducted on samples from the furnace itself. The presence or absence of certain minerals seen in the X-ray diffractograms yields additional information on the temperature reached at different spots.

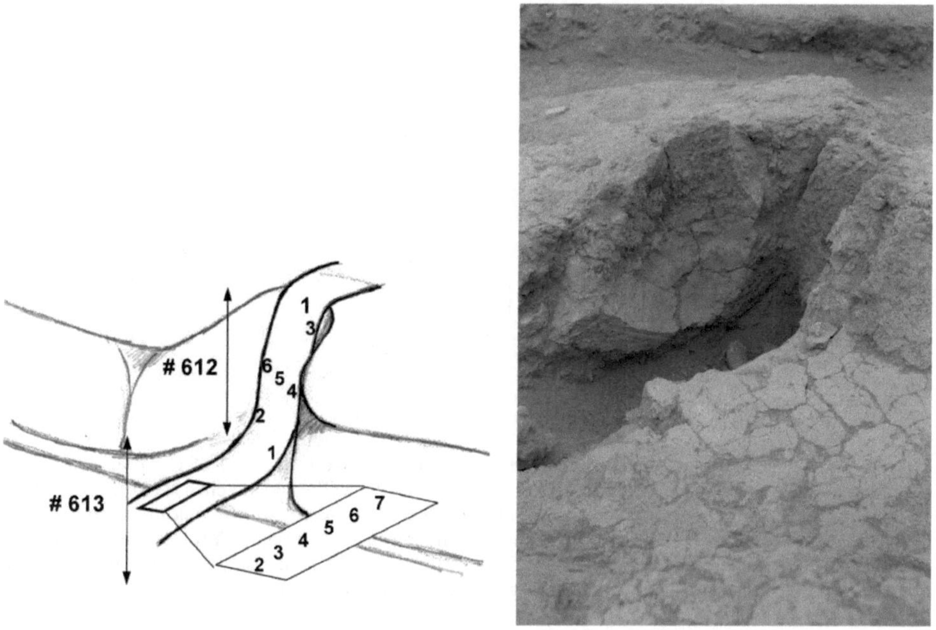

Fig. 2 Schematic view of the locations in the furnace where samples were taken and the numbering of these samples (*left*). The furnace after removal of the sample material is shown on the *right*

2 Samples and experiments

Figure 2 shows schematically at which locations inside the furnace samples were taken and how they were numbered. The specimens usually covered an area of several cm^2, and were between 30 and 55 mm thick, depending on what thickness of material was available at the respective position in the furnace. They were cut from the larger pieces with a diamond saw. Near the inner surface of the furnace the material was usually quite hard, in some places it appeared partly vitrified. The color was gray to black down to a depth of about 20 mm, but varying from spot to spot. Further below the surface the color changed to red, sometimes with a magenta hue. As an example, a part of piece 612 representing the middle of the rear wall down to the region of the bottom of the furnace is shown in Fig. 3.

The specimens from the individual positions were subdivided into about 5 mm thick layers reaching from the inner surface of the furnace down to the maximum depth for which material was available. The separation was made with a knife for the soft crumbling deeper layers or with a diamond circular saw for the hard material close to the surface. The samples were then ground in an agate mortar for Mössbauer, XRD and XRF measurements.

It turned out in the course of the studies that in the lowest available layers the temperature of the soil had reached about 400 °C. This is the temperature at which the soil was baked sufficiently to solidify somewhat, enough to stick to the pieces recovered from the furnace.

Of all 98 samples that were obtained as described above, ^{57}Fe Mössbauer spectra were measured at room temperature (RT), and at 4.2 K in a liquid He bath cryostat.

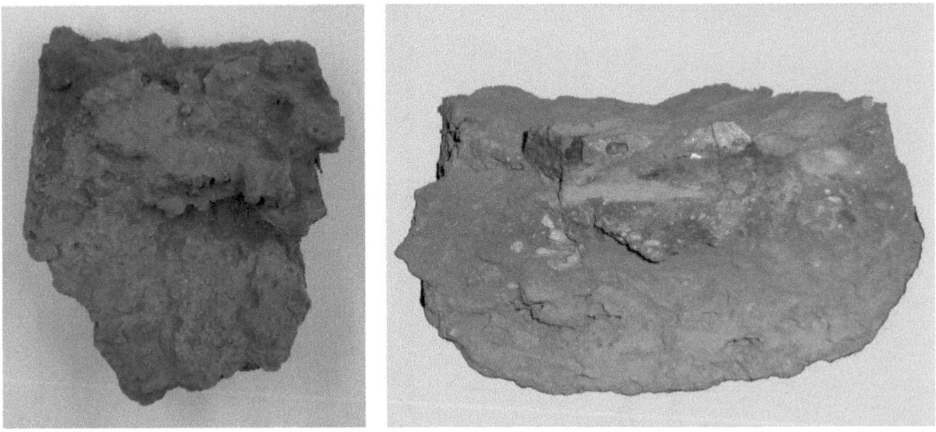

Fig. 3 On the left the uneven, partially vitrified inner surface of a piece of the rear wall of the furnace is shown. The piece is about 16 cm high and 13 cm wide and reaches from the middle to the lower end of the rear wall. From its upper rim about 3 cm were cut off, yielding samples 612/4, 612/5 and 612/6 (cf. Fig. 2). On the right, a view of the cut face of the piece left after removing these samples is shown. The piece reaches about 50 mm deep into the soil. The bright spot in the middle about 5 mm below the surface is a rather large metallic copper inclusion

The spectra were fitted with suitable superpositions of Lorentzian lines grouped into quadrupole doublets, magnetic sextets or patterns corresponding to (sometimes asymmetric) Gaussian distributions of electric quadrupole splittings or magnetic hyperfine fields. Isomer shifts (IS) will be given as measured with respect to the ^{57}Co:Rh source having the same temperature as the absorber.

X-ray powder diffraction (XRD) analyses were performed on selected samples with a Philips PW 1070 diffractometer equipped with a graphite monochromator and using CoK_{α} radiation (40 kV and 30 mA), in steps of 0.02 °2θ with a counting time of 5 s for each step. The positions of the X-ray reflexes were fitted using the Macclayfit routine [5]. For the XRF determinations of the element contents a portable Niton XL3t analyzer was used.

3 Results and discussion

The evaluation of the data is still in progress, and additional re-firing experiments are planned. In the following, data on only two positions inside the furnace will be discussed in some detail: specimen 612/6 taken near the middle of the rear wall, and specimen 612/3 taken about 6 cm higher up on the rear wall. These examples show what kind of information can be obtained and give a first impression of the conditions prevailing during the firing of the furnace.

Piece 612/6, being about 50 mm thick, was subdivided into a series of 10 layers, from sample 612/6-1 covering the topmost 5 mm to sample 612/6-10 at a depth of 45–50 mm in the soil. Many small copper inclusions several hundred micrometers in size were visible under the optical microscope or even with the bare eye. An example of a rather large inclusion of millimeter size is shown on the right side of Fig. 3. XRF showed that the uppermost 5 mm of 612/6 contain only about 0.25 % of copper, but

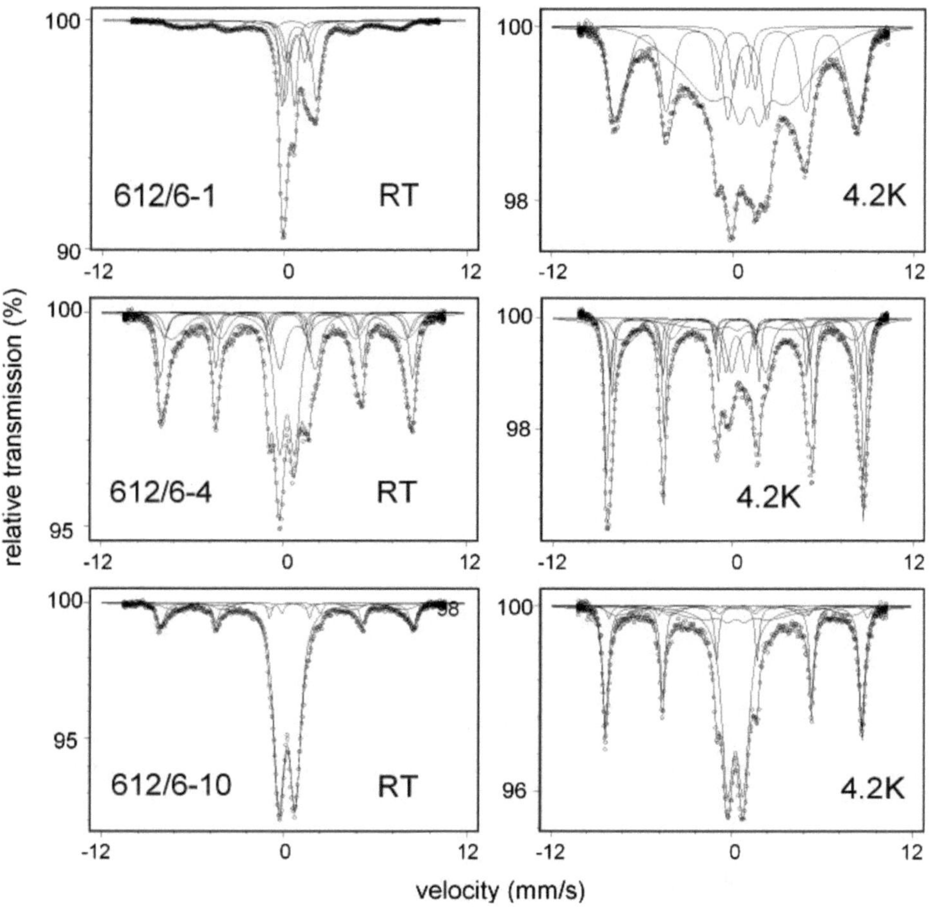

Fig. 4 Mössbauer spectra taken of specimen 612/6 at room temperature (*left*) and 4.2 K (*right*) for the topmost layer(612/6-1, 0–5 mm), an intermediate layer (612/6-4, 15–20 mm) and the lowest layer (612/6-10, 45–50 mm)

the neighboring piece 612/4 contains 4 % of copper in its uppermost layer. A slight arsenic enrichment is observed at some spots in the copper-rich surface regions.

Figure 4 shows some of the Mössbauer spectra of piece 612/6; the corresponding XRD patterns are shown in Fig. 5. In the uppermost layer (612/6-1) about half of the iron is divalent, showing that the conditions were reducing. Part of this is bound in hercynite ($FeAl_2O_4$) which forms in ceramics beginning at 750 °C under reducing conditions [4], yields a spectrum consisting of several quadrupole doublets at RT and splits into a complicated magnetic pattern at 4.2 K [6, 7]. The Fe^{2+} which still yields a quadrupole doublet at 4.2 K is presumably a vitreous phase but might also, at least in part, be an impure form of hercynite that is not ordered at 4.2 K. In the RT spectrum, trivalent iron is also present, both as a quadrupole doublet and as a phase that already splits magnetically at RT. This component is not a common iron oxide like hematite or maghemite or magnetite, but rather represents the $Mg(Fe,Al)_2O_4$ spinel phase, which at higher iron contents is already ordered at RT [8]. In fact, the surface layer

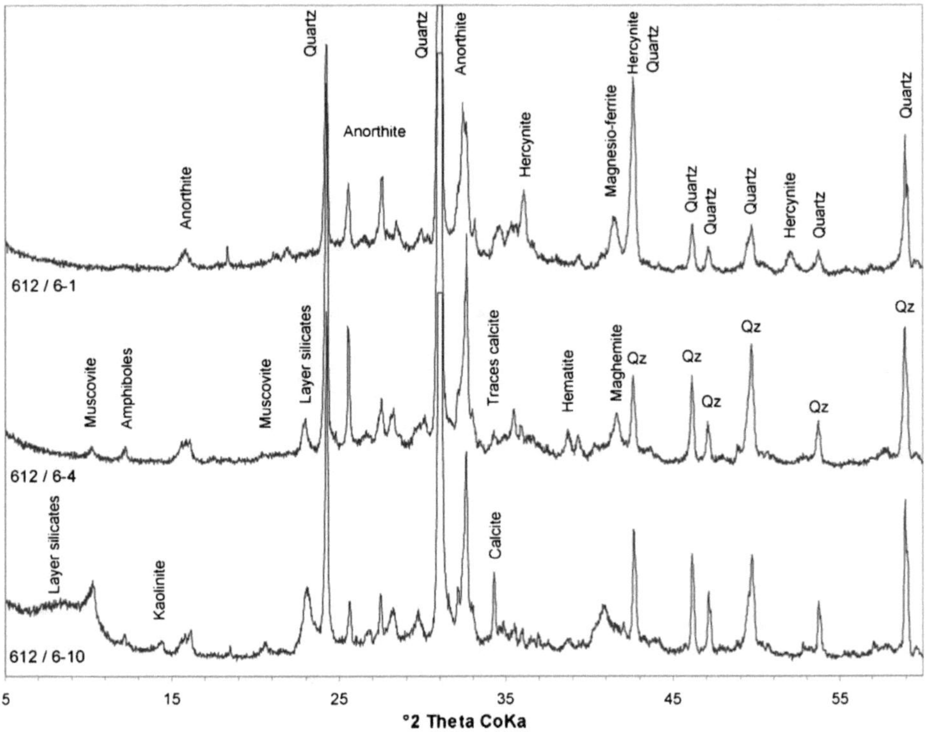

Fig. 5 XRD patterns of specimen 612/6 for the topmost layer (612/6-1, 0–5 mm), an intermediate layer (612/6-4, 15–20 mm) and the lowest layer (612/6-10, 45–50 mm)

of 612/6 and neighboring regions is rich in Mg (about 10 % according to XRF). The spinel phase and hercynite can also be identified in the X-ray diffractogram (Fig. 5). Mica and amphiboles are absent in the uppermost layer, though they are present in the lower ones. Since these minerals become unstable above about 900 °C, one may conclude that the temperature reached at least 900 °C.

Below the uppermost layer, the amount of divalent iron decreases until, at a depth of 15–20 mm (612/6-4) it amounts to no more than about 10 %. In this depth, the main Fe^{3+} components are hematite and maghemite, which split already at RT and dominate the 4.2 K spectrum. Since maghemite is unstable above about 750 °C, the temperature in a depth of 15–20 mm cannot have been higher. Hematite and maghemite are also visible in the XRD pattern, and so are peaks of mica and amphibole.

Finally, in the lowest layer, at 45–50 mm (612/6-10), the XRD pattern still shows the presence of clay minerals (chlorite, kaolinite), which means that the soil there cannot have been hotter than about 400 °C. This is confirmed by the Mössbauer patterns. The RT Mössbauer pattern is dominated by a quadrupole doublet attributable mainly to Fe^{3+} in the clay minerals present of the soil. Firing studies of a number of clays from the Lambayeque region have shown [3, 4] that the quadrupole splitting (QS) of this doublet changes due to the dehydroxylation in a typical manner as is shown in Fig. 6. The behavior of different clays was found to differ but very little; the curve shown in Fig. 6 is the average over 7 different clay samples from the

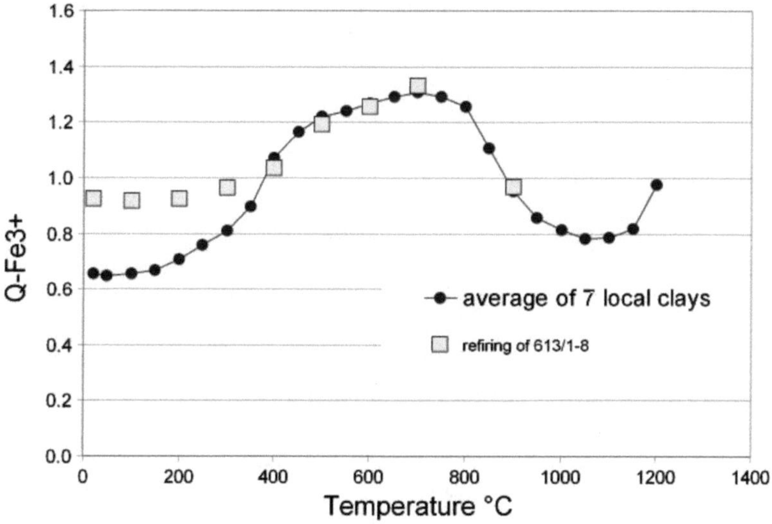

Fig. 6 Room temperature quadrupole splitting of the paramagnetic Fe^{3+} component of clays from the Lambayeque region as a function of the firing temperature. The filled dots and the curve are averages of data for 7 different clay samples. The squares are the results of re-firing samples from a deep layer at the bottom of the furnace, 613/1-8 (25–30 mm), in the laboratory. The points show that the original temperature reached there was 350 °C, and that the clay of the soil into which the furnace was dug behaves in the manner expected for clays from the Lambayeque region

Lambayeque region. The initial value is around QS = 0.6 mm/s, a value typical for ferric iron in clay minerals. The splitting begins to rise when dehydroxylation begins between 200 °C and 300 °C. It reaches QS = 1.20 mm/s on firing at 500 °C. After a broad maximum, it drops quite abruptly on firing above 800 °C. The initial rise is well suited for determining low firing temperatures. Assuming that the master curve of Fig. 6 is also valid for the soil into which the furnace was dug, the splitting of QS = 1.02 mm/s found in sample 612/6-10 corresponds to a firing temperature of 380 °C (Fig. 6). It is also worth mentioning that the diffraction pattern shows the presence of calcite. Expectedly, the layers closer to the surface reached higher temperatures. In 612/6-7 (30–35 mm), for instance, one finds QS = 1.19 mm/s corresponding to 500 °C. Figure 6 also shows QS values obtained by laboratory re-firing experiments on a sample from the bottom of the furnace (613/1-8; 25–30 mm below the surface), which by the same arguments has reached 400 °C in antiquity. The fact that the data points obtained by the re-firing experiments follow the master curve at higher temperatures shows that the soil into which the furnace was built behaves like other regional clays.

Piece 612/3 is remarkable since, although it was taken only about 6 cm above 612/6, it contains hardly any divalent iron. More than 97 % of the iron is trivalent in the uppermost layers, showing that the firing conditions were strongly oxidizing. Some typical Mössbauer spectra are shown in Fig. 7, the corresponding X-ray diffraction patterns in Fig. 8. The RT Mössbauer spectrum of the uppermost layer, 612/3-1 (0–5 mm), exhibits the magnetically split spinel phase (25%) that is also present in the surface of 612/6, and a minor amount of hematite (8%). The majority of the iron in paramagnetic and requires fitting with two quadrupole doublets. One of these, with QS = 0.69 mm/s, IS = 0.26 mm/s and an area of 35% has parameters that are

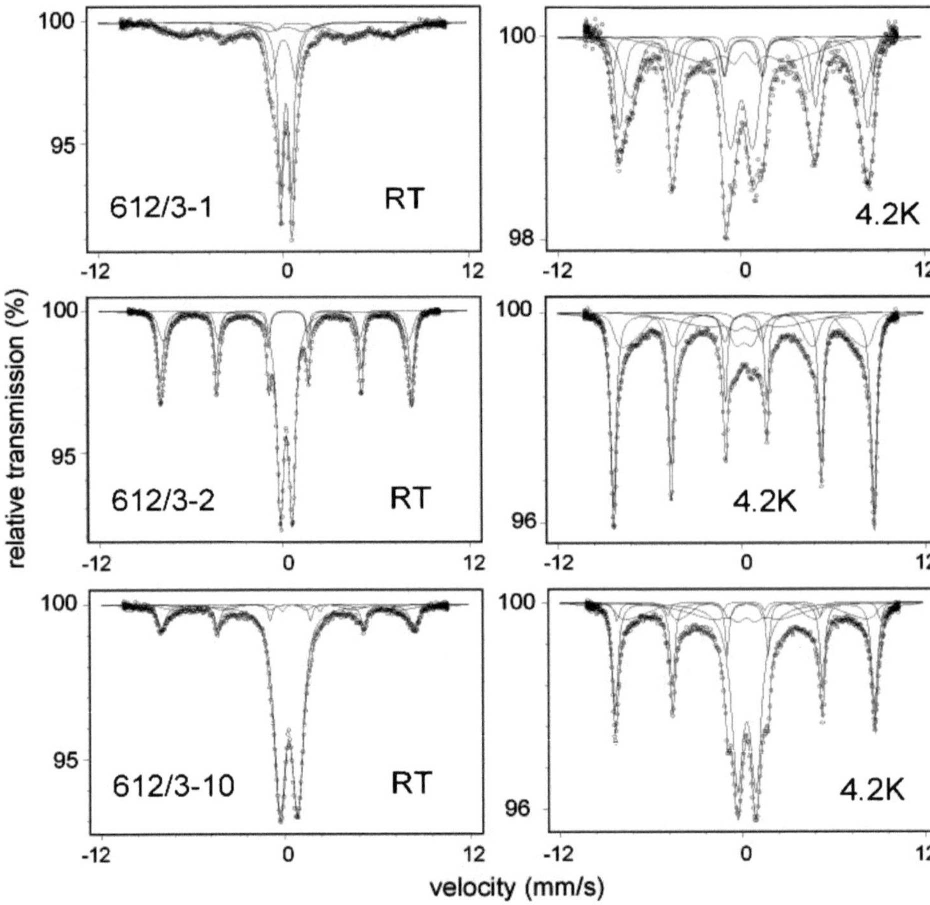

Fig. 7 Mössbauer spectra of piece 612/3 taken at room temperature (*left*) and 4.2 K (*right*) for the topmost layer (612/3-1, 0–5 mm), the layer just below (612/3-2, 5–10 mm) and the lowest layer (612/3-10, 45–50 mm)

often found in ceramics fired at high temperatures. The other (QS = 1.58 mm/s; IS = 0.12 mm/s; area 30%) has parameters that can be attributed to iron-bearing gehlenite ($Ca_2(Al,Fe)_2SiO_7$) if the two lattice sites T1 and T2 of gehlenite [9] cannot be resolved, perhaps due to bad crystallinity. The formation of gehlenite has been observed after firing calcareous clays at 850 or 900 °C [10–12]. The presence of gehlenite in 612/3-1is confirmed by the X-ray diffractogram (Fig. 7), as is the presence of a spinel phase, presumably $Mg(Fe,Al)_2O_4$. The absence of mica and amphiboles in the XRD pattern confirms the high firing temperature. XRF showed that the surface layer of 612/3 is not only rich in Mg (17 %) but also in Ca (17 %), while the layers below contain only about 3 % of Ca and very little Mg.

Surprisingly, the layer below the uppermost one, 612/3-2 (5–10 mm), has completely different Mössbauer spectra. Both the RT and 4.2 K spectra are like spectra of clays fired in air between 800 °C and 900 °C: The sample contains most of the iron as hematite (50%). The remainder is also magnetically split but with broad patterns

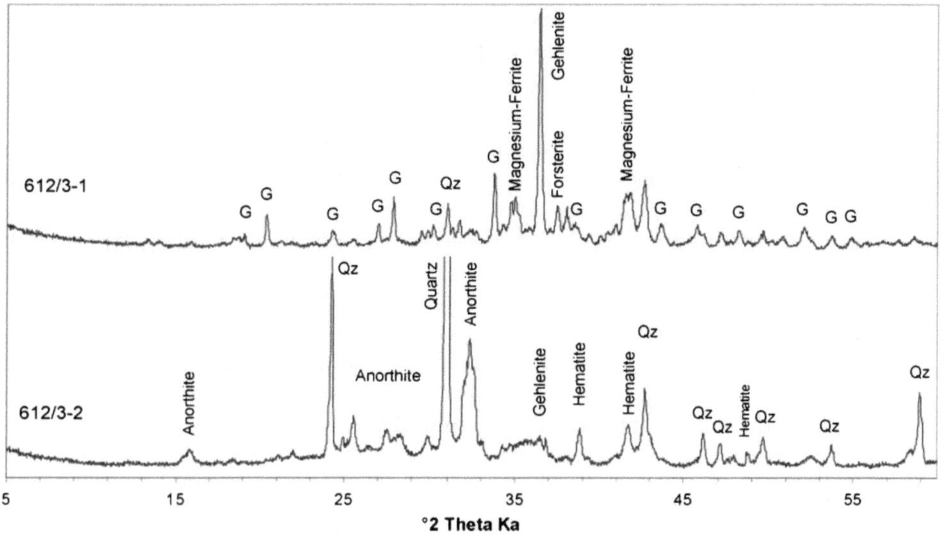

Fig. 8 XRD patterns of piece 612/3 for the topmost layer (612/3-1, 0–5 mm) and the layer immediately below (612/3-2, 5–10 mm). The peaks bearing the letter G in can all be attributed to gehlenite. In 612/3-2 gehlenite and magnesium ferrite is practically absent, but hematite is well visible

that required fitting with distributions of hyperfine fields. These are attributed to small oxidic clusters that order magnetically at low temperatures [13, 14]. Only 5 % of the iron is still paramagnetic at 4.2 K. The difference between 612/3-1 and 612/3-2 does not seem to be due to different temperatures or redox conditions. Re-firing experiments at 1000 °C in air did not change either spectrum decisively. XRF showed, however, that the Ca content in 612/3-2 is only about 3 % and the Mg content is even less and thus equal to the values also found deeper in the soil. The low Ca content favors the formation of hematite rather than gehlenite. The high Ca and Mg content in the surface layer may be due to repairs of the furnace walls with a material rich in these elements.

At a depth of 45–50 mm (612/3-10) one finds a spectrum with a RT quadrupole splitting of 1.14 mm/s, which according to Fig. 6 corresponds to a firing temperature of 400 °C, much like 612/6-10. Going towards the surface, the quadrupole splitting rises, for instance to QS = 1.29 mm/s for 612/3-7 (30–35 mm), which corresponds to about 600 °C. This behavior is very similar to that of 612/6, but the temperatures reached at the same depths are somewhat higher than in 612/6.

4 Conclusions

The investigation of the furnace from the Pampa de Chaparrí is still being continued; the present paper is intended to give only a first impression of the information that can be gleaned from these studies.

Substantial amounts of divalent iron are found in samples from the middle of the rear wall, but not at a spot about 5 cm higher up in the rear wall. This implies that the conditions in the middle of the rear wall were reducing not only during the smelting process but also during cooling. Possibly the furnace was filled with charcoal and smelting debris up to about half height of the rear wall, while the surface was uncovered farther up. In the bed of charcoal the conditions would have been reducing. In fact, at the bottom of the furnace (positions 612/2 and 613/1, cf. Fig. 2) the iron in the surface layers is also found to be largely divalent. The temperatures reached at the inner surface of the furnace at positions 612/6 and 612/3, i.e., near the middle of the rear wall, were around 900 °C or even higher. Generally, these findings are in agreement with observations during replicative smelting experiments conducted by Shimada and Merkel [2].

Presumably the furnace was used repeatedly. The influence of repeated heating of the soil into which the furnace was built has not yet been investigated in any detail. Supposedly, the Mössbauer patterns of the furnace material will largely retain a record of the last firing, or alternatively the highest temperature reached at any spot. We intend to study details of this by model experiments in the laboratory.

Acknowledgements Research at 256A01 was funded by a National Science Foundation grant (BCS-0838211) to Hayashida and carried out with the permission of the Peruvian Instituto Nacional de Cultura, and with the logistical support of the Sicán National Museum. Simonnet Pérez and Lewis Borck excavated the furnaces described in this article. The authors thank these individuals and institutions for their contributions.

References

1. Killick, I.D., Hayashida, F.: Lung-powered copper smelting of the Middle Sicán Period (ca. 900–1000 CE), North coastal Peru, 39th International Symposium in Archaeometry, Leuven, Belgium (2012)
2. Shimada, I., Merkel, J.F.: Sci. Am. **26**, 80 (1991)
3. Shimada, I., Häusler, W., Hutzelmann, T., Wagner, U.: Hyperfine interact. **150**, 73 (2003)
4. Wagner, F.E., Wagner, U.: Hyperfine interact. **154**, 35 (2004)
5. Stanjek, H., Häusler, W.: Ber. Dtsch. Ton- Tonminer. **7**, 256 (2000)
6. Dormann, J.L., Seqqat, M., Fiorani, D., Noguès, M., Soubeyroux, J.L., Bhargava, S.C., Renaudin, P.: Hyperfine interact. **54**, 503 (1990)
7. Kalvius, G.M., Krimmel, A., Hartmann, O., Litterst, F.J., Wäppling, R., Wagner, F.E., Tsurkan, V., Loidl, A.: Eur. Phys. J. **77**, 7 (2010)
8. Sundararajan, M.D., Narayanasamy, A., Nagarajan, T., Häggström, L., Swamy, C.S., Ramanujachary, K.V.: J. Phys. C: Solid State Physics **17**, 2953 (1984)
9. Akasaka, M., Ohashi, H.: Phys. Chem. Miner. **12**, 13 (1985)
10. Cultrone, G., Rodriguez-Navarro, C., Sebastian, E., Cazalla, O., De La Torre, M.J.: Eur. J. Miner. **13**, 621 (2001)
11. Maniatis, Y., Simopoulos, A., Kostikas, A.: J. Am. Ceram. Soc. **64**, 263 (1981)
12. Maniatis, Y., Simopoulos, A., Kostikas, A., Perdikatsis, V.: J. Am. Ceram. Soc. **66**, 773 (1983)
13. Friederich, J., Gebhard, R., Krause, R., Riederer, J., Wagner, F.E., Wagner, U.: J. Phys.: Conf. Ser. **217**, 102064 (2010)
14. Costa, B.F.O., Silva, A.J.M., Wagner, F.E., Wagner, U.: Hyperfine interact. **202**, 81 (2011)

DOI 10.1007/978-94-007-6482-8_20
Reprinted from *Hyperfine Interactions*, DOI 10.1007/s10751-013-0860-3

Structural, calorimetric and magnetic properties study of the $Cu_{0,91}Fe_{0,09}O$ system

H. D. Colorado · J. S. Trujillo Hernandez · G. A. Pérez Alcázar · Alberto Bolaños

Published online: 24 April 2013

Abstract In this work the $Cu_{0.91}Fe_{0.09}O$ nanocrystalline system was prepared via the co-precipitation method. Using Mössbauer Spectrometry, X-Ray Diffraction, Vibrating Sample Magnetometry, Thermogravimetry and Differential Scanning Calorimetry, we study the magnetic behavior, and the structural and calorimetric properties of this system. X-ray diffraction shows only the presence of the CuO structural monoclinic phase, suggesting that Cu atoms are substituted by Fe ones. This hypothesis was confirmed by Mössbauer spectrometry at room temperature, because it shows that the spectrum is formed by two doublets, which correspond to Fe^{+2} and Fe^{+3} sites. Hysteresis cycles obtained by vibrating sample magnetometry detect a soft ferromagnetic behavior at room temperature with coercive fields between 8 and 20 Oe. At T = 20 K the sample shows a hard-magnetic behavior. The thermogravimetry results show a Néel temperature ($T_N > 440$ °C). The differential scanning calorimetry curve show two endothermic peaks in the 90–120 °C range.

Keywords Mössbauer spectrometry · X-ray diffraction · Vibrating sample magnetometry · CuO

1 Introduction

Magnetism in nanoparticles is currently the subject of interest because of its wide application in industrial as well as research areas [1]. CuO, as a narrow band gap p-type semiconductor, has been recognized as an industrially important material for

Proceedings of the Thirteenth Latin American Conference on the Applications of the Mössbauer Effect, (LACAME 2012), Medellín, Columbia, 11–16 November 2012.

H. D. Colorado (✉) · J. S. T. Hernandez · G. A. P. Alcázar
Departamento de Física, Universidad del Valle, Meléndez, A. A 25360 Cali, Colombia
e-mail: hernan.colorado@correounivalle.edu.co

A. Bolaños
Departamento de Química, Universidad del Valle, Meléndez, A. A 25360 Cali, Colombia

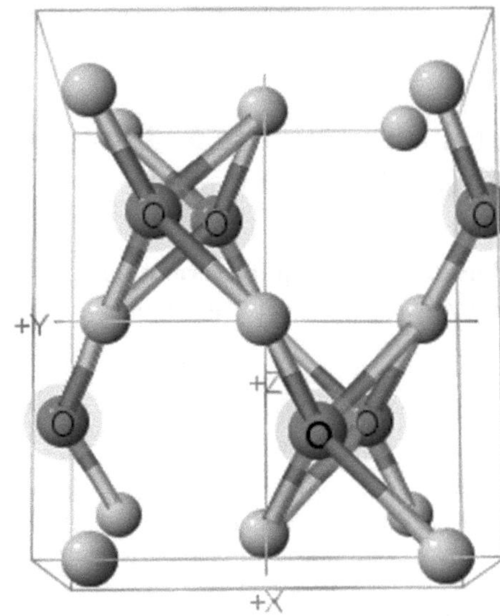

Fig. 1 Monoclinic structure of the CuO system. Image taked from the reference [13]

a variety of practical applications, such as gas sensors [2], catalysis [3], optics, solar energy conversion [4], and field emission [5]. CuO is antiferromagnetic bellow 230 K [6], but when it is doped at very low concentrations, it could exhibits anomalous magnetic properties [7, 8]. For this reason, it is interesting to know the response of this compound when it is doped with other materials, for example Fe. In recent years, ferromagnetism in semiconductors, including oxides, has received renewed attention, partly due to interest in spintronic device concepts, as spin valve for magnetic recording, sensor devices, spin transistors and nonvolatile storage devices, in which it can be controlled spin injection by weak magnetic field [9, 10]. CuO nanoparticles exhibit unique properties, such as ferromagnetic response due to the uncompensated surface spins, and an optical band gap (2.43 eV) which is larger than that of bulk CuO (1.85 eV). These surface spins are responsible for the observed anomalous magnetic properties of CuO nanoparticles [11]. For example, temperature dependent magnetic effects of the surface spin lead to several interesting phenomena such as enhanced magnetic moments, superparamagnetism, and magnetic hysteresis [12]. CuO presents a lattice with a monoclinic symmetry and a magnetic susceptibility temperature behavior that is unusual for 3d antiferromagnets. Each atom has four nearest neighbors of the other kind [13] (see Fig. 1). The bulk CuO is an antiferromagnetic material with Néel temperatures reported near 215 and 230 K. However, surface spins can dominate magnetization in nanostructures because of uncompensated exchange coupling. This may lead to ferromagnetic like behavior at low temperatures. In recent years, several studies on Fe-doped CuO nanomaterials have been reported [14–18].

In this work we study the structural, thermal, and magnetic behavior of Fe-doped CuO, taking into account its possible technological applications and the basic aspects of magnetism that are not understood in this system. In this work CuO samples doped with 9 % Fe are prepared using the co-precipitation method. This paper presents the magnetic behavior of the sample at 300 K and 20 K, studied by transmission Mössbauer spectrometry (MS) and hysteresis loops. The structural properties were studied by X-ray diffraction (XRD). The thermal properties were studied by the thermogravimetry (TGA) and the differential scanning calorimetry (DSC) techniques.

2 Experimental method

The production process of copper monoxide doped with Fe was arisen by the co-precipitation method. We used stoichiometric amounts of Cupper Nitrate, $Cu(NO_3)3H_2O$ and Iron Nitrate, $Fe(NO_3)_39H_2O$. The starting salts were put in an aqueous solution under magnetic stirring at 600 rpm for 40 min at 40 °C. Then with the addition of Sodium Hydroxide (NaOH), co-precipitation occur releasing heat to stabilize and eventually kill a temperature of 40 °C. NaOH is used to control the PH which remained at 12.53 in the aqueous solution on magnetic stirring for 1 h. The resulting precipitate, from the chemical reaction, is slowly cooled to room temperature, and then transferred to a centrifuge at 2000 rpm for 10 min in order to decant the $Cu(FeOH)_2$ and remove the $NaNO_3$ formed. This process was repeated four times until the complete removal of $NaNO_3$. Then the precipitate was dried for 192 h at 50 °C and calcined for 5 h at 260 °C. A pure CuO sample was also produced by the same method. Mössbauer measurements were performed in a conventional spectrometer at room temperature with transmission geometry using a ^{57}Co source of 25 mCi in a Rh matrix. All spectra were fitted with the Mosfit program [19], and the isomer shift values are referred to that of α-Fe. The X-ray measurements were obtained using Cu-Kα radiation and the patterns were refined using the GSAS program [20], which is based in the Rietvelt method combined with Fourier analysis to describe the broadening of the lines. With this refinement the average values of the lattice parameter, crystallite size, and the structural phase were obtained. The behavior of the magnetization as a function of the field was studied by using the vibrating sample magnetometer (VSM) making use of a Physics Properties Measurement System (PPMS) equipment of the Excellence Center for New Materials (CENM). Surface morphology of the sample was observed using a scanning electron microscope (SEM) of the same center. The thermogravimetry (TGA) measurements were performed with a small applied field ($\approx$ 20 Oe) to determine the T_N as a function of temperature. The measurement was performed in N_2 atmosphere with a heating rate of 10 °C/min.

3 Results and discussion

Figure 2 shows the XRD patterns of the CuO and $Cu_{0.91}Fe_{0.09}O$ samples. It can be seen that both samples present a single monoclinic structural phase, and this confirms that all Fe atoms are substituting Cu atoms in the CuO lattice [18]. From the refinement of the diffractogram of CuO it was obtained that: the perpendicular and parallel crystallite sizes are equal to 51,4 ± 0,1 and 25,7 ± 1,3 nm, respectively, indicating that the crystallites have an elongated shape; and that the lattice parameters of this structure are a = 4,68 Å, b = 3,42 Å and c = 5,13 Å. From the diffractrogram of CuO doped with Fe it was found that: the perpendicular and parallel crystallite sizes are equal to 36,8 ± 0,2 nm and 40,4 ± 0,5 nm, respectively, showing that the

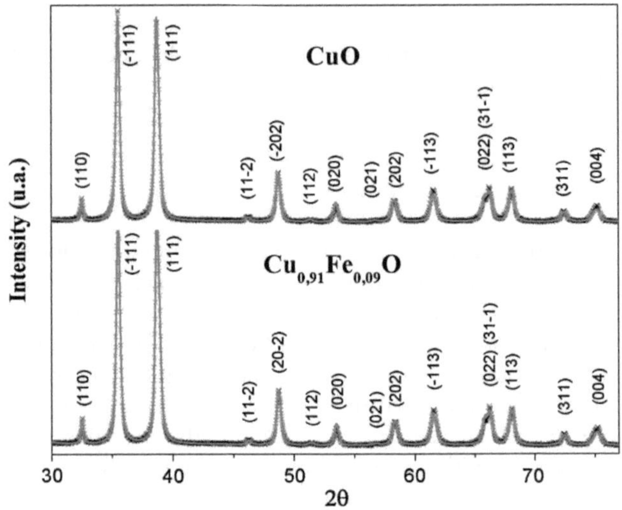

Fig. 2 XRD patterns of the $Cu_{0,91}Fe_{0,09}O$ system

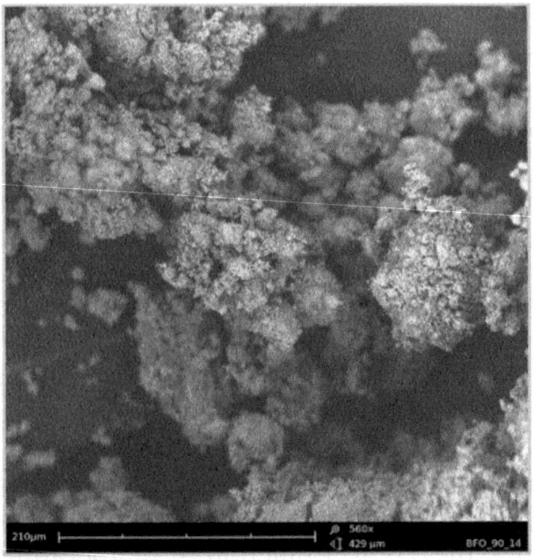

Fig. 3 SEM image (6100X) of the $Cu_{0,91}Fe_{0,09}O$ sample after calcination at 260 °C

crystallites have a nearly oblate shape; the lattice parameters of this structure are a = 4,69 Å, b = 3,41 Å and c = 5,13 Å. Comparing the lattice parameters of both samples it is observed that they do not vary significantly, indicating that there is no deformation of the lattice by the Fe doping. Additionally, it can be said that the reduction of crystallite size obtained due the replacement of Cu by Fe atoms produces a more fragile material.

According to the micrograph obtained by SEM, and shown in Fig. 3, it can be noted that the particles present an agglomeration tendency; this result is consistent with [11]. But the resolutions do not permit to obtain the mean particle size.

Figure 4 shows the Mössbauer transmission spectrum of the $Cu_{0,91}Fe_{0,09}O$ sample measured at room temperature. The Mössbauer spectrum was fitted with two doublets indicating the paramagnetic nature of the sample and the total dissolution of

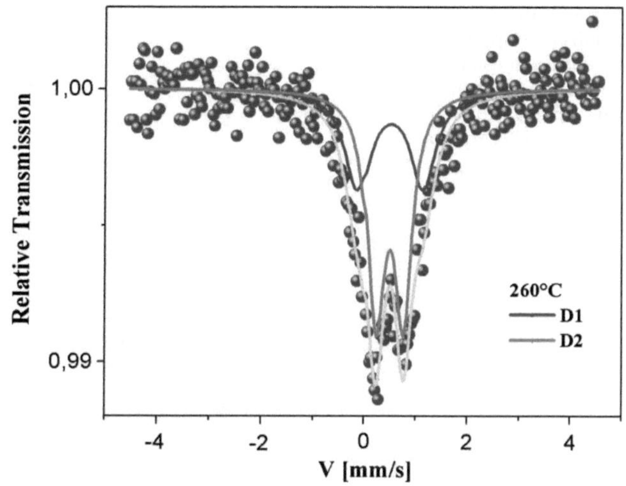

Fig. 4 Transmission Mössbauer spectrum of the $Cu_{0,91}Fe_{0,09}O$ system

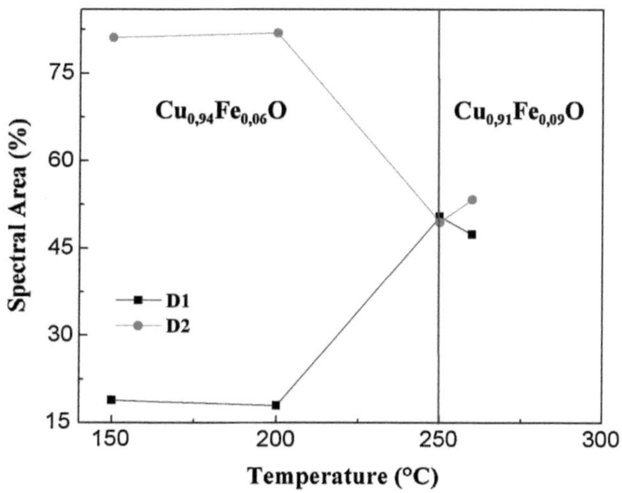

Fig. 5 Variation of the spectral area with the different calcinations temperatures of the $Cu_{0,94}Fe_{0,06}O$, results from reference [17] and compared with the current $Cu_{0,91}Fe_{0,09}O$ system

the Fe atoms, substituting Cu atoms. The doublets have a line half width of 0,50 mm/s showing a disordered paramagnetic behavior of the system. The two doublets present basically the isomer shift of the Fe^{+3} (δ = 0,35 mm/s), but different quadrupole splitting ($\Delta Q_1 \cong$ 1,10 mm/s and $\Delta Q_2 \cong$ 0,49 mm/s, respectively). Then it is proposed that the second site (D_2) corresponds to a Fe^{+3} site and the first one (D_1) to the Fe^{+3} site but with a more asymmetric charge distribution around it.

Figure 5 presents the behavior of the spectral area of each doublet, with the calcination temperature; these data were compared with those of Ref. [17]. These authors reported the variation of the spectral area and of the quadrupole splitting with the different calcination temperatures of the $Cu_{0,94}Fe_{0,06}O$ system obtained at temperatures of 150, 200 and 250 °C. The fourth points plotted in Fig. 5 are the current data reported for the $Cu_{0,91}Fe_{0,09}O$ at a calcination temperature of 260 °C. It is observed that the D_2 area tends to increase with temperature while that of D1

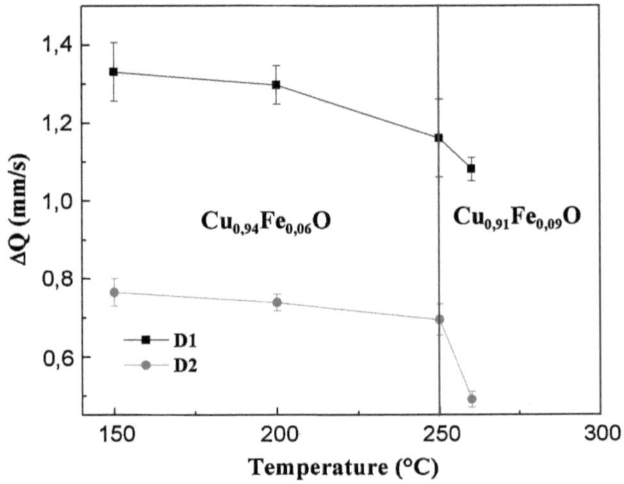

Fig. 6 Variation of the quadrupole splitting with the different calcinations temperatures of the $Cu_{0,94}Fe_{0,06}O$, results from reference [17] and for the present $Cu_{0,91}Fe_{0,09}O$ system are compared

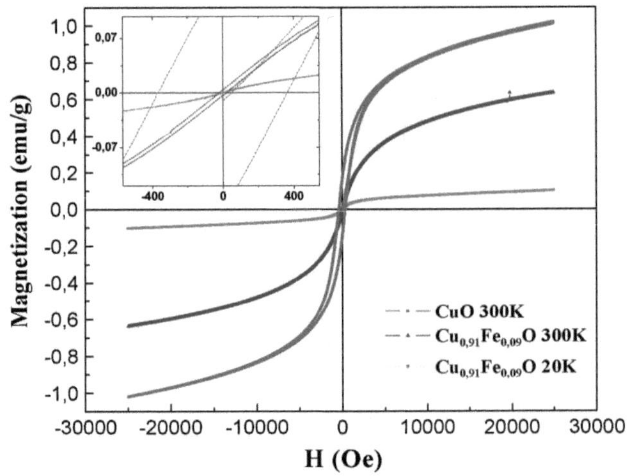

Fig. 7 Hysteresis loop of the $Cu_{0,91}Fe_{0,09}O$ system

decreases. Between 250 °C and 260 °C the diffusion of atoms due to the calcination temperature favors the Fe^{+3} site with the highest charge asymmetric than that of the other site.

In Fig. 6 the quadrupole splitting of both sites, obtained for different temperatures, are compared. It is observed that between 250 °C and 260 °C both values decrease, indicating that by increasing the calcination temperature and the Fe doping, a reduction of the asymmetry is obtained.

Magnetization measurements of both samples show a ferromagnetic behavior with coercive field of 8 ± 1 Oe for the CuO and 20 ± 1 Oe for the Fe-doped sample, as seen in Fig. 7. These magnetization curves were taken at T = 300 K and indicates a soft magnetic behavior of the material [21]. The Fe-doped sample measured at T = 20 K shows a coercive field of 372 ± 5 Oe indicating that there is a bigger ordering of magnetic moments due the temperature decrease. The increase in the field shows that the material behaves as magnetically hard at low temperatures.

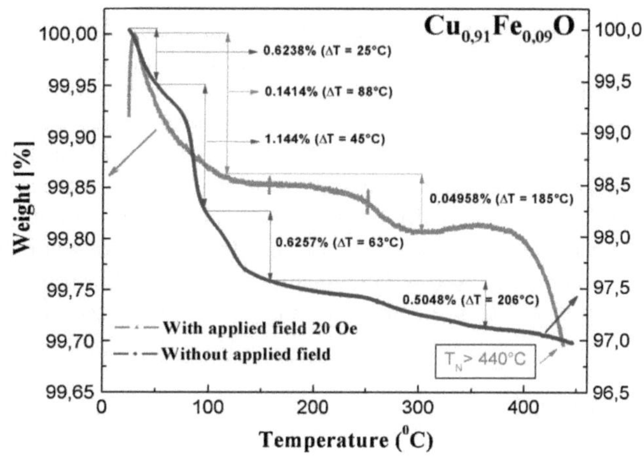

Fig. 8 Termogravimetry without applied field and with applied field of (≈ 20 Oe) of the $Cu_{0,91}Fe_{0,09}O$ system

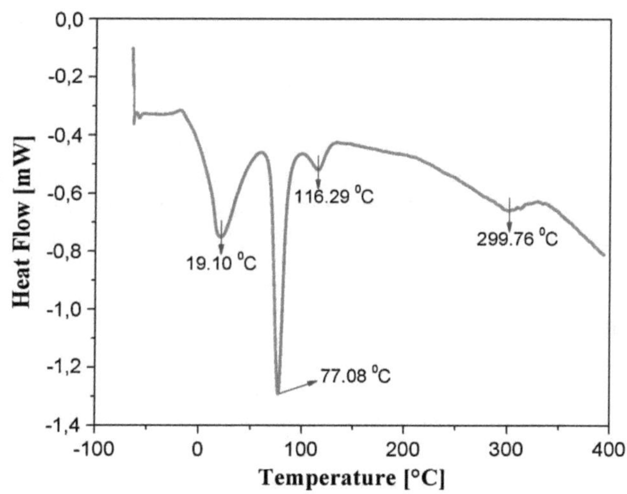

Fig. 9 Differential scanning calorimetry of the $Cu_{0,91}Fe_{0,09}O$ system

The different behavior shown by the Mossbauer (paramagnetic) and by hysteresis cycles (soft ferromagnetic) is a consequence of the different measurement times of both techniques. The hysteresis cycles in these samples are part of the anomalous behaviors which were also reported by other authors [12] and they attribute this behavior to the superficial spins.

In Fig. 8 the TGA measurement of the doped sample is shown. A weight loss of 3 % was obtained between room temperature and nearly 120 °C. Besides the maximum at 20 °C, small anomalies can be noted near 80 and 120 °C and these may be due to the evaporation of water and/or gases which are present at the surface of the sample. The water presence is a consequence of the hygroscopic character of the CuO. Two small anomalies can also be observed at around 250 and 400 °C.

Figure 8 also shows the effect of an applied field (≈ 20 Oe) on the thermogravimetric answer of the doped sample and it can be noted that the anomalies at 300 and 400 °C are enhanced, especially that at around 400 °C. We associated this anomaly corresponds to a structural phase transition. This results show a Néel temperature (T_N > 440 °C), this result is consistent with [18].

In Fig. 9 the DSC measurement is shown and three endothermic peaks around 20, 80 and 120 °C are detected. They correspond to irreversible processes and agree with the anomalies detected for the loss of weight of Fig. 8 and correspond to the energy adsorbed by the sample to remove impurities, water or gas, on the sample surface. Additionally between 300 and 350 °C a change of curvature corresponding to a reversible processes can be attributed. However, this still needs to be checked with magnetic and XRD measurements at higher temperatures, in order to prove our proposal or to see if this anomaly corresponds to a structural phase transition.

4 Conclusions

XRD results indicate that the sample has a characteristic pattern of CuO, when Fe atoms replace Cu atoms. A decrease of the crystallite size is obtained with the addition of Fe. These DRX results indicate that the material is nanostructured and SEM micrograph proves that the particles are agglomerated. Transmission Mössbauer spectrometry results show that there is a disordered paramagnetic behavior due to the large value of the line half-width of the obtained doublets. The results obtained for the hysteresis loop indicate that the material at room temperature (300 K) shows a soft magnetic behavior, but by making measurements at 20 K this presents an ordering of magnetic moments which increases the coercive field indicating that the material has a hard magnetic behavior at this temperature.

References

1. Batlle, X., Labarta, A.: J. Phys. D **35**, R15 (2012)
2. Zhang, J., Liu, J., Peng, Q., Wang, X., Li, Y.: Chem. Mater. **18**(4), 867–871 (2006)
3. Teng, F., Yao, W., Zheng, Y., Ma, Y., Teng, Y., et al.: Sens. Actuators B, Chem. **134**, 761–768 (2008)
4. Anandan, S., Wen, X., Yang, S.: Mater. Chem. Phys. **93**, 35–40 (2005)
5. Hsieh, C.-T., Chen, J.-M., Lin, H.-H., Shih, H.-C.: Appl. Phys. Lett. **83**, 3383–3385 (2003)
6. Roden, B., Braun, E., Freimuth, A.: Solid State Commun. **64**, 1051–1052 (1987)
7. Borzi, R.A., Stewart, S.J., Punte, G., et al.: J. Appl. Phys. **87**, 4870–4872 (2000)
8. Bishta, V., Rajeeva, K.P., Banerjee, S.: arXiv:0911.1838v2 (2011)
9. Wolf, S.A., Awschalom, D.D., Buhrman, R.A., et al.: Sci., **294**, 1488–1495 (2001)
10. Fukimura, T., Yamada, Y., Toyosaki, H., et al.: Appl. Surf. Sci. **223**, 62–67 (2004)
11. Punnoose, A., Magnone, H., Seehra, M.S., Bonevich, J.: Phys. Rev. B. **64**, 174420 (2001)
12. Rao, G.N., Yao, Y.D., Chen, J.W.: IEEE Trans. Magn. **41**, 3409–3411 (2005)
13. Crystal lattice structures [On-Line]. http://cst-www.nrl.navy.mil/lattice/spcgrp/index.html. Accessed 5 Jan 2012
14. Li, Y., Xu, M., Pan, L., Zhang, Y., Guo, Z., Bi, C.: J. Appl. Phys. **107**, 113908 (2010)
15. Yin, S.Y., Yuan, S.L., Tian, Z.M., Liu, L., Wang, C.H., Zheng, X.F., Duan, H.N., Huo, S.X.: J. Appl. Phys. **107**, 043909 (2010)
16. Jing, Z., Qinglin, Z., Jinmin, L.: **33**(013001), 1–3 (2012)
17. Colorado, H.D., Pérez Alcázar, G.A.: Hyperfine Interact. **202**, 139–144 (2011)
18. Joseph, D.P., Venkateswaran, C., Vennila, R.S.: Adv. Mater. Sci. Eng. **2010**, 1–4 (2010)
19. Teillet, J., Varret, F.: MosfitProgramm, University du Maine, unpublished, (1976)
20. Larson, A.C., Von Dreele, R.B.: General Structure Analysis System (GSAS), Los Alamos National Laboratory Report LAUR, pp. 86–748 (1994)
21. O'Handley, R.C.: Modern Magnetic Materials. Wiley, New York (1999)

DOI 10.1007/978-94-007-6482-8_21
Reprinted from *Hyperfine Interactions*, DOI 10.1007/s10751-013-0857-y

Synthesis and characterization of uncoated and gold-coated magnetite nanoparticles

L. León-Félix · J. Chaker · M. Parise · J. A. H. Coaquira · L. De Los Santos Valladares · A. Bustamante · V. K. Garg · A. C. Oliveira · P. C. Morais

Published online: 22 March 2013

Abstract We report on the synthesis and characterization of uncoated and gold coated magnetite nanoparticles. Structural characterizations, carried out using X-ray diffraction, confirm the formation of magnetite phase with a mean size of ~7 and ~8 nm for the uncoated and gold covered magnetite nanoparticles, respectively. The value of the gold coated Fe_3O_4 nanoparticles is consistent with the mean physical size determined from transmission electron microscopy images. Mössbauer spectra at room temperature are consistent with the thermal relaxation of magnetic moments mediated by particle-particle interactions. The 77 K Mössbauer spectra are modeled with four sextets. Those sextets are assigned to the signal of iron ions occupying the tetrahedral and octahedral sites in the core and shell parts of the particle. The room-temperature saturation magnetization value determined for the uncoated Fe_3O_4 nanoparticles is roughly ~60 emu/g and suggests the occurrence of surface effects such as magnetic disorder or the partial surface oxidation. These surface effects are reduced in the gold-coated Fe_3O_4 nanoparticles. Zero-field–cooled and

Proceedings of the Thirteenth Latin American Conference on the Applications of the Mössbauer Effect, (LACAME 2012), Medellín, Colombia, 11–16 November 2012.

L. León-Félix · M. Parise · J. A. H. Coaquira (✉) · V. K. Garg · A. C. Oliveira · P. C. Morais
Núcleo de Física Aplicada, Institute of Physics, University of Brasília, Brasília, DF 70910-900, Brazil
e-mail: coaquira.ja@gmail.com

J. Chaker
Faculdadede Ceilândia, University of Brasília, Brasília, DF 72220-140, Brazil

L. De Los Santos Valladares
Cavendish Laboratory, Department of Physics, University of Cambridge, J.J. Thomson Avenue, Cambridge CB3 0HE, UK

A. Bustamante
Laboratorio de Cerámicos y Nanomateriales, Facultad de Ciencias Físicas, Universidad Nacional Mayor de San Marcos, Ap. Postal 14-0149, Lima, Perú

field-cooled curves of both samples show irreversibilities which are consistent with a superparamagnetic behavior of interacting nanoparticles.

Keywords Au-coated magnetite nanoparticles · Mössbauer spectroscopy · Superparamagnetism

1 Introduction

Magnetite (Fe_3O_4) is one of the iron oxide phases of great interest for technological applications due to its high saturation magnetization, the highest ordering temperature among spinel ferrites and its low toxicity [1]. Magnetite nanoparticles (NPs) are interesting systems to be used in biomedical applications, bioseparation, magnetic recording media, target drug delivery, biochemical sensors, among others [2]. Therefore, the stability of iron oxide particles is impeded notably by the occurrence of spontaneous oxidation surface [3]. Recently, it has been revealed the possibility of surface passivation of magnetic nanoparticles (γ-Fe_2O_3, Fe_3O_4, Co, Fe, etc.) by a nonmagnetic shell (SiO_2, gold, silver, polymer, etc.) [4]. Core-shell structure of magnetic NPs is an attractive way to fabricate systems possessing unique physical and chemical properties [5]. Special attention has been given to gold (Au) NPs, since they show large surface area and suitable surface chemistry to be functionalized with thiolated molecules via covalent bonds [3]. Recently, core-shell gold coated magnetite NPs are reported to exhibit interesting optical properties (due to Au) and magnetic properties (due to Fe_3O_4) [5]. Therefore, Au-coated Fe_3O_4 NPs can provide a suitable platform for combine the magnetic properties of the iron oxide core with the gold shell which can be conjugated with biocompatible molecules in order to be used in biotechnological and biomedical applications.

It is known that bulk magnetite crystallizes in the cubic inverse spinel structure. The oxygen ions form a closed packed cubic structure with Fe ions localized in two different sites, octahedral and tetrahedral. Fe^{3+} ions occupy the tetrahedral sites (A) and Fe^{3+} and Fe^{2+} occupy octahedral sites (B) and are randomly arranged at room temperature because of an electron transfer process between these Fe^{3+} and Fe^{2+} ions. Below the $T_V \sim 125$ K, the electron hopping freezes out leading to an ordered array of Fe^{2+} and Fe^{3+} ions with static charge. This effect is manifested as a metal-insulator transition, which is known as the Verwey transition [6, 7]. That transition also modifies the magnetic properties and hyperfine interactions. Reports indicate that for untwined single crystals of magnetite, the low-temperature spectrum is composed of five sextets: one corresponds to Fe^{3+} ions at tetrahedral sites, two to Fe^{3+} ions at octahedral sites and two to Fe^{2+} ions at octahedral sites [8].

On the other hand, non-stoichiometric magnetite shows a reduced content of Fe^{2+} in octahedral sites. This non-stoichiometric magnetite stabilizes by the generation of cation vacancies at B sites. It is known that these vacancies screen the charge transfer and isolate the electron hopping process, since for each vacancy five Fe^{3+} ions in octahedral sites are trapped [6]. In Mössbauer spectrum these trapped Fe^{3+} ions at the octahedral sites and Fe^{3+} ions at the tetrahedral sites are indistinguishable without applying an external magnetic field [6]. Therefore, in a room-temperature Mössbauer spectrum of highly non-stoichiometric magnetite, the intensity transfer from $Fe_B^{2.5+}$ to Fe_A^{3+}-like component is observed. In this case, below

T_V the Mössbauer spectrum becomes strongly complicated. It is reported that the spectrum comprises a superposition of at least five components. For a small density of cation vacancies at B sites, the Verwey transition is shifted to lower temperatures and becomes less sharp (smoother) than in stoichiometric magnetite [6].

In the case of magnetite with reduced dimensionality, such as in nanoparticles, the surface or interface effects and particle-particle interactions strongly affect their hyperfine properties. It is difficult to distinguish between intrinsic size effects and effects related to deviation from the stoichiometry at the surface, driving to complicated Mössbauer spectra. Since at room temperature, Mössbauer spectrum shows features related to thermal relaxation, the models to fit the spectrum in order to determine the hyperfine parameters are not unique. A report suggests the use of single sextet to fit the room-temperature Mössbauer spectrum of magnetite NPs of ~10 nm [9]. More recently publications on Mössbauer and magnetic study of magnetite NPs with a mean size of ~8 nm reported the use up to 16 sextets in order to improve the fit quality of room-temperature Mössbauer spectra [10, 11].

Therefore, better knowledge of the magnetic, morphological and structural properties of coated magnetite nanoparticles is needed yet. In this work, we report the synthesis and characterization of uncoated and gold-coated Fe_3O_4 nanoparticles.

2 Experimental

The Fe_3O_4 nanoparticles were synthesized by the coprecipitation of Fe(II) and Fe(III) chlorides in aqueous solution. The solution was rinsing several times with distilled water and the black colored precipitation was collected by using a small magnet. Subsequently, 8 mL of Fe_3O_4 nanoparticles solution was added to 50 mL of 0.1 M sodium citrate and stirred at room temperature by ~30 min to get the exchange absorbed OH^- with citrate anions [3]. In order to obtain the gold coated Fe_3O_4 nanoparticles, the latter solution was heated up to the boiling point by stirring vigorously. At this point, 2 μM of $HAuCl_4$ was slowly spilled into the nanoparticles solution and cooled by stirring [12]. The resultant material was washed with water and separated by high-speed ultra-centrifugation.

In order to determine the crystalline structure and estimate the mean crystallite size X-ray powder diffraction (XRD) was performed using a commercial diffractometer (Bruker, D8 advanced) with CuK_α radiation. Transmission electron microscopy (TEM) images were obtained by using a microscope (JOEL, model 1011) to determine the morphology, the mean particles size and the size distribution. Mössbauer spectroscopy (MS) measurements were carried out using a computer-controlled Wissel spectrometer with a triangular velocity mode and using a ^{57}Co(Rh) source with an activity of ~20 mCi. A multichannel platform of 512 channels (before folding) has been used to register the spectra. 297 and 77 K spectra were obtained by cooling the sample in a commercial cryostat. A thin α-Fe foil has been used for calibration and a calibration error less than ~0.5 % was estimated. The systematic error of unfolded spectrum is estimated to be ±0.02 mm/s. Thus, the velocity resolution and systematic error for hyperfine parameters should be ±0.04 mm/s for the 512-channels spectra and ±0.08 mm/s for 256-channels spectra. Isomer shifts are given related to α-Fe, at room temperature. All spectra were computer fitted with the least-square fitting routine using the NORMOS program and assuming

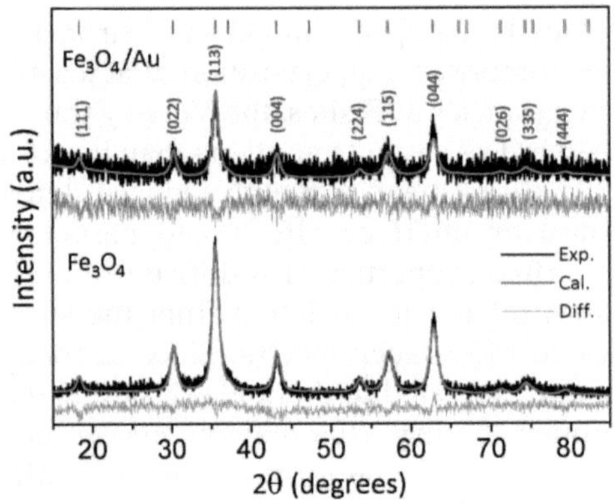

Fig. 1 Room temperature XRD patterns of the uncoated and gold-coated Fe_3O_4 nanoparticles. The observed and calculated data are represented by the *points* and *solid line*, respectively. The Bragg reflections are also indicated

Lorentzian lineshapes. The fit of Mösbauer spectra with a distribution of hyperfine fields has been carried out using the histogram method of the DIST subroutine of the NORMOS program, where the linewidth of each subspectrum is fixed to 0.3 mm/s. The magnetic properties were studied by magnetization measurements using the vibrating sample magnetometer (VSM) modulus of a physical property measurement system (PPMS) in a wide range of temperatures (from 5 to 340 K) and applying magnetic fields up to 90 kOe.

3 Results and discussion

Figure 1 shows the X-ray diffraction patterns of the uncoated and gold-coated Fe_3O_4 nanoparticles. As observed in the bottom part of Fig. 1, the Bragg reflections are consistent with the spinel structure of bulk magnetite (space group, $Fd\bar{3}m$). Although lower signal-to-noise ratio is obtained for the gold-coated Fe_3O_4 nanoparticles, no Bragg reflections corresponding to the Au crystalline phase have been determined and we assumed that the gold is located at the particle surface.

In order to estimate the particle size, XRD patterns have been refined by using the Rietveld method and modeling the peaks profile with a Pseudo-Voigt function. The particle size (D) is related to the linewidth (Γ) by the Scherrer equation: $D = 0.916\ \lambda/\Gamma \cdot \cos\theta_{hkl}$, where λ is the wavelength of the X-rays (0.15424 nm), and θ_{hkl} is the Bragg's angle [13]. The average crystallite sizes of $\langle D \rangle_{XRD} = (7.4 \pm 0.2)$ nm and (8.2 ± 0.2) nm are estimated for the uncoated and Au-coated NPs, respectively. The larger value obtained for the Au-coated NPs suggests that the gold covering process helps with the recrystallization of disordered regions likely present at the particle surface.

In order to corroborate these XRD results, TEM images were obtained. Figure 2 shows one of the TEM images obtained for the uncoated Fe_3O_4 NPs. With the help of the Imag-J software, several images have been used to count N $\sim$ 1,000 particles. Subsequently, a particle size histogram has been mounted using the Sturges method [14]. The bin-width (W) is obtained from the relation: $W = (D_{max} - D_{min})/k$, where $k = 1 + 3.322 \log(N)$. That particle size histogram (see inset in Fig. 2) was modeled with a lognormal distribution: $f(D) = (1/\sqrt{2\pi}\ \sigma D) \exp\left[-\ln^2(D/D_0)/2\sigma^2\right]$, where D_0 is the median diameter and σ represents the degree of polydispersion. The mean value

Fig. 2 TEM images of uncoated Fe_3O_4 NPs. In the inset is shown the histogram mounted as described in the text and modeled by a lognormal function

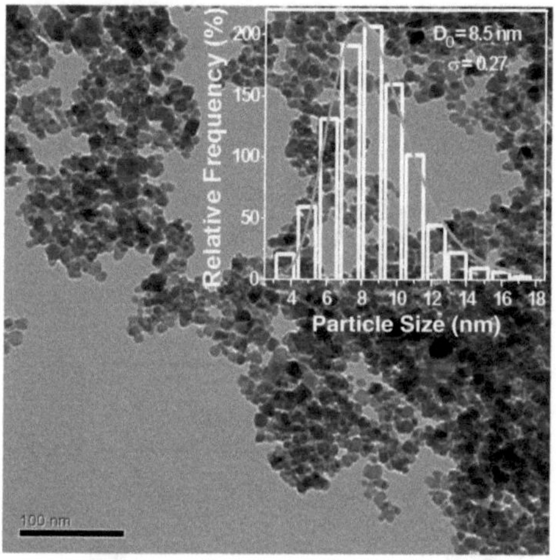

and its standard deviation are given by $< D >= D_0 e^{\sigma^2/2}$ and $\sigma_D =< D > \sqrt{e^{\sigma^2} - 1}$, respectively. Using those values, a mean particle size of $\langle D \rangle_{TEM} = (8.8 \pm 2.0)$ nm is estimated. TEM images (not shown here) obtained for the Au-coated Fe_3O_4 NPs show particle size consistent with the value obtained for the uncoated Fe_3O_4 NPs. Despite the large uncertainty, that mean value is in consistency to that obtained from XRD data of Au-coated NPs. This result implies that, after the gold surface covering, the crystalline size of the particle becomes comparable to the physical size. This result reinforces the finding that gold covering helps with the recrystallization of disordered regions likely formed at the particle surface.

Mössbauer spectra measured at room temperature and 77 K for the uncoated and Au-coated Fe_3O_4 NPs are shown in Figs. 3 and 4. At room temperature, thermal relaxation features related to the small particles size are observed. This relaxation drive to the starting collapse of magnetic sextets observed on both samples. These results are in agreement with magnetization data obtained near to 300 K (see below). In order to model the experimental data, the fit using a few number of sextets (for two sextets, one obtains a $\chi^2 \sim 1.81$) provides reasonable fits, but large linewidths ($\Gamma \sim 1.0$–2.0 mm/s). More plausible fits are obtained by using a distribution of hyperfine fields in which case a $\chi^2 \sim 1.03$ is obtained. The resultant histograms of the hyperfine field distribution are shown in Fig. 3. The broad distribution of hyperfine fields (from ~0 to 55 T) is in consistent with recent publications of Mössbauer spectroscopy results of magnetite nanoparticles with a mean size of ~8 nm, where the authors used a large number of sextets with hyperfine fields distributed in a broad range [11, 12].

The absence of a central doublet in the room temperature spectra of both spectra must be related to the strong particle-particle interaction and to the occurrence of magnetic correlations among the smallest particles. In order to compare the results obtained for both samples, the most probable component of the distribution is used. A most probable sextet with a hyperfine field of $B_{hf}^{max} = (44.5 \pm 0.4)$ T, quadrupole shift of $2\varepsilon = (-0.02 \pm 0.08)$ mm/s and IS $= (0.36 \pm 0.08)$ mm/s are obtained for the uncoated NPs. Slightly different values are obtained for the Au-coated NPs ($B_{hf}^{max} = (46.0 \pm 0.4)$ T, $2\varepsilon = (-0.00 \pm 0.08)$ mm/s and IS $= (0.38 \pm 0.08)$ mm/s).

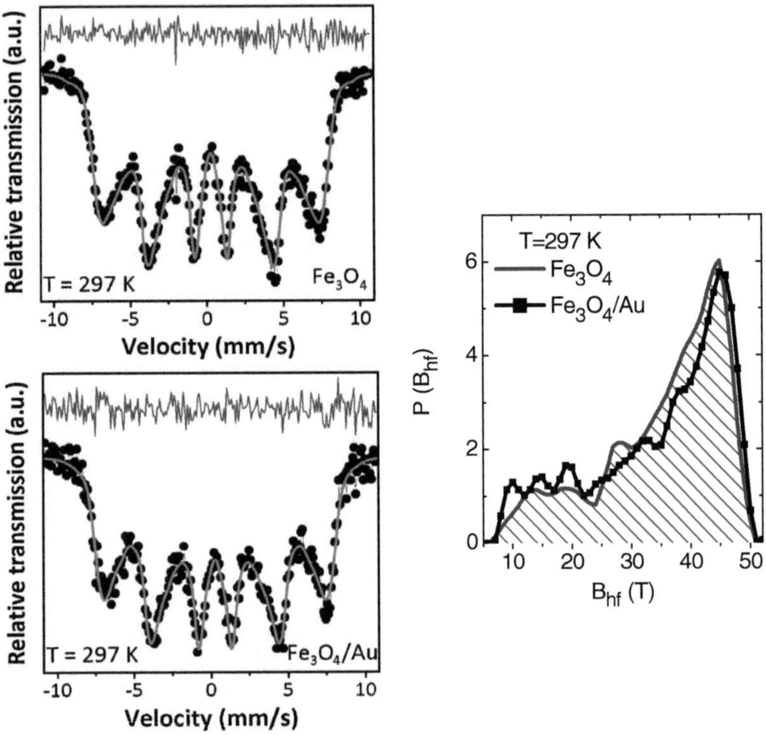

Fig. 3 Mössbauer spectra of Fe_3O_4 and Fe_3O_4/Au NPs obtained at room temperature. The points represent the experimental data and the *red line*, the calculated data. The difference between experimental and calculated data is shown at the top part of the plots. These spectra have been fitted using a distribution function of the hyperfine magnetic fields and the histograms are shown at the right-hand side

It suggests that the presence of gold ions at the particle surface slightly affects the particle-particle interaction, likely because of the partial surface coating.

At 77 K, the magnetic collapse is not else observed and the spectra of uncoated and Au-coated Fe_3O_4 NPs are well fitted by considering three sextets. $\chi^2 = 1.8$ and 2.5 are obtained for the Au-coated and the uncoated NPs, respectively. However, lower χ^2 are obtained (1.6 and 2.1 for the Au-coated and uncoated NPs, respectively) by considering four sextets (see Fig. 4). Fits considering more than five sextets provide lower χ^2, but unphysical values for the linewidth are obtained (<0.1 mm/s). As mentioned above, below 125 K, the electron transfer between Fe^{2+} and Fe^{3+} ions in B sites of bulk magnetite becomes slow and the presence of ferrous and ferric cations can be differentiated in the Mössbauer spectrum. Since it is difficult to distinguish between iron ions occupying tetrahedral or octahedral sites [8, 15], the origin of these sextets (S1, S2 and S3 components) are assigned to Fe^{3+} ions occupying both the tetrahedral and octahedral sites. Although the IS's for all sextets for uncoated NPs are the same within the experimental error the same, the lower hyperfine field and the larger linewidth of the S3 component suggest that this signal could come from the defective sites. The lowest hyperfine magnetic field of the S4 component could be assigned to iron ions located in the surface region of the

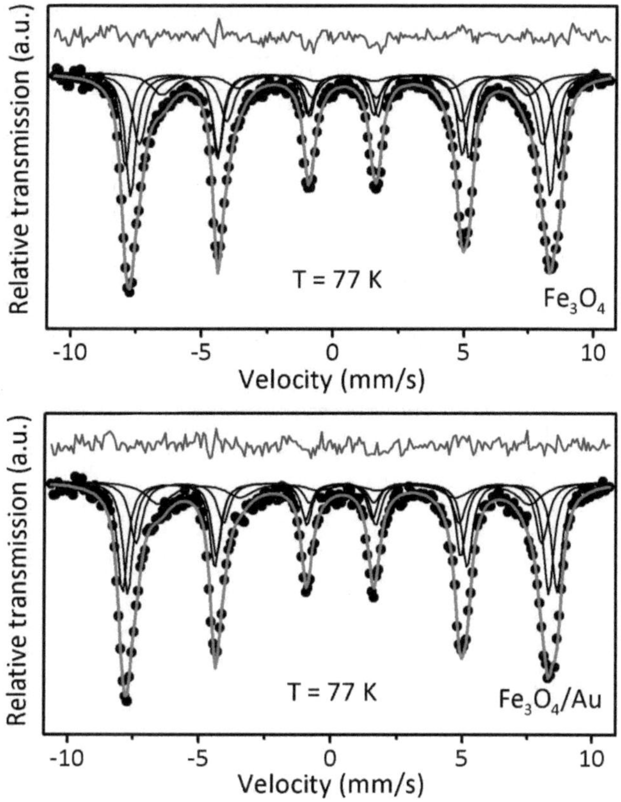

Fig. 4 Mössbauer spectra of Fe_3O_4 and Fe_3O_4/Au NPs obtained at 77 K. The *points* represent the experimental data and the *red line* represents the calculated data. The subspectra are also plotted as *continuous lines*. The difference between experimental and calculated data is shown at the top part of the plots

particles. Within the experimental resolution, the spectral areas of all components (see Table 1) seem to remain the same in both samples. Although the spectral area of the S4 component seems to remain constant, the isomer shift shows a larger value for the Au-coated NPs. This increase in the IS could be assigned to the surface coating effects, since it seems that Au coating drives to the partial recrystallization of the particle surface. These results are preliminary and further study with more resolved Mössbauer spectroscopy measurements will permit to clarify this issue better.

Magnetizations (M) vs. magnetic field (H) curves have been obtained for both Fe_3O_4 and Fe_3O_4/Au NPs at 300 and 5 K. The M vs. H curves of uncoated and Au-coated NPs are shown in Fig. 5a. At 300 K, the saturation magnetization (Ms), determined by using the law of approach to saturation (a linear term is included in order to account for the linear increase of M at high fields), is found to be ~61 emu/g for the uncoated Fe_3O_4 NPs. This saturation magnetization value is consistent with the value reported in the literature for uncoated magnetite NPs of around the same mean size (~8 nm) [11]. Slightly larger value is obtained for the Fe_3O_4/Au NPs (~63 emu/g), which is consistent with the Mössbauer results as previously discussed. The increase of the saturation magnetization indicates that the Au covering of the NPs helps to reduce magnetically disordered regions, likely at the particle surface. Similar effects have been reported for surface-passivated magnetite nanoparticles which have been assigned to the improvement of the crystalline quality [11]. Moreover, those saturation magnetization values are significantly lower than the value

Table 1 List of hyperfine parameters, linewidths (Γ) and spectral areas (A) obtained from the fit of 77-K Mössbauer spectra of Fe_3O_4 and Au-coated Fe_3O_4

Sample	Site	IS (mm/s)	2ε (mm/s)	Bhf (T)	Γ (mm/s)	A (%)
Fe_3O_4	Fe^{3+} (S1)	0.48	−0.01	51.3	0.42	28
	Fe^{3+} (S2)	0.40	0.01	49.3	0.48	33
	Fe^{3+} (S3)	0.48	−0.02	47.8	0.66	27
	Fe′ (S4)	0.56	0.01	43.3	0.74	12
Fe_3O_4/Au	Fe^{3+} (S1)	0.47	−0.01	51.3	0.48	33
	Fe^{3+} (S2)	0.41	0.01	49.7	0.47	31
	Fe^{3+} (S3)	0.47	−0.05	47.9	0.56	21
	Fe″ (S4)	0.67	−0.10	43.8	0.70	15

The error of the spectral areas is estimated to be not more than 10 %

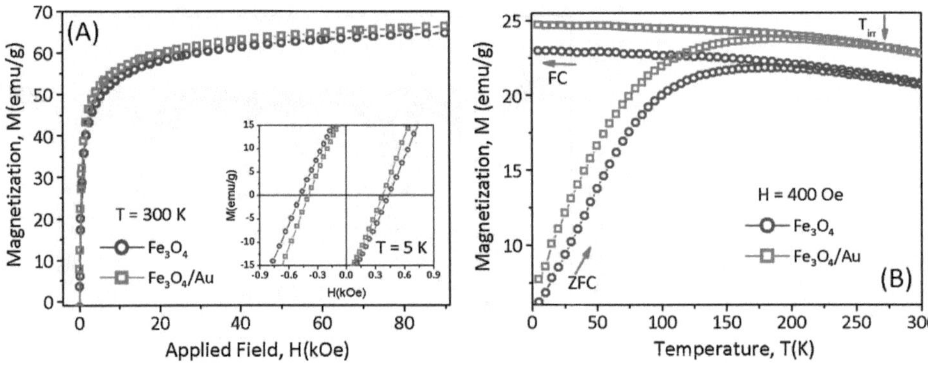

Fig. 5 **a** Hysteresis loops of Fe_3O_4/Au NPs at $T = 5$ K and room temperature. **b** Temperature dependence of zero-field-cooled (ZFC) and field-cooled (FC) curves of the Fe_3O_4 and Fe_3O_4/Au NPs

reported for bulk Fe_3O_4 (Ms ∼ 82 emu/g). These lower saturation magnetizations must be related to the surface effects [16, 17] or surface oxidation [6]. Those effects must be the responsible for the increasing tendency of the M vs. H curve observed in the high-field region.

At $T = 5$ K, the coercive field (H_C) determined for the uncoated Fe_3O_4 NPs is ∼460 while that determined for the Fe_3O_4/Au NPs is ∼390 Oe. The lower coercive field of the Fe_3O_4/Au NPs can be related to the recrystallization process of amorphous regions fulfilled by the gold shell covering the particles surface. Moreover, at 300 K, the coercive field of both samples decreases to $H_C \sim 12$ Oe. Since a zero coercive field is expected for the superparamagnetic behavior of small particles, the nonzero coercive field must be associated to the particle-particle interactions which delay the thermal relaxation of the magnetic moments of the particles [18].

Zero-field cooled (ZFC) and field cooled (FC) curves obtained with $H = 400$ Oe for the Fe_3O_4 and Fe_3O_4/Au NPs are shown in Fig. 5b. No evidences for the occurrence of the Verwey transition are observed in the temperature range from 5 to 300 K. The irreversible behavior of ZFC and FC curves reveals a typical thermal blocking process of superparamagnetic NPs. The ZFC curve shows a broad maximum at $T_m = 179 \pm 3$ K and 189 ± 3 K for the Fe_3O_4 and Fe_3O_4/Au NPs, respectively. In order to determine whether the studied samples can behave as superparamagnetic systems, the magnetic particle size has been estimated. According to the Ref. [19] the mean magnetic sizes is given by: $D_{MAG} = \left(18 k_B T \chi_i / \pi\, \rho\, M_s^2\right)^{1/3}$, where $\chi_i = (dM/dH)_{H \to 0}$ is the initial susceptibility, k_B is the Boltzmann constant, Ms is the saturation magnetization and ρ is the mass density of magnetite. A value of $D_{MAG} \sim$ 9 nm, is obtained for both studied samples, which is in good agreement with the value obtained from XRD and TEM experiments. This result strongly suggests that the Fe_3O_4 NPs (uncoated and Au-coated) behave as weakly interacting entities. Therefore, the maximum of the ZFC curve can be related to the mean value of the blocking temperature (T_B). Considering a hypothetical system of monodispersed particles, then the relationship between these two temperatures is given by: $T_m = \beta \langle T_B \rangle$, where β is a parameter whose value is in the range of 1.4–1.6 [20]. Using $\beta = 1.5$ one can obtain $\langle T_B \rangle = 119 \pm 3$ K and $\langle T_B \rangle = 126 \pm 3$ K for the Fe_3O_4 and Fe_3O_4/Au NPs, respectively. The slightly larger T_B obtained for the Au-coated NPs, must be related to the larger particle size obtained from XRD and TEM data analysis.

One can estimate the anisotropy constant by using the relation $K_{an} V \gamma = 25\, k_B T_B$, which is valid for a random distribution of a single domain magnetic particle. Here γ is a constant expected to be depend on the crystal size distribution ($\gamma = 9.2$) [21], K_{an} is the effective anisotropy constant and V is the volume the particle. Considering a $T_B \sim 120$ K and $D = 88$ nm for the uncoated NPs, an affective anisotropy constant of $K_{an} \sim 1.6 \times 10^5$ erg/cm^3 is determined. This value is close to that one reported in the literature for Fe_3O_4 NPs [22].

4 Conclusions

Structural characterization confirms the successful production of ~8-nm gold-coated Fe_3O_4 NPs. This mean particle size is corroborated by transmission electron microscopy. Room-temperature Mössbauer spectra of the uncoated and Au-coated Fe_3O_4 NPs are consistent with a distribution of magnetic particles which shows thermal relaxation of magnetic moments mediated by particle-particle interactions. However, at 77 K the Mössbauer spectra are well-modeled by four sextets, which are assigned to the signal of iron ions occupying tetrahedral and octahedral sites in the core and shell regions of the particles. The presence of gold on the particle surface produces effects which seem to be associated with the recrystallization of the shell region. Saturation magnetization values are lower than that of bulk magnetite and suggest the occurrence of surface effects which become important due to the large surface-to-volume ratio. The non-zero coercive field determined at room temperature indicates that the superparamagnetic behavior of NPs is mediated by the occurrence of particle-particle interactions.

Acknowledgements This work was financially supported by the Brazilian agencies CNPq and CAPES. JAHC thanks to DPP/UnB for the partial support.

References

1. Lima, Jr. E., Brandl, A.L., Arelaro, A.D., Goya, G.F.: J. Appl. Phys. **99**, 083908 (2006)

2. Zhou, X., Xu, W., Wang, Y., Kuang, Q., Shi, Y., Zhong, L., Zhang, Q.: J. Phys. Chem. C **114**, 19607 (2010)
3. Lyon, J.L., Fleming, D.A., Stone, M.B., Schiffer, P., Williams, M.E.: Nano Lett. **4**, 719 (2004)
4. Wu, W., He, Q., Chen, H., Tang, J., Nie, L.: Nanotechnology **18**, 145609 (2007)
5. Pal, S., Morales, M., Mukherjee, P., Srikanth, H.: J. Appl. Phys. **105**, 07B504 (2009)
6. Korecki, J., Handke, B., Spiridis, N., Slezak, T., Flis-Kabulska, I., Haber, J.: Thin Solid Films **412**, 14 (2002)
7. Voogt, F.C., Fujii, T., Smulders, P.J.M., Niesen, L., James, M.A., Hibma, T.: Phys. Rev B **60**, 11193 (1999)
8. Hargrove, R.S., Kündig, W.: Solid State Comm. **8**, 303 (1970)
9. Goya, G.F., Berquó, T.S., Fonseca, F.C., Morales, M.P.: J. Appl. Phys. **94**, 3520 (2003)
10. Oshtrakh, M.I., Sepelák, V., Rodriguez, A.F.R., Semionkin, V.A., Ushakov, M.V., Santos, J.G., Silveira, L.B., Marmolejo, E.M., De Souza Parise, M., Morais, P.C.: Spectrochim. Acta, Part A **100**, 94 (2013)
11. Oshtrakh, M.I., Ushakov, M.V., Semenova, A.S., Kellerman, D.G., Šepelák, V., Rodriguez, A.F.R., Semionkin, V.A., Morais, P.C.: Hyperfine Interact (2013). doi:10.1007/s10751-012-0666-8
12. Loaiza, Ó.A., Jubete, E., Ochoteco, E., Cabañero, G., Grande, H., Rodríguez, J.: Biosens. Biolectron. **26**, 2194 (2011)
13. Cullity, B.D., Stock, S.R.: Elements of X-ray Diffraction, 3rd edn. Prentice-Hall, Upper Saddle River, NJ (2001)
14. Aragón, F.H., de Souza, P.E.N., Coaquira, J.A.H., Hidalgo, P., Gouvêa, D.: Physica B **407**, 2601 (2012)
15. Presniakov, I.A., Sobolev, A.V., Baranov, A.V., Demazeau, G., Rusakov, V.S.: J. Phys., Condens. Matter **18**, 8943 (2006)
16. Gee, S.H., Hong, Y.K., Erickson, D.W., Park, M.H., Sur, J.C.: J. Appl. Phys. **93**, 7560 (2003)
17. Morales, M.P., Veintemillas Verdaguer, S., Montero, M.I., Serna, C.J., Roig, A., Casas, Ll., Martínez, B., Sandiumenge, F.: Chem. Mater. **11**, 3058 (1999)
18. Cho, S.-J., Idrobo, J.-C., Olamit, J., Liu, K., Browning, N.D., Kauzlarich, S.M.: Chem. Mater. **17**, 3181 (2005)
19. Carpenter, E.E.: J. Magn. Magn. Mater. **225**, 17 (2001)
20. Coaquira, J.A.H., Vaccari, C.B., Tedesco, A.C., Morais, P.C.: IEEE Trans. Magn. **45**, 4059 (2009)
21. Mikhaylova, M., Kim, D.K., Bobrysheva, N., Osmolowsky, M., Semenov, V., Tsakalakos, T., Muhammed, M.: Langmuir **20**, 2472 (2004)
22. Jonsson, T., Mattsson, J., Djurberg, C., Khan, F.A., Nordblad, P., Svedlindh, P.: Phys. Rev. Lett. **75**, 4138 (1995)

DOI 10.1007/978-94-007-6482-8_22
Reprinted from *Hyperfine Interactions*, DOI 10.1007/s10751-013-0850-5

Size dependence of the magnetic and hyperfine properties of nanostructured hematite (α-Fe_2O_3) powders prepared by the ball milling technique

J. André-Filho · L. León-Félix · J. A. H. Coaquira · V. K. Garg · A. C. Oliveira

Published online: 22 March 2013

Abstract In this work we present the study of hematite (α-Fe_2O_3) nanostructures synthesized by the ball milling technique. The structural characterization and the crystallite size estimation have been carried out using the X-ray diffraction (XRD) technique. Data analyses indicate that the hematite phase (space group, R-3C) is preserved after the milling process. As the milling time is increased, a second phase (α-Fe) appears. The mean crystallite size shows a decreasing tendency as the milling time is increased. High-resolution transmission electron microscopy (HRTEM) images show the formation of grains composed of crystallites with irregular shapes. Mössbauer spectra of milled powders carried out at 297 and 77 K are well modeled with a histogram distribution of hyperfine fields. The presence of one additional sextet which corresponds to the α-Fe phase is also determined in agreement with XRD data analysis. Magnetic measurements suggest the suppression of the Morin transition in the milled samples and the absence of thermal relaxation effects in agreement with the Mössbauer spectroscopy results.

Keywords Hematite nanoparticles · Ball-milling technique · Mössbauer spectroscopy

1 Introduction

The need to detect leaks from hydrogen-powered machines in order to prevent explosion hazards makes it necessary to develop improved hydrogen sensors. The most important requirement for a practical hydrogen sensor is the ability to quickly

Proceedings of the Thirteenth Latin American Conference on the Applications of the Mössbauer Effect, (LACAME 2012), Medellín, Colombia, 11–16 November 2012.

J. André-Filho · L. León-Félix · J. A. H. Coaquira (✉) · V. K. Garg · A. C. Oliveira
Núcleo de Física Aplicada, Instituto de Física, Universidade de Brasília,
Brasília, DF 70910-900, Brazil
e-mail: coaquira.ja@gmail.com

respond to the presence of hydrogen before the explosive limit (~4% H_2 in air) is attained. Canonical sensors employ the systematic changes in the electrical properties of oxide semiconductors to detect hydrogen gas [1]. However, they suffer from low sensitivity, serious difficulties associated with the complex electrical contacts, especially for uses in mobile environments and reactive atmospheres. The use of nanoparticles with very large surface-to-volume ratios can significantly improve the gas sensitivity because the interaction between the gas and the sensing material mainly occurs at the particles surface. Therefore, the development of new gas sensing methods, which can effectively utilize the large surface-to-volume ratio of nanoparticulated systems, is nowadays needed. This will allow the rapid miniaturization of current bulky gas sensors, allowing them to be used in mobile micro and nanoscale devices, for instance in hydrogen fueled automobiles [1]. Bulk antiferromagnets are magnetically inactive materials due to the antiparallel arrangement of spins. Antiferromagnetic (AF) order in metal oxides, such as hematite (α-Fe_2O_3), results from the superexchange interaction between the Fe^{3+} spins mediated by neighboring oxygen ions. A considerable fraction of the spins on the surface are uncompensated; however, their contribution is insignificant in bulk materials. When AF materials are prepared in nanoparticle form, the role of the uncompensated spins at the particle surface becomes the main magnetic contribution. It is known that the exact magnitudes of the magnetic parameters could vary with the oxygen stoichiometry and/or other defects [1–3]. Therefore, it is expected that the oxygen stoichiometry and/or surface defect density of hematite can be altered by an oxidizing or reducing atmosphere.

Bulk hematite present the Morin transition ($T_M \sim 260$ K), which is the magnetic change from a high-temperature weakly ferromagnetic phase to a low-temperature antiferromagnetic phase caused by spin canting. The crystal structure of hematite is similar to that of corundum (Al_2O_3) and it can be described as a rhombohedral structure with R-3c space group. Moreover, it has been reported that the Morin transition depends on crystallite size, impurities, strain and defects [4–7].

In this work, we study the effects of the grinding, in air atmosphere, on the structural and magnetic properties of hematite powders ground by the ball milling technique. The structural and microstructural parameters of hematite and the particle size dependence on the milling time have been analyzed by X-ray diffraction (XRD) technique and Transmission Electron Microscopy (TEM). The magnetic properties of the samples have been studied by magnetic and Mössbauer spectroscopy measurements.

2 Experimental

Bulk polycrystalline *hematite* in the form of a *commercial hematite* (α-Fe_2O_3) powder has been used to prepare the samples.. The milling process has been carrying out using a high energy ball mill (SPEX 8000). Hardened steel vial and balls have been used and a ball-to-powder weight ratio of ~ 4:1. In order to determine the particle size dependence on the milling time, five samples with different milling times have been prepared using t = 0, 1, 2, 5 and 10 h as the milling times. The structural properties of the resultant powders have been studied by using a commercial X-ray diffraction (XRD) diffractometer (Bruker D8 Focus) with Cu-Kα radiation.

The diffractograms were recorded in the 2θ range from 15° to 90° and a step size of 0.05°. The corroboration of the particle size and the morphology has been obtained by transmission electron microscopy (TEM) images. Mössbauer spectroscopy (MS) measurements were performed at room temperature and at 77 K, using a constant acceleration spectrometer with a triangular waveform using a $^{57}Co/Rh$ source. Isomer shifts are quoted relative to α-Fe. Magnetic measurements were carried out using the vibrating sample magnetometer (VSM) unit of a physical property measurement system (PPMS, from Quantum Design) in abroad range of temperature (from 5 to 300 K) and applying magnetic field up to ±90 kOe.

3 Results and discussion

In Fig. 1 is shown the XRD patterns of three samples. The observed Bragg reflections are consistent with the hematite phase (space group, R-3C) for all samples. Two main features are observed as the milling time is increased. The Bragg reflections become broader as the milling time is increased and some extra reflections arise after the milling process. In order to further analyze the patterns, the Rietveld refinement method has been used. During the refinement the R_{WP}/R_{EXP} (the R-weighted to R-expected ratio) was used to observe the convergence of the fitting parameters and to obtain a good fitting [8]. It has been determined that the extra reflections correspond to the metallic iron phase, which mass concentration is not more than ~2%. That impurity phase concentration does not show a clear dependence on the milling time. The origin of that extra phase has been assigned to the release of iron from the vial set. The Bragg reflection broadening has been associated to the reduction of the coherent diffraction domains (crystallites).

Using a Lorentzian function as the profile of Bragg reflection peaks and the Scherrer equation [9]: $D = 0.916\,\lambda/\Gamma \cdot \cos\theta_{hkl}$, where λ is the wavelength of the X-rays (0.15418 nm), Γ is the diffraction line broadening (in radians), and θ_{hkl} is the Bragg's angle, the mean value of the crystallites ($<D>$) can be estimated. Lattice constants and $<D>$ size obtained from the Rietveld refinement are listed in Table 1. As can be inferred from Table 1, the particle size shows a rapid decrease as the milling time is increased and becomes ~10 nm after 10 h of milling time. This mean value of D has been corroborated by high resolution TEM (HRTEM) images. In Fig. 2 is shown one HRTEM image for the hematite powder milled during 10 h. As observed, there are clustcred nanostructures of irregular shapes which sizes are not more than ~16–17 nm. Those nanostructures form clusters (grains) with size up ~80 nm. The nanostructures show larger size when compared to the mean size of the crystallites estimated from XRD data and it suggests that the hematite nanostructures contain noncrystalline portions, likely located at the interface of the crystallites.

In Fig. 3 is shown the Mössbauer spectra of three hematite powders obtained at room temperature. The spectra show well-resolved magnetic sextets at room temperature and at 77 K (not shown here). This implies that the thermal relaxation effects associated to the nanostructured samples are not present. However, due to the asymmetric shape and the peaks broadening with the milling time, the fits of the spectra are carried out using a histogram distribution of sextets for the milled samples. For the latter samples, the inclusion of a sextet consistent with the

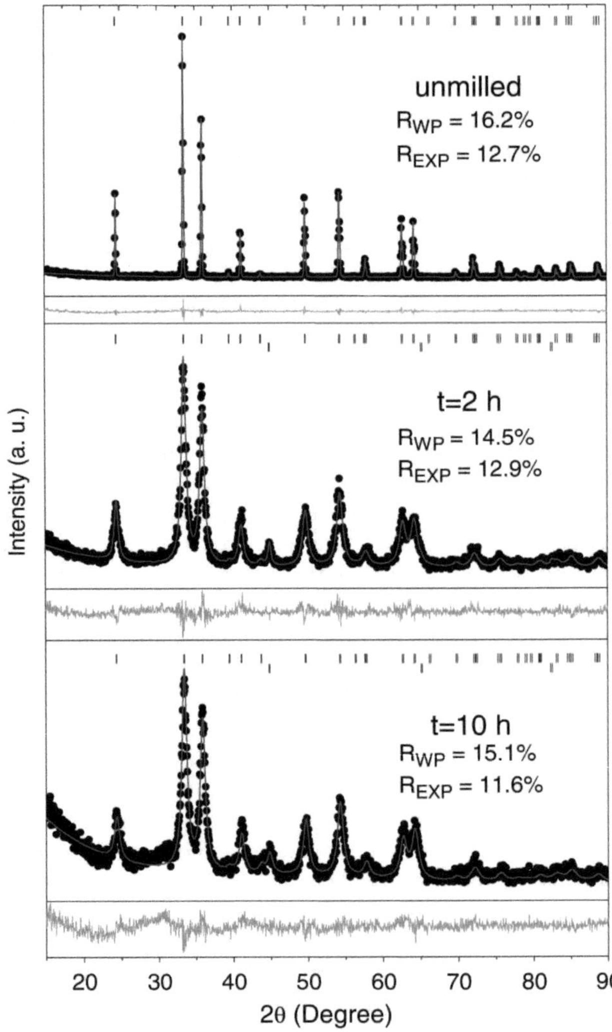

Fig. 1 X-ray diffraction patterns of the hematite powders milled at different times. The R-factors which represent the refinement quality are included

Table 1 Mean size of the coherent diffraction domains and lattice constants and volume of the unit cell obtained from the Rietveld refinement are listed for the milled hematite powders

Milling time (h)	<*D*> (nm)	a = b (Å)	c (Å)	Volume (Å^3)	R_{WP}/R_{Exp}
unmilled	few μm	5.0348	13.7483	301.8	1.3
1	13	5.0379	13.7634	302.5	1.2
2	12	5.0392	13.7561	302.5	1.1
5	9	5.0410	13.7677	303.0	1.2
10	10	5.0405	13.7659	302.9	1.3

The quality factor ratios are also included

presence of α-Fe has been needed, in agreement with the structural analysis. In Table 2 are listed the hyperfine parameters, the distribution width and spectral areas determined from the fits. In agreement with the structural study, a small percentage

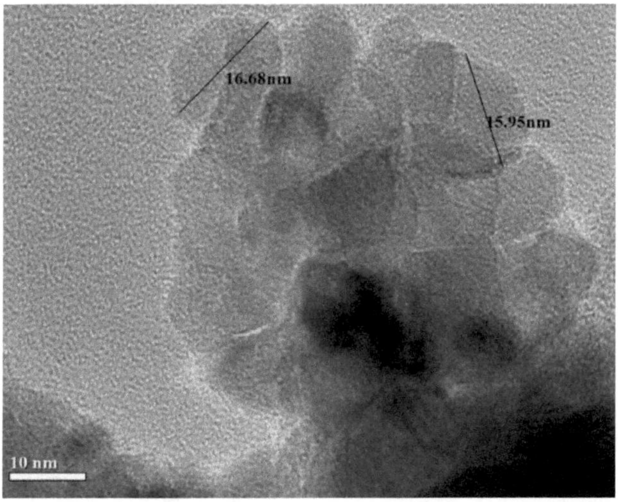

Fig. 2 High resolution transmission electron microscopy (HRTEM) image of the hematite powder milled during 10 h

of metallic iron has been detected as shown in Table 2. The room temperature and 77 K Mössbauer spectra of the un-milled powder were fitted with only one sextet. The isomer shifts values suggest the presence of only Fe^{3+} ions in all samples in accordance with the reported values of a typical α-Fe_2O_3 phase [10]. For the unmilled samples, the change in signal of the quadrupole shift from room temperature to 77 K (from $2\varepsilon = -0.199$ mm/s to 0.383 mm/s) evidences the occurrence of the Morin transition, which separates a weakly ferromagnetic phase at room temperature from an antiferromagnetic phase at 77 K [11, 12]. Since one expects that the crystalline grains are not homogeneous, attempts to fit the Mössbauer spectra of the milled samples with a sextet, which represents the non-defective hematite component, and the histogram distribution, which represents the disordered hematite regions, fail. Therefore, fits obtained with only a histogram distribution are considered for the analysis.

Meanwhile, the δ of the milled samples shows values similar to the unmilled sample, at the respective temperatures, the 2ε of the histogram distribution shows negative values at both temperatures (~ -0.14 mm/s at 77 K and -0.16 mm/s at 297 K). This implies that the structural disorder introduced by the milling process is large enough to inhibit the Morin transition observed in the unmilled sample. Moreover, the increase of the lattice constants with increasing the milling time, as inferred from Table 1, must be related to those structural modifications. The increase of lattice constants of hematite powders with the milling time has been reported in the literature [14]. Other effect of the milling process must be the evolution of the average value of the hyperfine field, since smaller values are determined as the milling time is increased (see Table 2). The latter dependence of the hyperfine field on the mean crystallite size (i.e., on the milling time) is in agreement with other reports on ball-milled hematite powders [15].

No evidence for the transition from the high-temperature weak ferromagnetic phase to the low-temperature antiferromagnetic phase is observed for the milled samples in comparison to the un-milled one as shown in Fig. 4. This is in agreement with the absence of signal change in the 2ε obtained from Mössbauer spectra of

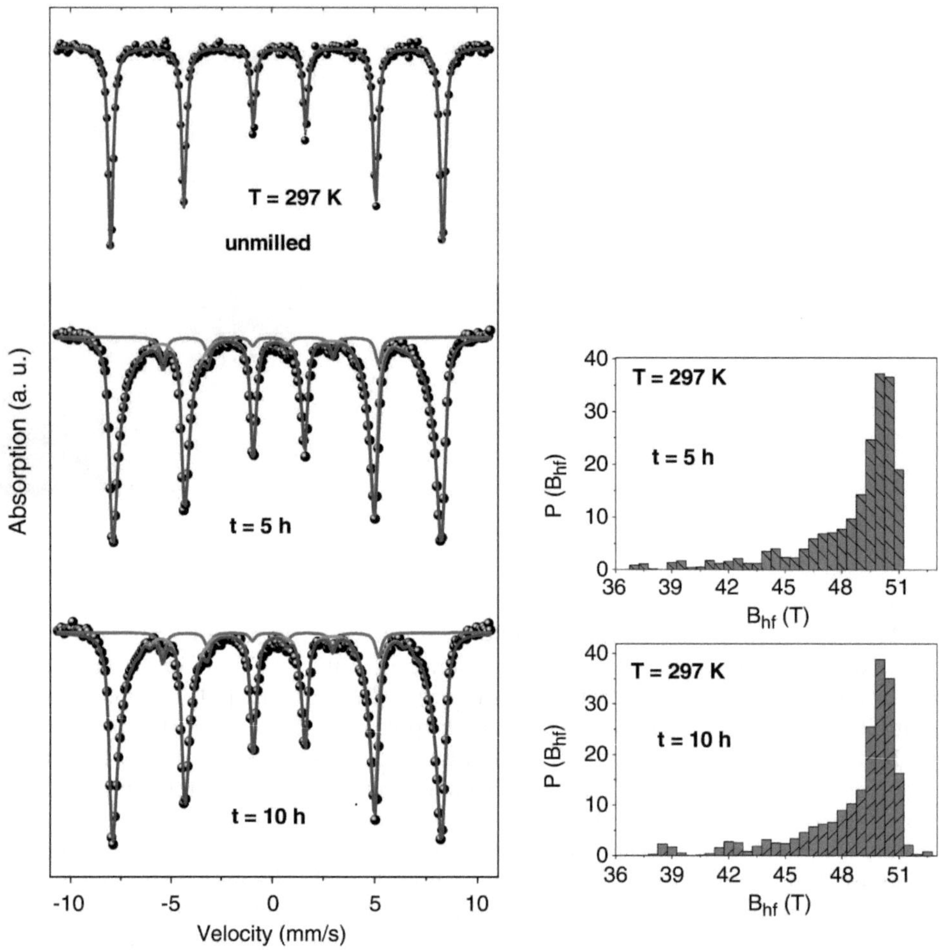

Fig. 3 Mössbauer spectra of un-milled and milled hematite powders obtained at room temperature

Table 2 Mössbauer parameters and spectral areas (area) determined from the spectra fit of the unmilled and milled powders

Milling time (h)	Hematite T (K)	δ (mm/s)	2ε (mm/s)	$<B_{hf}>$ (T)	W (T)	Area (%)	α-Fe Area (%)
Unmilled[a]	77	0.46	0.38	53.0	–	100	0
	297	0.36	−0.20	50.7	–	100	0
1	77	–	–	–	–	–	–
	297	0.36	−0.18	49.3	2.5	96	4
5	77	0.47	−0.13	51.6	1.6	94	6
	297	0.36	−0.16	48.5	2.8	93	7
10	77	0.47	−0.14	51.2	2.4	95	5
	297	0.35	−0.16	48.6	2.9	93	7

$<B_{hf}>$ and W are the average values of the hyperfine field and the width of the histogram distribution, respectively. The isomer shift (δ) is given with respect to metallic iron. Errors: ±0.01 mm/s, ±0.1 T and ±2 %

[a]The fit of unmilled sample was carried out with only one sextet

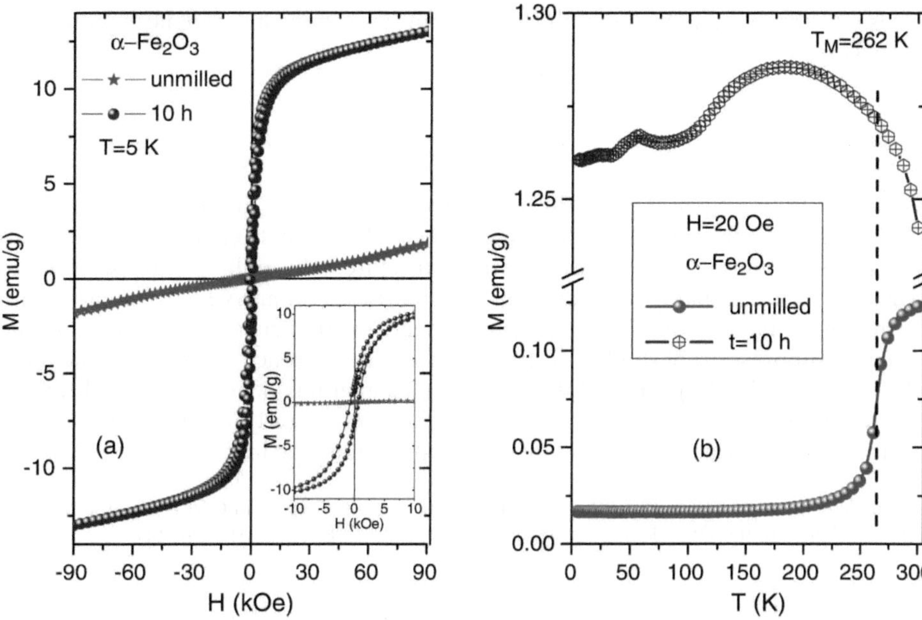

Fig. 4 **a** Magnetization (M) vs. magnetic field (H) curves of the un-milled and milled hematite powder is shown for comparison. **b** Magnetization (M) vs. temperature (T) curves are shown for comparison

milled powders. Magnetization (M) vs. magnetic field (H) curve of the milled sample obtained at 5 K (see Fig. 4a) shows ferromagnetic contribution with a saturation magnetization of $M_S = 11$ emu/g and a coercive field of $H_C \sim 760$ Oe. At room temperature, those values are $M_S = 10$ emu/g and $H_C \sim 240$ Oe. That ferromagnetic contribution is accompanied by a linear increase of M at high fields. These features suggest that the milling process leads to the rearrangement of magnetic moments of iron ions and drives to the suppression of the Morin transition, likely associated to the structural disorder. These results are in agreement with reports on the literature which suggest that the Morin transition is suppressed in hematite nanoparticles with a size below ~20 nm [13]. The ferromagnetic signal could be assigned to two sources: i) the population of spins located in the disordered regions, likely at the crystallite interfaces, which are not perfectly antiparallel to each other and becomes stronger as the disordered regions grows and ii) the small percentage of nanostructured metallic iron which is expected to show ferromagnetic order. On the other hand, despite the nanostructured nature of the milled powders, no evidences of thermal relaxation have determined from both magnetization and Mössbauer spectroscopy measurements up to 300 K. Reports on hematite powders synthetized by the ball milling technique indicate that the blocking temperature strongly depends on the interparticle interactions [16]. In that work, it is shown that the room-temperature Mössbasuer spectra do not exhibit features related to the thermal relaxation. However, after an ultrasonic separation of the clustered crystallites, the thermal relaxation effects appear [16]. Therefore, the results obtained for the milled powders studied in this work suggest that the interactions of crystallites inside the grains are strong enough to inhibit the thermal relaxations in agreement with other works [13].

4 Conclusions

We successfully prepared hematite nanoparticles by a ball milling technique. The XRD data analysis and HRTEM images show the formation of inhomogeneous crystallites forming clusters with hematite structure. The milling process leads to structural disorder, likely developed in the interface regions, which increases with the milling time. Mössbauer spectra and magnetization measurements show the complete suppression of the Morin transition and the absence of thermal relaxation effects in the milled samples. A ferromagnetic signal has been determined for the milled samples, which has been assigned to non-collinearly oriented spins likely located in the interface and to the small fraction of metallic iron.

Acknowledgements The authors thank the Brazilian agencies CNPq and CAPES for the financial support. Thanks to the Dr. J. Dias for carrying out the X-ray diffraction experiments.

References

1. Punnoose, A., Reddy, K.M., Thurber, A., Hays, J., Engelhard, M.H.: Nanotechnology **18**, 165502 (2007)
2. Dong, Q., Kumada, N., Yonesaki, Y., Takei, T., Kinomura, N., Wang, D.: J. Mater. Sci. **45**, 5685 (2010)
3. Cregan, R.F., Mangan, B.J., Knight, J.C., Birks, T.A., Russell, P.St.J., Roberts, P.J., Allan, D.C.: Science **285**, 1537 (1999)
4. Tadić, M., Čitaković, N., Panjan, M., Stojanović, Z., Marković, D., Spasojević, V.: J. Alloys Compd. **509**, 7639 (2011)
5. Williamson, D.L., Venturini, E.L., Graham, R.A., Morosin, B.: Phys. Rev. B **34**, 1899 (1986)
6. Cinader, G., Flanders, P.J., Shtrikman, S.: Phys. Rev. **162**, 419 (1967)
7. Bødker, F., Hansen, M.F., Koch, C.B., Lefmann, K., Mørup, S.: Phys. Rev. B **61**, 6826 (2000)
8. Rietveld, H.M.: J. Appl. Crystallogr. **2**, 65 (1969)
9. Cullity, B.D., Stock, S.R.: Elements of X-ray Diffraction, 3rd edn. Prentice-Hall, Upper Saddle River, NJ (2001)
10. Randrianantoandro, N., Mercier, A.M., Hervieu, M., Grenèche, J.M.: Mater. Lett. **47**, 150 (2001)
11. Van San, E., De Grave, E., Vandenberghe, R.E.: J. Magn. Magn. Mater. **269**, 54 (2004)
12. Gee, S.H., Hong, Y.-K., Sur, J.C., Erickson, D.W., Park, M.H., Jeffers, F.: IEEE Trans. Magn. **40**, 2691 (2004)
13. Sánchez, L.C., Arboleda, J.D., Saragovi, C., Zysler, R.D., Barrero, C.A.: Physica B **389**, 145 (2007)
14. Florez, J.M., Mazo-Zuluaga, J., Restrepo, J.: Hyperfine Interact **165**, 253 (2005)
15. Bhowmik, R.N., Saravanan, A.: J. Appl. Phys. **107**, 053916 (2010)
16. Borzi, R.A., Stewart, S.J., Punte, G., Mercader, R.C., Vasquez-Mansilla, M., Zysler, R.D., Cabanillas, E.D.: J. Magn. Magn. Mater. **205**, 234 (1999)

DOI 10.1007/978-94-007-6482-8_23
Reprinted from *Hyperfine Interactions*, DOI 10.1007/s10751-013-0858-x

Magnetic composites from minerals: study of the iron phases in clay and diatomite using Mössbauer spectroscopy, magnetic measurements and XRD

M. Cabrera · J. C. Maciel · J. Quispe-Marcatoma · B. Pandey · D. F. M. Neri · F. Soria · E. Baggio-Saitovitch · L. B. de Carvalho Jr

Published online: 28 March 2013

Abstract Magnetic particles as matrix for enzyme immobilization have been used and due to the enzymatic derivative can be easily removed from the reaction mixture by a magnetic field. This work presents a study about the synthesis and characterization of iron phases into magnetic montmorillonite clay (mMMT) and magnetic diatomaceous earth (mDE) by ^{57}Fe Mössbauer spectroscopy (MS), magnetic measurements and X-ray diffraction (XRD). Also these magnetic materials were assessed as matrices for the immobilization of invertase via covalent binding. Mössbauer spectra of the magnetic composites performed at 4.2 K showed a mixture

Proceedings of the Thirteenth Latin American Conference on the Applications of the Mössbauer Effect, (LACAME 2012), Medellín, Columbia, 11 16 November 2012.

M. Cabrera · J. C. Maciel · L. B. de Carvalho Jr (✉)
Laboratório de Imunopatologia Keizo Asami (LIKA), Universidade Federal de Pernambuco, Cidade Universitária, PE CEP 50670-901, Recife, PE, Brazil
e-mail: lbcj@hotlink.com.br

M. Cabrera · F. Soria
Instituto de Investigaciones para la Industria Química, Universidad Nacional de Salta - CONICET, Buenos Aires N° 177, 4400, Salta, Argentina

J. Quispe-Marcatoma · B. Pandey · E. Baggio-Saitovitch
Centro Brasileiro de Pesquisas Físicas, Urca, 22290-180, Rio de Janeiro, RJ, Brazil

B. Pandey
Department of Applied Science, Symbiosis Institute of Technology, Mulsi, Pune 412 115, India

D. F. M. Neri
Universidade Federal do Vale de São Francisco, Campus Petrolina, 56304-917, Petrolina, PE, Brazil

L. B. de Carvalho Jr
Departamento de Bioquímica, Centro de Ciências Biológicas, Universidade Federal de Pernambuco, Cidade Universitária, 50670-901, Recife, PE, Brazil

of magnetite and maghemite about equal proportion in the mMMT, and a pure magnetite phase in the sample mDE. These results were verified using XRD. The residual specific activity of the immobilized invertase on mMMT and mDE were 83 % and 92.5 %, respectively. Thus, both magnetic composites showed to be promising matrices for covalent immobilization of invertase.

Keywords Magnetic particles · Montmorillonite · Diatomite · Immobilization · Invertase

1 Introduction

Inorganic materials have been widely used as carriers for enzyme immobilization. Their advantages are rigid structure, durability, high mechanical strength and relatively low cost [1]. Montmorillonite belongs to the smectite clays and its crystal structure consists of two tetrahedral silicate layers with an edge-shared octahedral layer of either alumina or magnesia [2]. On the other hand, diatomaceous earths or diatomite are mineral deposits of diatomaceous algae and are the major silica source on earth [3]. These minerals have many properties that make them interesting matrices for the immobilization of proteins. Some of them are chemical inertness, large surface area, high porosity and mechanical strength, besides being readily available mineral in nature. It is advantageous to use magnetic particles as matrix for enzyme immobilization because the enzymatic derivatives are insoluble in water and can be easily removed from the reaction mixture by an external magnetic field. Our group has reported several works related to biomolecules immobilized on magnetite and different magnetic composites [4–10]. The objective of the present work is to study the different iron phases in the magnetic montmorillonite clay (mMMT) and magnetic diatomaceous earth (mDE) by ^{57}Fe Mössbauer spectroscopy (MS), magnetic measurements and X-ray diffraction (XRD). It is also the objective to propose and assess the mMMT and mDE as matrices for the immobilization of invertase via covalent binding.

2 Materials and methods

2.1 Magnetization of clays

Montmorillonite (MMT) and diatomaceous earth (DE) were kindly supplied by Minarmco S.A. (Neuquén, Argentina) and TAMER S.A. (Salta, Argentina), respectively. The synthesis of magnetic composites was performed according to Amaral et al. [11] The magnetic particles obtained were washed with distilled water and recovered by a magnetic field (Ciba Corning; 0.6 T). The mMMT and mDE were dried at 50 °C overnight.

2.2 Characterization of magnetic composites

The phases of iron present in the resulting mMMT and mDE were investigated by X-ray diffraction and Mössbauer spectroscopy. X-ray diffraction patterns were

Fig. 1 XRD patterns of (**a**) magnetite, (**b**) mMMT and (**c**) mDE. M = magnetite; C = montmorillonite clay; Q = quartz; DE = diatomaceous earth

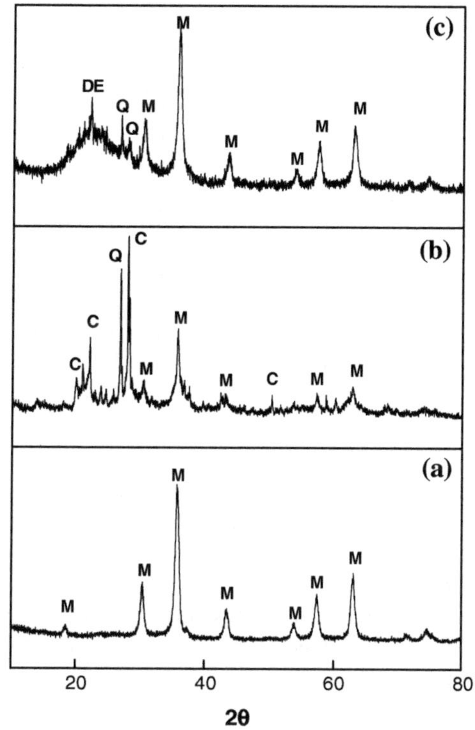

measured at room temperature in a Siemens D5000 X-ray diffractometer, using CuKα radiation ($\lambda = 1.5406$ Å). Mössbauer spectra were recorded at 4.2 K in a transmission geometry using a conventional ^{57}Fe Mössbauer spectrometer employing a 50 mCi ^{57}Co/Rh source. The spectra were analyzed using the least squares method assuming Lorenzian line shapes and a hyperfine field distribution. The isomer shift (δ) values are relative to α-Fe at room temperature. Magnetization measurements were performed at 298 K in magnetic fields varying from 0 to 50 kOe (5.0 T) using a SQUID magnetometer (Quantum Design Model MPMS-5 S).

2.3 Immobilization process

The magnetic composites were silanized with aminopropyltriethoxysilane (APTES, 2.5 % v/v) stirring at 25 °C. The activation of the silanized mMMT and mDE with glutaraldehyde (10 % v/v) also was carried out stirring at 25 °C. Functionalized materials were washed several times with distilled water. The invertase from Baker's yeast (1 mL, prepared in 0.2 M sodium acetate buffer, pH 5.0) was incubated with mMMT and mDE (0.01 g) 4 °C under mild stirring. Afterwards the material was washed five times with 0.2 M sodium acetate buffer, pH 5.0. The invertase immobilized on mMMT (mMMT-invertase) and mDE (mDE-invertase) were collected by the magnetic field and the supernatants including the first two washings were used for protein determination according to Lowry et al. [12] using bovine serum albumin as the standard protein. The immobilized derivatives were stored in sodium acetate

Table 1 Mössbauer parameters

Sample	Grain size XRD (nm)	Component	δ (mm/s)	Δ (mm/s)	Γ (mm/s)	B (T)	Area %
Magnetite	11	Sextet 1	0.31	−0.01	0.65	47.5	49.0
(RT = 300 K)		Sextet 2	0.33	−0.03	1.20	43.2	51.0
mMMT	25	Sextet 1	0.33	0.01	0.45	52.3	30.0
(4.2 K)		Sextet 2	0.38	−0.12	0.60	50.0	62.5
		Doublet	0.33	0.53	0.30	–	7.5
mDE	12	Sextet 1	0.33	0.01	0.47	52.3	49.5
(4.2 K)		Sextet 2	0.28	−0.03	0.64	50.3	50.5

Isomer shift (δ), Quadrupole splitting (Δ), Line width (Γ), Relative area (Area) and Hyperfine field (B). Uncertainty in particle size is 0.5 nm while that the uncertainty in δ and B are 0.02 mm/s and 0.5 T, respectively. Areas are accurate within 2%

buffer at 4 °C for further use. Invertase activity was determined by using 0.15 M sucrose (10 mL) prepared in sodium acetate buffer (0.2 M, pH 5.0). After exactly 15 min of incubation at 25 °C, 20 μL the sample was withdrawn and added to 2.0 mL of working solution in order to measure released glucose using a glucose oxidase-peroxidase (GOD/POD) enzymatic kit (Doles, Goiás, Brazil). The enzyme activity unit (U) was defined as the amount of enzyme releasing 1 μmol of glucose per minute under the assay conditions.

3 Results and discussion

The XRD patterns of the magnetic particles are presented in Fig. 1. The 2θ peaks at 18.44°, 30.32°, 35.75°, 43.32°, 53.89°, 57.34° and 62.96° are attributed to the crystal planes of magnetite at (111), (220), (311), (400), (422), (511) and (440) respectively [13]. The characteristic peaks of magnetite and quartz were observed in all magnetic composites. By analyzing the XRD patterns it is observed that the magnetite is the most predominant crystalline phase in mDE, while that in mMMT the aluminosilicates are the predominant crystalline phase. The mMMT and mDE exhibited broad and low intensity peaks in the base line (Fig. 1b and c). This broad X-ray structure suggests an amorphous component in the prepared composite [14]. The grain size estimated from the main reflections of each diffractogram, by using the Scherrer formula, is shown in Table 1. In this calculus we are not considering possible contributions of crystal stress. The XRD patterns of magnetite and maghemite are very similar. The main difference consists of a few low-intensity diffractions (<5 %) which are only present for the maghemite structure [15]. These diffraction lines are present in the patterns presented in mMMT and mDE (Fig. 1b and c), but this does not confirm or exclude the presence of maghemite in the composites produced. Furthermore, the fact that the low intensity peaks become visible in the XDR pattern does not prove that the transition from magnetite to maghemite took place; it could simply be due to an increase of the particle size [15]. However, in a more recent report [16], the differentiation between magnetite and maghemite was made on the basis of high angle peaks corresponding to plane (511) and (440) peak-heights and its resolution through the deconvolution.

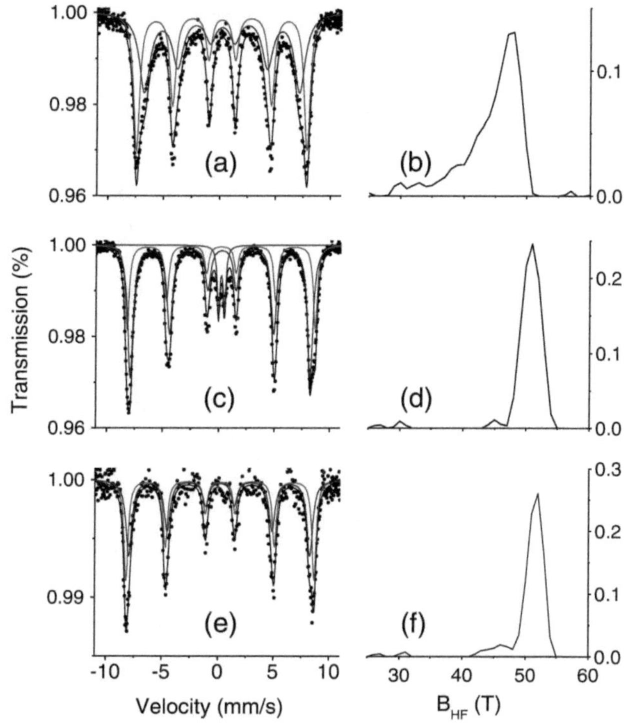

Fig. 2 Mössbauer spectra and their corresponding p-B distribution (**a**) and (**b**) magnetite at room temperature (**c**) and (**d**) mMMT at 4.2 K and (**e**) and (**f**) mDE at 4.2 K. Scattered points are data point and the fitted spectrum is shown in *black line*. The subspectra shown in *red* and *blue lines* are the component subspectra corresponding to A-site and B-site iron respectively, whereas in (**c**) the subspectrum shown in *dark line* is showing doublet

The hyperfine parameters obtained by Mössbauer Spectroscopy (MS) at an appropriate temperature can be used to identify the magnetic signal of the iron oxide and to obtain information about the Fe^{3+} linking the components of the material [17]. The measurements were carried out at 4.2 K to check any superparamagnetic state present in the samples. Only sample of pure magnetite may be analyzed at room temperature (300 K). ^{57}Fe Mössbauer spectrum of pure magnetite shows two sextets (Fig. 2a). The first one (A sites) has a hyperfine magnetic field, B = 47.5 T, and an isomer shift, $\delta = 0.31$ mm/s; assigned to Fe^{3+} ions; the second sextet (B sites) has B = 43.2 T and $\delta = 0.33$ mm/s; this sextet corresponds to mixed Fe^{2+}–Fe^{3+} ions [18]. The line width (Γ) of the second sextet corresponding to the B-site is quite high. This could be because of defects in the sample, particle size distribution and presence of different iron environments which leads to intermediate iron oxidation states. Due to the high linewidth the two components in the spectrum are not fully resolved. As it is also clear from the broad hyperfine field distribution (Fig. 2b) that there may be more than two sextets present in the sample corresponding to different iron oxidation states or different iron minerals. These values are similar to the bulk material (sextet 1: B = 49.0 T and $\delta = 0.26$ mm/s and sextet 2: B = 46.0 T and $\delta = 0.67$ mm/s) [19], but the δ for second component is significantly lower, this might indicate the presence of some other iron mineral such as maghemite. The deviation in the ideal area ratio (1:2) of the iron in tetrahedral and octahedral position obtained from the subspectra area is due to the smaller particle size compared to their bulk counterpart [20].

The hyperfine magnetic fields for mMMT (sextet 1 equal to 52.3 T and sextet 2 equal to 50.0 T) and mDE (sextet 1 equal to 52.3 T and sextet 2 equal to 50.3 T)

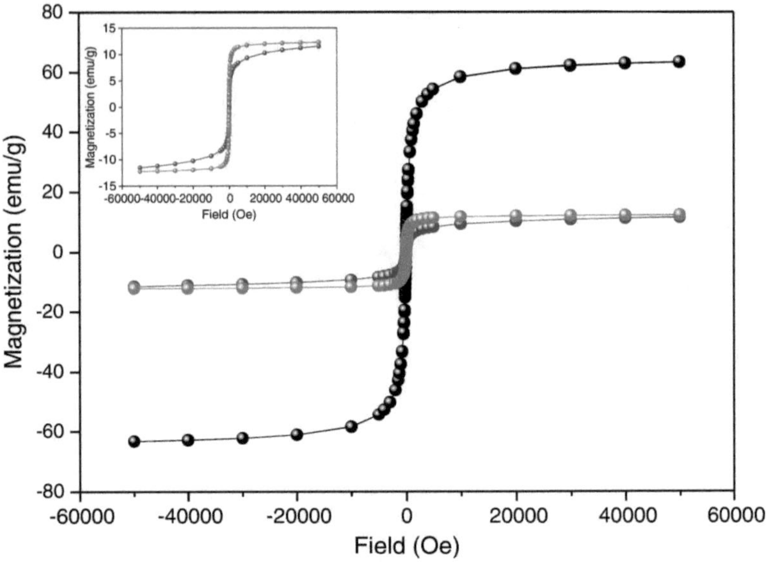

Fig. 3 Magnetization measurements for the magnetite (*black*), mMMT (*red*) and mDE (*gray*). The inset shows a magnified view of the magnetization curves of the mMMT and mDE

showed slightly higher values than pure magnetite (Fig. 2c, e and Table 1). In addition to the signals relating to magnetite, Fig.2c (mMMT) also shows a doublet having an isomer shift equal to 0.33 mm/s and an area equal to 7.5 %. This doublet emanates from ferric iron in a non-spherical local surrounding, maybe coming from the rim of the iron oxide core, i.e., the magnetic relaxation effect which is attributed to the presence of superparamagnetism as well as the ferromagnetic nanoparticles [20]. The absence of a doublet in the Mössbauer spectrum of mDE suggests that there is no non-magnetic and non-spherical iron surrounding present in the mDE sample. The mMMT and mDE spectra (Fig. 2c and e) show an increase in the B sites compared to the spectrum for pure magnetite, whereas the intensity of the A sites (Fe^{3+}) decreases. Table 1 shows the hyperfine parameters obtained from fitting the Mössbauer spectra. The percentage area reported under the curve of the lines of best fit is related to the composition of the obtained materials. According to these spectra and the hyperfine parameters, it is evident that the mMMT showed a higher increase in the B sites than mDE as compared to the pure magnetite, i.e., the presence of the clay caused more modifications in the magnetite compared to diatomaceous earth.

The magnetic properties of magnetite, mMMT and mDE particles were measured by applying an external magnetic field at 298 K. The saturation magnetization for magnetite, mMMT and mDE was determined by the magnetization curve at maximum magnetic field. As shown in Fig. 3 the saturation magnetization of mMMT and mDE was around 10 emu g^{-1} lower than the value of 60 emu g^{-1} found for the magnetite particles at 298 K. The decreased saturation magnetization can be attributed to surface effects, such as a magnetically inactive layer produced by the disordered surface [21]. In addition, the magnetite, mMMT and mDE particles exhibited superparamagnetic behavior.

Important parameters in the immobilization process such as reaction conditions of enzyme, solid support and linker determine the biochemical, mechanical and kinetic properties of the immobilized enzyme. The magnetic composites proposed as matrices for the immobilization of invertase via covalent binding showed excellent results. Thus, the residual specific activity of mMMT-invertase and mDE-invertase was 83 % and 92.5 %, respectively. No decrease in specific activity was observed, suggesting the potential application of these magnetic composites from minerals of low cost as matrices for the immobilization of invertase or other enzymes.

4 Conclusion

The synthesis of magnetic composites from two minerals of low cost as well as the study of the iron phases in mMMT and mDE was successfully performed. X-ray diffraction measurements of mMMT and mDE exhibited similar peak compared to those of magnetite and showed that the montmorillonite clay and diatomaceous earth minerals does not significantly interfere with the structure of synthesized magnetite. Through Mössbauer spectra we observed that the nanoparticles of mMMT are composed by a mixture of magnetite and maghemite whereas the mDE showed a pure magnetite phase. All magnetic particles displayed superparamagnetic behavior in agreement with particles size distribution at the nano scale. The residual specific activity of the mMMT-invertase and mDE-invertase was 83 % and 92.5 % respectively. Finally, these results suggest that mMMT and mDE are promising matrices for covalent immobilization of invertase and could be used for the immobilization or purification of other enzymes of industrial interest for biotechnological applications.

Acknowledgements This work was financially supported by the Brazilian Agencies CAPES and CNPq. The authors are grateful to Dr. José Albino Oliveira de Aguiar for XRD analyses and Dr. Adilson Jesus Aparecido de Oliveira for magnetization measurements.

References

1. Sanjay, G., Sugunan, S.: Glucoamylase immobilized on montmorillonite: synthesis, characterization and starch hydrolysis activity in a fixed bed reactor. Catal. Commun. **6**(8), 525–530 (2005). doi:10.1016/j.catcom.2005.04.016
2. Lagaly, G., Ziesmer, S.: Colloid chemistry of clay minerals: the coagulation of montmorillonite dispersions. Adv. Colloid Interface Sci. **100–102**(0), 105–128 (2003). doi:10.1016/s0001-8686(02)00064-7
3. Ghiazza, M., Gazzano, E., Bonelli, B., Fenoglio, I., Polimeni, M., Ghigo, D., Garrone, E., Fubini, B.: Formation of a vitreous phase at the surface of some commercial diatomaceous earth prevents the onset of oxidative stress effects. Chem. Res. Toxicol. **22**(1), 136–145 (2008). doi:10.1021/tx800270g
4. Maciel, J.C., Andrad, P.L., Neri, D.F.M., Carvalho, L.B.Jr., Cardoso, C.A., Calazans, G.M.T., Albino Aguiar, J., Silva, M.P.C.: Preparation and characterization of magnetic levan particles as matrix for trypsin immobilization. J. Magn. Magn. Mater. **324**(7), 1312–1316 (2012)
5. Neri, D.F.M., Balcão, V.M., Cardoso, S.M., Silva, A.M.S., Domingues, M.d.R.M., Torres, D.P.M., Rodrigues, L.R.M., Carvalho, L.B.Jr., Teixeira, J.A.C.: Characterization of galactooligosaccharides produced by β-galactosidase immobilized onto magnetized Dacron. Int. Dairy J. **21**(3), 172–178 (2011). doi:10.1016/j.idairyj.2010.10.009
6. Neri, D.F.M., Balcão, V.M., Carneiro-da-Cunha, M.G., Carvalho Jr, L.B., Teixeira, J.A.: Immobilization of β-galactosidase from Kluyveromyces lactis onto a polysiloxane–polyvinyl alcohol

magnetic (mPOS–PVA) composite for lactose hydrolysis. Catal. Commun. **9**(14), 2334–2339 (2008). doi:10.1016/j.catcom.2008.05.022
7. Neri, D.F.M., Balcão, V.M., Costa, R.S., Rocha, I.C.A.P., Ferreira, E.M.F.C., Torres, D.P.M., Rodrigues, L.R.M., Carvalho, L.B.Jr., Teixeira, J.A.: Galacto-oligosaccharides production during lactose hydrolysis by free Aspergillus oryzae β-galactosidase and immobilized on magnetic polysiloxane-polyvinyl alcohol. Food Chem. **115**(1), 92–99 (2009). doi:10.1016/j.foodchem.2008.11.068
8. Neri, D.F.M., Balcão, V.M., Dourado, F.O.Q., Oliveira, J.M.B., Carvalho, L.B.Jr., Teixeira, J.A.: Immobilized β-galactosidase onto magnetic particles coated with polyaniline: support characterization and galactooligosaccharides production. J. Mol. Catal., B Enzym. **70**(1–2), 74–80 (2011)
9. Neri, D.F.M., Bernardino, D.P.B., Beltrão, E.I.C., Carvalho, L.B.Jr.: Purines oxidation by immobilized xanthine oxidase on magnetic polysiloxane–polyvinyl alcohol composite. Appl. Catal. A Gen. **401**(1–2), 210–214 (2011). doi:10.1016/j.apcata.2011.05.026
10. Soria, F., Ellenrieder, G., Oliveira, G., Cabrera, M., Carvalho L.: α-L-Rhamnosidase of Aspergillus terreus immobilized on ferromagnetic supports. Appl. Microbiol. Biotechnol. **93**(3), 1127–1134 (2012). doi:10.1007/s00253-011-3469-y
11. Amaral, I.P.G., Carneiro-da-Cunha, M.G., Carvalho, L.B.Jr., Bezerra, R.S.: Fish trypsin immobilized on ferromagnetic Dacron. Process Biochem. **41**(5), 1213–1216 (2006). doi:10.1016/j.procbio.2005.11.023
12. Lowry, O.H., Rosebrough, N.J., Farr, A.L., Randall, R.J.: Protein measurement with the folin phenol reagent. J. Biol. Chem. **193**(1), 265–275 (1951)
13. Wen, X., Yang, J., He, B., Gu, Z.: Preparation of monodisperse magnetite nanoparticles under mild conditions. Curr. Appl. Phys. **8**(5), 535–541 (2008). doi:10.1016/j.cap.2007.09.003
14. Fang, F.F., Kim, J.H., Choi, H.J.: Synthesis of core–shell structured PS/Fe3O4 microbeads and their magnetorheology. Polymer **50**(10), 2290–2293 (2009)
15. Pinna, N., Grancharov, S., Beato, P., Bonville, P., Antonietti, M., Niederberger, M.: Magnetite Nanocrystals: nonaqueous synthesis, characterization, and solubility†. Chem. Mater. **17**(11), 3044–3049 (2005). doi:10.1021/cm050060+
16. Kim, W., Suh, C.-Y., Cho, S.-W., Roh, K.-M., Kwon, H., Song, K., Shon, I.-J.: A new method for the identification and quantification of magnetite–maghemite mixture using conventional X-ray diffraction technique. Talanta **94**(0), 348–352 (2012). doi:10.1016/j.talanta.2012.03.001
17. Wang, J., Wu, H.-Y., Yang, C.-Q., Lin, Y.-L.: Room temperature Mössbauer characterization of ferrites with spinel structure. Mater. Charact. **59**(12), 1716–1720 (2008). doi:10.1016/j.matchar.2008.03.013
18. Korecki, J., Handke, B., Spiridis, N., Slezak, T., Flis-Kabulska, I., Haber, J.: Size effects in epitaxial films of magnetite. Thin Solid Films **412**, 14–23 (2002)
19. Dyar, M.D., Agresti, D.G., Schaefer, M.W., Grant, C.A., Sklute, E.C.: Mössbauer spectroscopy of earth and planetary materials. Ann. Rev. Earth Planet. Sci. **34**, 83–125 (2006)
20. Cabrera, L., Gutierrez, S., Menendez, N., Morales, M.P., Herrasti, P.: Magnetite nanoparticles: electrochemical synthesis and characterization. Electrochim. Acta **53**(8), 3436–3441 (2008). doi:10.1016/j.electacta.2007.12.006
21. Zhang, L.-Y., Gu, H.-C., Wang, X.-M.: Magnetite ferrofluid with high specific absorption rate for application in hyperthermia. J. Magn. Magn. Mater. **311**(1), 228–233 (2007)

DOI 10.1007/978-94-007-6482-8_24
Reprinted from *Hyperfine Interactions*, DOI 10.1007/s10751-013-0844-3

Enhancing vibration measurements by Mössbauer effect

G. A. Pasquevich · A. Veiga · P. Mendoza Zélis · N. Martínez · M. Fernández van Raap · F. H. Sánchez

Published online: 13 March 2013

Abstract The measurement of the Mössbauer effect in a system excited with a periodic perturbation can provide information about it. For that purpose, the Mössbauer absorption of a source-absorber set which hyperfine parameters are well known, is measured at a constant relative velocity (i.e. at a defined spectral energy). The resulting Mössbauer absorption periodic signal provides information of the sample *ac* perturbation response. This approach has been used time ago to measure small tympanic vibrations (mechanical perturbations). In this work we present an extension of the vibration experiments, by measuring them at various absorber-source relative velocities within a constant-velocity strategy. As a demonstration test, the frequency response of a piezoelectric diaphragm in the 100 Hz–5 kHz range is obtained with a custom electronic counter. The experiments are performed using a ^{57}Co(*Rh*) source and a 25-μm-thick stainless-steel absorber fixed to a piezoelectric diaphragm. Phase shifts and amplitude vibrations with velocities in the range from 1.5 μm/s to 20 mm/s are well characterized, extending the linearity limit well beyond the earlier suggested one of 1 mm/s.

Keywords Vibrations · Mössbauer · Velocity

Proceedings of the thirteenth Latin American Conference on the Applications of the Mössbauer Effect, (LACAME 2012), Medellín, Columbia, 11–16 November 2012.

G. A. Pasquevich (✉) · A. Veiga
Facultad de Ingeniería, UNLP, La Plata, Argentina
e-mail: gpasquev@fisica.unlp.edu.ar

G. A. Pasquevich · A. Veiga · P. Mendoza Zélis · M. Fernández van Raap · F. H. Sánchez
Instituto de Física La Plata, CONICET, La Plata, Argentina

N. Martínez · P. Mendoza Zélis · M. Fernández van Raap · F. H. Sánchez
Facultad de Ciencias Exactas, UNLP, La Plata, Argentina

N. Martínez
CICpBA, La Plata, Argentina

1 Introduction

The measurement of Mössbauer effect synchronously with a periodic excitation at a defined constant Doppler energy can be used to obtain the dynamic response of a perturbed sample. Variations on the intensity, position, or other parameters of the Mössabuer spectral lines can be interpreted as a characterization of the perturbation and its effect on the sample.

Relative movements induced by an external perturbation can be detected by Mössbauer effect. In fact, Mössbauer spectroscopy is based on relative motion through Doppler Effect: a Mössbauer spectrum display absorption versus relative velocity. Therefore, if the spectrum is well known, changes in the absorption can be interpreted as a measurement of the relative velocity.

This approach has been widely used to characterize basilar-membrane frequency responses of different species [1–6]. The pioneering work in this line is from 1967 by P. Gilad et al. [1] in which a single line absorbent and a counter synchronized with a sound generator is used to study the tympanic membrane response of guinea pigs in the range from 250 Hz to 9 kHz. All those works have been of major importance in its discipline, they are still relevant and are being cited up to day.

In addition to biologic applications, mechanical vibration studied by Mössbauer effect has some characteristics that makes it powerful and unique. Hence should not be let aside. Accordingly, we can mention: the ability of the gamma ray to pass through opaque materials, and the atomic selectivity of the Mössbauer effect. The former allows the study of internal parts in macroscopic complex systems such as instruments, machines, etc.; the latter (the ability to distinguish different structural phases of the same atom) let the study of particular locations into micro-scale complex systems (e.g. soft condensed matter materials).

The procedure used to detect and quantify effects of vibrations by Mössbauer effect consists on measuring the Mössbauer absorption synchronously with the vibration source at a chosen position (velocity) in the spectrum. It means that the source-absorber relative velocity has a steady (*dc*) component, to adjust that point (constant velocity) and a superimposed harmonic wave (*ac*) producing the perturbation. The sensitivity of the method depends strongly on the constant velocity selection. For low amplitude perturbations, the most sensitivity spectral position corresponds to the one of maximal slope. Relative movements induced in the absorbent appear as changes in the relative velocity that produce variations in the transmission rate which are proportional to the spectrum slope. Note that in all the previous works (related to tympanic response) the radiation source was attached rigidly to the membrane to be characterized and the absorber was at rest in the laboratory frame. Therefore, the *dc* constant velocity was zero. This velocity election is the simplest way to perform this kind of experiments, since no electronic velocity branch is needed. In those cases, the slope at the zero velocity spectral position is just determined by the relative isomer shifts of the source and absorber. In what follows, we will assume a geometry in which the absorber is fixed to the sample (which will be perturbed) while the radioactive source is used to select the constant velocity relative to the laboratory frame. Although both geometries are equivalent, the latter is the one used in the experiments of this work.

Let us assume a sinusoidal time perturbation ε of frequency f and amplitude ε_0,

$$\varepsilon(t) = \varepsilon_0 \sin(2\pi f t) \tag{1}$$

under which the absorber oscillates. Let us also assume a linear absorber movement response, in such a way that the response of the absorber-velocity V is only defined by an amplitude V_0 and a phase-shift φ,

$$V(t) = V_0 \sin(2\pi f t + \varphi) \tag{2}$$

If the source is moved at a constant velocity v, corresponding to a flank of a spectral absorption line, and if the resulting V_0 is low enough, then the transmission rate r will harmonically oscillate retaining information about the amplitude and phase-shift of the velocity response:

$$r(t) = r_0 + \alpha V_0 \sin(2\pi f t + \varphi), \tag{3}$$

where r_0 is the rate in absence of perturbation and α is the spectrum slope ($\partial r/\partial v$) at the constant velocity position. For higher amplitudes the shape of the absorption line introduces non linearities and the last equation is not valid any more. This nonlinearity puts a limit to the maximum amplitudes than can be retrieved through this methodology. This limit was estimated around 1 mm/s, a rough evaluation of the effective velocity wideness of an absorption line [4, 7].

The Mössbauer characterization of vibrations at different constant velocities improves considerably the limitations previously mentioned. Besides, the ability of measuring synchronized Mössbauer signals allows the incursion in other kind of perturbation studies as the recently proposed magnetic susceptibility study by Mössabuer effect [8], which is a frequency extension of the Mössbauer magnetic scans [9]. Therefore, the development of equipments with this facilities is of interest. In this work we present a complete device with the ability of performing synchronous Mössbauer experiments. The test of the equipment is carried out characterizing a commercial piezoelectric diaphragm, by studying oscillations up to 5 kHz induced on a rigidly attached stainless steel sheet.

2 Experimental details

A piezoelectric diaphragm (PD) 7BB-35-3 from muRata Manufacturing was used for the experimental set-up. A 5 mm hole was perforated in its center to act as a window for the gamma ray. A 25 μm thick stainless-steel sheet (the absorber), mounted between two acrylic plates of 1 mm thick, was fixed to the PD covering the hole. The PD was fed with an harmonic *ac* signal generated by a driver as is shown in the Fig. 1. The stainless steel foil was carefully fasten to assure its motion represents the PD motion.

A multichannel scaler (MCS) designed in our laboratory [10] was used as wave reference generator for the PD excitation and as a multiscaler for the data acquisition. With this custom equipment, frequencies up to 75 kHz can be reached when only four acquisition channels are used. The constant velocity module (CVM), also designed in our laboratory [11], generates a rectangular asymmetric wave with zero mean value as a velocity reference wave for the velocity drive. The generated wave was carefully designed, considering the driver and transducer transfer function, in order to achieve a high quality constant velocity state as well a good temporal efficiency [11]. A logic signal synchronized with the velocity wave was also generated to enable the MCS to count only at the constant part of the velocity wave.

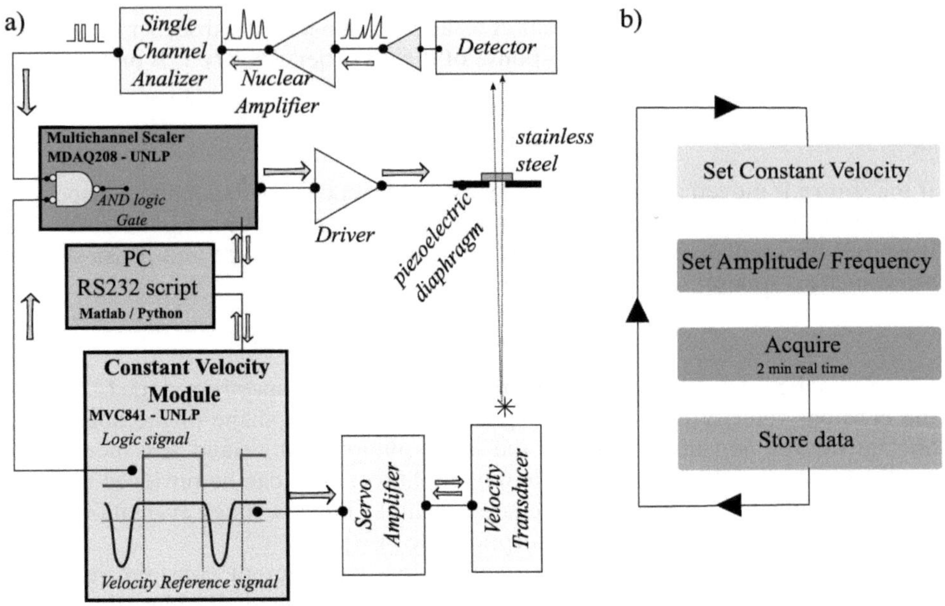

Fig. 1 **a** Experimental layout used to count 14.4 keV events in a synchronous way with a harmonic perturbation. The synchronous counting is carried out while the source is moved at constant velocity relative to the laboratory frame. **b** Flow diagram of the protocol used to measure different perturbation frequencies and amplitudes as well as different velocities during experiments

Both modules are connected to a personal computer using an RS232 serial protocol. Amplitude and frequency of the excitation wave are managed externally through the RS232 protocol. A Python 2.6 script on the computer was used to simultaneously drive both modules. A flow diagram of the script is shown at Fig. 1b.

The velocity calibration was performed by full spectra acquisition of three different calibration samples: alpha iron, stainless-steel and bulk magnetite. All the velocities given in this work are given relative to the Laboratory Frame, i.e. they are referenced to isomer shift of ^{57}Fe(Rh).

Two kinds of experiments were performed: frequency characterization of the piezoelectric response in the range 100 Hz to 5 kHz, and amplitude response at a fixed frequency (near resonance).

3 Frequency domain

For frequency characterization of the piezoelectric response, three velocities values were selected. Two at the highest slope values of the left and right flanks of the absorption lines (−0.49 and 0.03 mm/s respectively). The third at −3.52 mm/s, where resonant absorption in absence of perturbation is zero, i.e background velocity. The position are shown with vertical arrows in Fig. 2.

The range from 100 Hz to 4.4 kHz was sampled at two different PD voltage amplitudes: 90 mV and 150 mV. The corresponding results are shown in Fig. 3a and

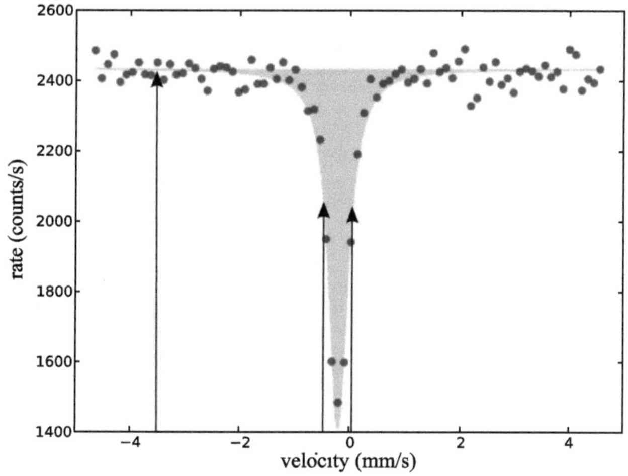

Fig. 2 Stainless-steel Mössbauer spectrum. Fixed velocities (−3.52 mm/s, −0,49 mm/s and 0.03 mm/s) chosen to perform the frequency characterization are indicated with *arrows*. The velocities are expressed relative to ^{57}Fe(*Rh*) isomer shift (i.e. to the laboratory frame)

b respectively. In each graph transmission vs. phase ($2\pi t/T$) of each of the three velocities are plotted. The L and R curves correspond to left and right flanks velocity positions (−0.49 and 0.03 mm/s respectively) while the B curve corresponds to the high velocity position (background).

At low and high frequencies transmission rate oscillations are harmonic enough as to follow (3). For frequencies between 2 kHz and 3 kHz, around resonance, the transmission response becomes nonlinear and it can be seen as a double peak response. The figure also shows a relative phase shift of 180° for all frequencies between R and L curves. It is apparent since the curves represent the behavior at each side of the absorption line. For the higher amplitudes near 2.4 kHz the transit of the whole absorption line can be appreciated over both flanks positions. This manifestations are more evident in the experiment at higher voltage amplitude (Fig. 3b).

In order to retrieve information on the amplitude V_0 and phase-shift φ from the data, the temporal transmission rate dependence was modeled as the absorption of a Lorentzian line whose center oscillates sinusoidally,

$$r(t; v_i, V_0, \varphi) = 1 - \frac{\Gamma^2}{4} \frac{h}{(\delta_0 - v_i - V_0 \sin(2\pi f t + \varphi))^2 + \frac{\Gamma^2}{4}} \tag{4}$$

where δ_0 and Γ are the isomeric shift and linewidth, h is the line depth, v_i with $i = 1, 2, 3$ are the three constant velocities. A multiple fit at the selected velocities was performed over all the frequencies range. Γ, h and δ_0 were common parameters to all the velocities and frequencies. V_0 and φ parameters were different for each frequency, but common for the three velocities. The three v_i were fixed parameters. The resulting fitted curve for each velocity is shown in Fig. 3 (solid black curve). Note that the explicit Lorentzian shape in the model incorporates the nonlinearity of the absorption profile to the analysis. On the other hand the explicit harmonic temporal dependence of the line center restricts the model to situations in which the PD responds linearly to the *ac* input voltage.

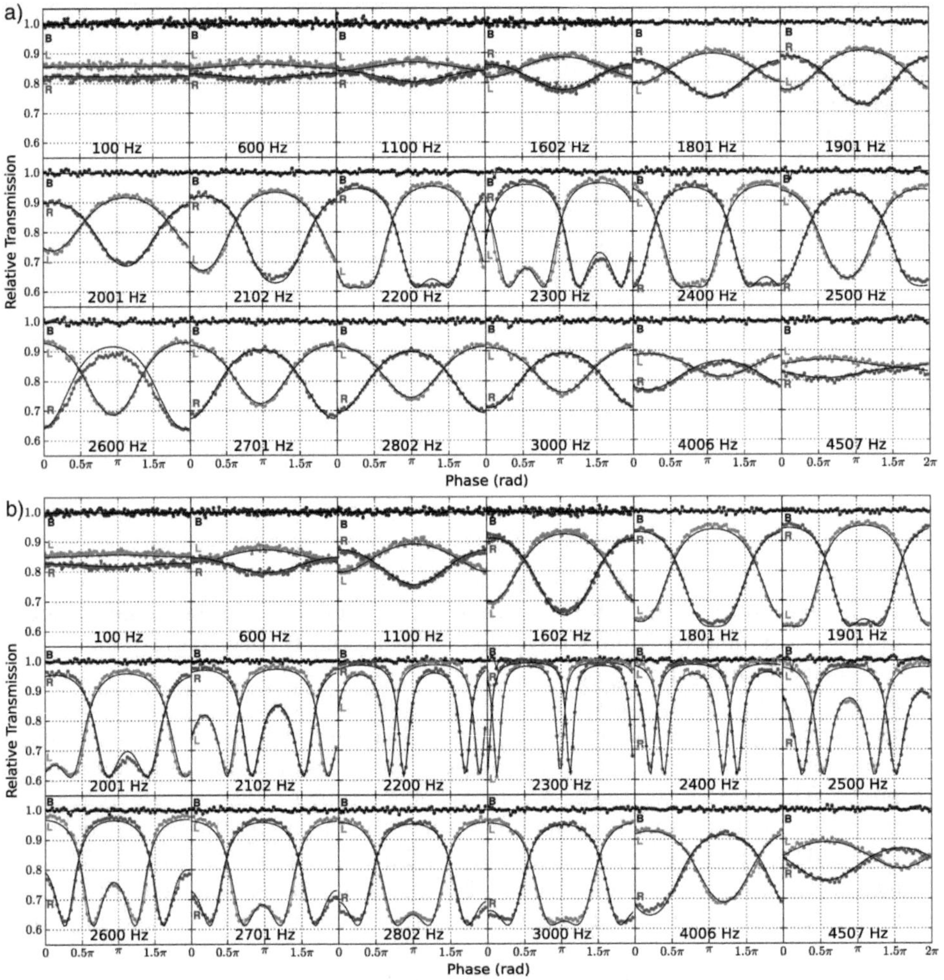

Fig. 3 Transmission rate at the three velocities indicated in Fig. 2. The curves indicated by R, L and B correspond to the fixed velocities at −3.52, −0.49 and 0.03 mm/s. They are related to the left and right flank of the absorption line and a background position, respectively. **a** Transmission response to 90 mV *ac* input voltage. **b** Response to 150 mV *ac* input voltage

From the fit procedure, amplitude and phase-shift of the PD response were determined in the whole frequency range (Fig. 4a and b). A resonance peak is observed at 2.3 kHz for both perturbation amplitudes. At this frequency a $\pi/2$ phase-shift change respect to low frequencies value is achieved, and completes to a net relative change π as the frequency increases. Moreover, a further change in curve's slopes can be noted at a frequency around 4 kHz. In terms of pole-zero diagrams this behavior corresponds to a linear third order system with a zero at the origin. The change in the PD input amplitude was reflected only as an increase in the velocity amplitude. The phase-shift was not affected by this parameter.

The above experiment is a good example of a system characterization using the method proposed in the present work.

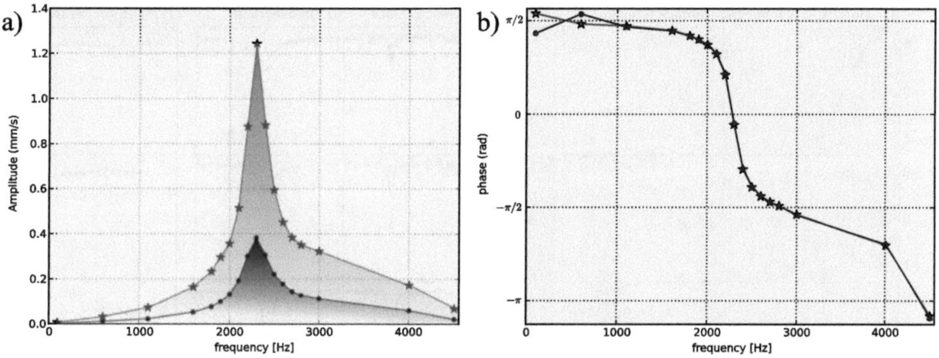

Fig. 4 Results of the multiple fit: **a** amplitude and **b** phase-shift as a function of the frequency for each PD ac-voltage amplitude (90 mV: *blue dots*, 150 mV: *red stars*)

On this set of experiments the extreme velocity amplitudes were 1.5 μm/s ($f = 100$ Hz, $\varepsilon_0 = 90$ mV) and 1.24 mm/s ($f = 2.3$ kHz, $\varepsilon_0 = 150$ mV). Since the movement is harmonic, the position amplitude can be deduced through $\Delta x = \frac{\Delta v}{2\pi f}$. The smallest position amplitude was 0.7 nm ($f = 4.5$ kHz, $\varepsilon_0 = 90$ mV) and the highest was 85 μm at ($f = 2.3$ kHz, $\varepsilon_0 = 150$ mV).

4 Amplitude domain

In order to characterize high amplitude vibrations, a large number of velocities covering a wider range is needed. A uniformly distributed set of twenty one velocities was chosen to characterize vibrations in the range from −10 mm/s to 10 mm/s. The transmission at each constant velocity was recorded at four voltage amplitudes of 1, 2, 4 and 8 V. The experiments were carried out at 2.4 kHz; a near resonance frequency selected to achieve high amplitudes with the available voltages. The phase was sampled in $N_{CH} = 64$ uniformly distributed channels.

The measured transmission as a function of phase and velocity is shown as 2D contour graphs in Fig. 5a. The Mössbauer transmission defines a surface in the *phase-velocity* space. The transmission at each point of this space is given by the color scale at the right side of the contour-graphs: light colors correspond to high transmission and dark colors to low transmission.

The velocity trace of the transmission surface at the velocity −0.2 mm/s is shown in Fig. 5b. This velocity corresponds to the center of the absorption line in absence of perturbation (i.e. the stainless-steel isomer shift). It is marked over the contour graphs with a dotted horizontal line. These traces show the two instants at which the absorption-line center coincides with −0.2 m/s, i.e. the moments at which the absorber is at rest relative to the laboratory. Note that these traces are equivalent to the plots shown in Fig. 3. Similarly, 0.15 rad phase traces are plotted in Fig. 5c. Again, the 0.15 rad phase is marked as a dotted vertical line on the contour graphs. This phase corresponds to one of the instant at which the absorber passes through its rest point.

To quantify φ and V_0 for each case a multiple fit of each data set was performed using (4). The velocity center of the absorption line, according to the resulting fitted

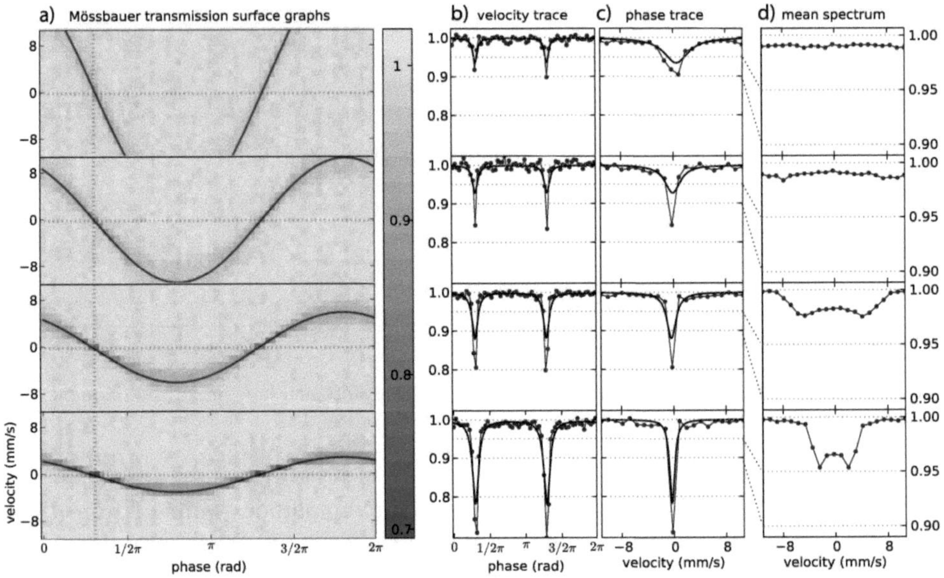

Fig. 5 **a** Contour graphs of Mössbauer transmission surface as a function of phase (abscissa) and constant velocity (ordinate). From bottom to top each graph corresponds to 1, 2, 4 and 8 V amplitude of the ac PD voltage input. The transmission intensity is given by right color scale. The *solid black line* indicates the center of the absorption line according to the least-square-fit result. **b** Velocity trace of the transmission surface at the velocity −0.2 mm/s (indicated as a *horizontal dotted line* in the contour graph). The *solid black line* is the velocity trace of the fitted surface. **c** Phase trace at 0.15 rad (indicated in the contour graph as a *vertical dotted line*). The *solid black line* is the surface trace. **d** Mean Mössbauer spectrum: average on phase of the transmission surface

surface, is plotted vs. phase over the contour graphs (solid black curve). In the same way, the fitted-surface velocity and phase traces are compared with the experimental traces in the corresponding axes.

From the fit procedure it is deduced that the phase shift ($\varphi = -0.8\pi$) remains quite independent of the input voltage amplitude. This conclusion was easily anticipated by a raw inspection of the contour graphs. The velocity amplitude displayed a linear dependence on the voltage amplitude. The amplitudes obtained from this fit analysis, together with the ones obtained in the previous section (for 2.4 kHz), are plotted in Fig. 6 evidencing the mentioned linearity. The maximum sensed amplitude is 20.6 mm/s.

Some cautionary remarks about experimental and fitting results shown in Fig. 5. First, when the line is passing near −0.2 mm/s (i.e. the rest point) the depth of the absorption-line decreases as V_0 increases (see Fig. 5a, b and c). This is an average effect related to the finite acquisition time of each MCS-channel (dwell-time t_d) and the movement of the absorption line. The quickest the absorption line moves through a constant velocity value, the lesser the time it will be contributing to the measured transmission of the corresponding channel. In other words, if the absorber velocity changes during t_d an amount ΔV of the order or larger than the line halfwidth, the recorded transmission at the channel will be related to the absorption time average of the moving line. As a consequence, the measured depth of the absorption line seems to be lower than the real one. In such cases, the time fraction that the line

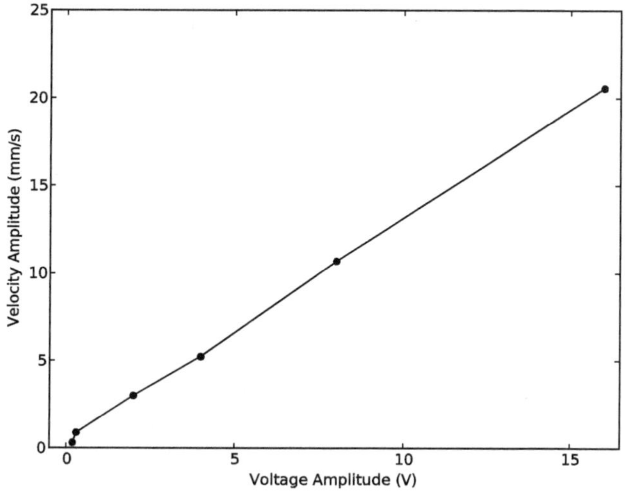

Fig. 6 Velocity amplitude vs. PD voltage amplitude obtained from fit analysis shown in Figs. 3 and 5

is not contributing to this velocity channel, it will be doing so at the neighboring channels. Hence, this line movement is also manifested as an increase of the observed linewidth in constant phase traces (Fig. 5c). Note that since it is an effect related with the instantaneous velocity change of the absorber, it will be more evident when the line moves faster, which occurs around the rest point zones.

Do not confuse this quick movement effect with the vibration effect over a Mössbauer spectrum. Although both consequences are related to vibrations, the technical origin is different as well as the magnitude of its effects. A harmonic vibration is manifested in a Mössbauer spectrum as a broadening of the line profile and a decreasing of the line depth. For high amplitude vibrations the Mössbauer-spectral-absorption lines take a double peak profile. In Fig. 5d the Mössbauer spectrum of the vibrating stainless-steel foil for each one of the voltage amplitudes is shown (the spectra were obtained by time averaging each transmission surface). Note that the magnitude of the vibration effects on the spectra are higher than the quick line-motion effects on the transmission surface. Take, for instance, the lowest V_0 case from Fig. 5. The spectrum shows a well resolved two peak profile; with a 5 % relative transmission (without vibration 40 % relative transmission was recorded—see Fig. 2). Whereas the absorption line observed in the surface graph is evidently narrower and deeper, as can be directly checked in the phase trace curve shown in Fig. 5c, in which a 30 % effect is evidencing. The point is that what is spurious for the spectrum analysis (the whole vibration) is not for the self vibration characterization. Instead, the quick line-motion effect is an spurious effect for this characterization.

Finally, the fitted surface does not reproduce exactly the experimental transmission surface. By raw inspection of the transmission surfaces is evident that the absorption line does not behave like a rigid Lorentzian profile moving sinusoidally, as the fitting model propose. Conversely, the absorption line is thinner and deeper when it pass through the rest points than when it reaches the equilibrium points (maximum velocities). On other words, the absorption line spread and shrink periodically as it passes by these points. This behavior is more important as V_0 increases being negligible at low amplitudes (as in the used in Section 3). The proposed fitting model does not cover this phase dependence behavior, therefore the depth and width fitting parameters should fall in the middle of the experimental extreme values. The spread of the absorption at high velocities might be originated on a dispersion of the PD velocity amplitudes or a chaotic noise over the PD movement. The quality of the experimental data allows a study of this features. However the proposed model allows a good determination of amplitude and phase-shift of the average sinusoidal movement, which was what we were looking for.

5 Discussion

The use of multiple velocities in synchronous Mössbauer acquisition increase considerably the experiment efficiency. Although, at low amplitude (see Section 3) a unique velocity position was enough to retrieve information of the oscillatory response. The use of more velocities is well justified due to the non-linearity of the velocity-transmission transfer curve given by the absorption line shape. In that sense, when the transmission rate response at a given velocity loses the pure harmonic behavior, the signal at different velocities allows to discern if the non linearities become from the absorption line shape or are originated in the voltage-velocity characteristic curve of the system under study. The choice of two velocities at both flanks of the absorption line improves considerably the characterization of low amplitude oscillations. Besides, a third velocity position at high energies allows a good quantification at a background energy, which is important to ensure a good estimation on the relative to background transmission rate at each velocity, which helps to dismiss spurious fluctuations or frequency dependent factors in the acquisition dead time.

In summary the use of multiple velocities allows to extend the maximum velocity limit previously indicated as 1 mm/s. In this work we proved that almost a maximum amplitude of 20.6 mm/s is attainable. Even more relevant is that this limit is not more determined by the velocity range of the sensitive absorption region of the absorption-line, but by ability to perform mechanics constant velocities movements. Instead of being an intrinsic limitation is a technical one.

Other interesting feature related to uniformly dispersed sampling velocities, is removal of nonlinearity, which was intrinsically presented in data interpretation. Indeed, single velocity data interpretation is restricted to the velocity-transmission transfer function (i.e the absorption line profile). However, the multiple velocity experiment allows to generate the surface graphs, in which a direct observation of the line velocity-position as a function of time (phase) is possible, even without considering the absorption line profile. The path that follows the absorption line in the velocity-time space is directly observed from these graphs.

Although amplitude and phase shift can be estimated from those surface graphs, a least square fit was used to quantify these values from the data. This procedure assures more precise values of those parameters. However, from the fit procedure not only the oscillatory movement is characterized, but also intrinsic parameters of the absorption line, i.e. the isomer shift, linewidth and depth. Therefore the proposed methodology serves as a tool to deconvolute those spectral parameters from unwanted vibrations. For this purpose the MCS must be triggered by the perturbation signal. However this practical application will be developed and discussed in a further work.

6 Conclusions

Vibration measurements using Mössbauer Effect Spectrometry was revisited. The technique capability is demonstrated by characterizing frequency and amplitude response of a commercial piezoelectric diaphragm using the constant-velocity mode. The smallest measured velocity-amplitude was 1.52 μm/s while the highest was 20.6 mm/s. In the position domain the highest position-amplitude was 1.3 μm at 2.4 kHz while the smallest amplitude was 0.7 nm at 4.5 kHz.

Acknowledgement We appreciate financial support to ANPCyT, Argentina (PICT 2785-2010).

References

1. Gilad, P., et al.: Application of the Mössbauer method to ear vibrations. J. Acoust. Soc. Am. **41**, 1232–1236 (1967)
2. Brafman, H., et al.: Use of the Mössbauer effect to measure small vibrations. Nucl. Intrum. Methods. **53**, 13–21 (1967)
3. Robles, L., et al.: Transient response of the basilar membrane measured in squirrel monkey using the Mössbauer effect. J. Acoust. Soc. Am. **59**, 926–939 (1976)
4. Lynch, T., et al.: Input impedance of the cochlea in cat. J. Acoust. Soc. Am. **72**, 108 (1982)
5. Robles, L., et al.: Basilar membrane mechanics at the base of the chinchilla cochlea. J. Acoust. Soc. Am. **80**, 1364–1374 (1986)
6. Ruggero, M.A., et al.: Middle-ear response in the chinchilla and its relationship to mechanics at the base of the cochlea. J. Acoust. Soc. Am. **87**, 1612–1629 (1990)
7. Ruggero, M.A., Rich, N.C.: Application of a commercially-manufactured Doppler-shift laser velocimeter to the measurement of basilar-membrane vibration. Hear. Res. **51**, 215–230 (1991)
8. Pasquevich, G.A., et al.: Local magnetic susceptibility by Mössbauer effect. In: CONFERENCIA Latinoamericana Sobre las Aplicaciones del Efecto Mössbauer (12 : 2010 : Lima), pp. 51–52. Abstracts book. Lima: LACAME, UNI, Fondo Editorial, UNMSM, Ediciones del Vicerrectorado Académico (2010)
9. Pasquevich, G.A., et al.: Determination of the iron atomic magnetic moments dynamics in the nanocrystalline ribbons Fe90Zr7B3 by Mössbauer magnetic scans. Phys., B Condens. Matter **384**, 348–350 (2006)
10. Veiga, A., et al.: Smooth driving of Mössbauer electromechanical transducers. Hyperfine Interact. **202**, 107–115 (2011)
11. Veiga, A., et al.: Advances in constant-velocity Mössbauer instrumentation. Hyperfine Interact. **167**, 905 (2006)

DOI 10.1007/978-94-007-6482-8_25
Reprinted from *Hyperfine Interactions*, DOI 10.1007/s10751-013-0835-4

Preparation and characterization of cobalt ferrite nanoparticles coated with fucan and oleic acid

P. L. Andrade · V. A. J. Silva · J. C. Maciel · M. M. Santillan · N. O. Moreno · L. De Los Santos Valladares · Angel Bustamante · S. M. B. Pereira · M. P. C. Silva · J. Albino Aguiar

Published online: 20 March 2013

Abstract Cobalt ferrite has attracted considerable attention in recent years due to its unique physical properties such as high Curie temperature, large magnetocrystalline anisotropy, moderate saturation magnetization, large magnetostrictive coefficient,

Proceedings of the thirteenth Latin American Conference on the Applications of the Mössbauer Effect, (LACAME 2012), Medellín, Columbia, 11–16 November 2012.

P. L. Andrade (✉) · J. Albino Aguiar
Programa de Pós-Graduação em Ciências de Materiais, CCEN,
Universidade Federal de Pernambuco, 50670-901, Recife PE, Brazil
e-mail: priscylandrade@df.ufpe.br

P. L. Andrade · V. A. J. Silva · J. C. Maciel · M. P. C. Silva
Laboratório de Imunopatologia Keizo Asami and Departamento de Bioquímica,
Universidade Federal de Pernambuco, Recife PE, Brazil

M. M. Santillan
Facultad de Ciencias Físicas, Universidad Nacional Mayor de San Marcos,
Ap. Postal 14-0149, Lima, Peru

N. O. Moreno
Departamento de Física, Universidade Federal de Sergipe, São Cristovão, Brazil

L. De Los Santos Valladares
Cavendish Laboratory, Department of Physics, University of Cambridge,
J. J. Thomson Avenue, Cambridge CB3 0HE, UK

L. De Los Santos Valladares · J. Albino Aguiar
Departamento de Física, Universidade Federal de Pernambuco, CEP 50670-901,
Recife PE, Brazil

A. Bustamante
Laboratorio de Cerámicos y Nanomateriales, Facultad de Ciencias Físicas,
Universidad Nacional Mayor de San Marcos, Apartado Postal 14-0149, Lima 14, Perú

S. M. B. Pereira
Departamento de Biologia, Universidade Federal Rural de Pernambuco, Recife PE, Brazil

excellent chemical stability and mechanical hardness. In this work we present the preparation, of fucan coated cobalt ferrite nanoparticles by a modified co-precipitation method and the study of their structural, microstructural and magnetic characteristics for their application as a solid support for enzymes immobilization and other biotechnology applications. Aqueous suspensions of magnetic particles were prepared by coprecipitation of Fe(III) and Co(II) in the presence of NaOH, acid oleic and fucan polymer. The X-ray diffraction indicates that the funtionalization does not degrade the core cobalt ferrite. The infrared (FTIR) bands, indicate the functional characteristics of the coating on the cobalt ferrite. Mössbauer spectra at room temperature indicate the presence of a broadened sextet plus a doublet which is typical of superparamagnetic relaxation. For the Co-ferrite uncoated and coated with fucan the doublets have areas of 36.1 % and 40.3 % respectively, indicating the presence of non-interacting particles and faster relaxation time. The Co-ferrite coated with oleic acid and oleic acid plus fucan have areas around 17.5 % and 17.1 % respectively which indicate a weak superparamagnetic relaxation due to a slow relaxation time. The magnetization measurements of the cobalt ferrite nanoparticles with and without coating confirm that they are superparamagnetic and this behavior is produced by the core nanoparticles rather than the coatings. The cobalt ferrite nanoparticles coated with oleic acid presented the highest magnetization than when coating with fucan.

Keywords Cobalt ferrite · Fucan · Magnetic nanoparticles · Mössbauer spectroscopy · Polysaccharides

1 Introduction

Nanomaterials based on ferrites have already numerous applications, including photocatalysis, adsorption technologies, gas sensor, microwave devices and others [1]. In particular, cobalt ferrite materials have attracted considerable attention in recent years due to their unique physical properties such as, high Curie temperature, moderate saturation magnetization, excellent chemical stability and mechanical hardness [2]. They can be used in a large number of biotechnological applications, such as DNA and RNA purification, cell separation, drug delivery, magnetic resonance imaging [3], magnetic hyperthermia for cancer treatments [4], biosensors [5] and others.

Fucans are highly heterogeneous sulfated polysaccharides composed of different sugar residues, such as fucose, galactose, mannose and uronic acid. It is made up of a-L-fucose units linked by (1–4) and (1–3) glycosidic bands and sulphate at positions 2 and/or 3 and/or 4 depending on the algal species. Fucans and fucoidans are distributed in the intercellular matrix in brown algae and have a wide spectrum of activity in biological systems of potential medicinal value, such as anticoagulant, antitumour, anti-inflammatory, antiviral and antioxidant activities [6, 7].

Various preparation techniques, such as sol-gel methods, citrate precursor techniques [8], electrochemical synthesis [9], combustion methods [10], solid state reaction [11], and mechanical alloying [12] are commonly used to produce ferrite nanoparticles. Among them, the chemical precipitation is an economical way to produce fine powders [13, 14]. This article describes the synthesis of cobalt ferrite ($CoFe_2O_4$) nanoparticles by the precipitation method, and their coating with oleic acid and fucan polysaccharide. The uncoated and fucan-coated $CoFe_2O_4$ nanoparticles were characterized structurally, microstructurally and magnetically by X-ray diffractometry, FTIR, Mössbauer spectroscopy and magnetization measurements.

2 Experimental

An aqueous mixture containing 50 mL of 1 M $FeCl_3.6H_2O$ and 50 mL of 0.5 M $CoCl_2.6H_2O$ was added to 0.5 mL of oleic acid [15] and sodium hydroxide to obtain a solution (pH 13). The mixture was heated at 100 ± 3 °C for 1h with vigorous stirring to obtain in the magnetic particles. The sample was then thoroughly washed with distilled water until it reached a neutral pH following by drying it at 60 °C. The preparation followed the same procedure described by Carneiro Leão et al. [16], except by modifications done during the incubation, which was performed at 85 °C during 30 min, and final pH 13. The nanoparticles were then added to 2 % of fucan polysaccharide solution in distilled water. The final step included drying the sample in an oven and the obtained powder were collected for characterization. The X-ray diffractogram (XRD) of the samples were taken in a Siemens model D5000 diffractometer. The samples were analyzed in the scale of the range $10° < 2\theta < 80°$ using CuKα radiation (λ = 1.5406 Å), in steps of 0.02° and the counting time of 1 second per step, with acquisition time of three hours. The Scherrer equation: $D = 0.916\lambda/\Gamma \cdot \cos\theta_{\mathrm{hkl}}$ was used to calculate the average grain size; where λ is the wavelength of the X-rays (0.15418 nm), β_{hkl} is the pure diffraction line broadening (in radians), which can be easily found by measuring the full-width-at-half-maximum (FWHM) of the hkl reflection and θ is the Bragg's angle [17]. The latest give a mean grain size of around 20 nm. The Fourier transform infrared (FTIR) was performed by the KBr tablet method in the range 4000–400 cm^{-1}, in a BRUKER model IFS 66 instrument. Two milligrams of samples were mixed with 200 mg of KBr and then were pressed at 490 atm and transformed into disks. The magnetization measuraments were realized in a SQUID MPMS magnetometer of Quantum Design. The uncoated cobalt ferrite nanoparticles were measured at different temperatures whereas the coated nanoparticles were measured at room temperature. The Mössbauer characterization was performed with a conventional transmission Mössbauer spectrometer, operating with 1024 channels and a Wissel INC. velocity module with a sinusoidal signal. The measurements were taken at room temperature (RT) and the obtained data were adjusted with the help of the program NORMOS [18] which does not present the residual data. In this program, the good fitting is controlled by the value of the chi^2. The source employed was a ^{57}Co in rhodium matrix with strength 25 mCi. The Isomer shifts and the velocity scale were calibrated with respect to a α-Fe film at RT. The sample holder used has a diameter 1 cm (0.7854 cm^2) and 24 mg mass which permitted to ascertain and quantify the small systematic effects of cosine smearing which usually occur in the folded Mössbauer spectra when relatively large collection solid angles are used.

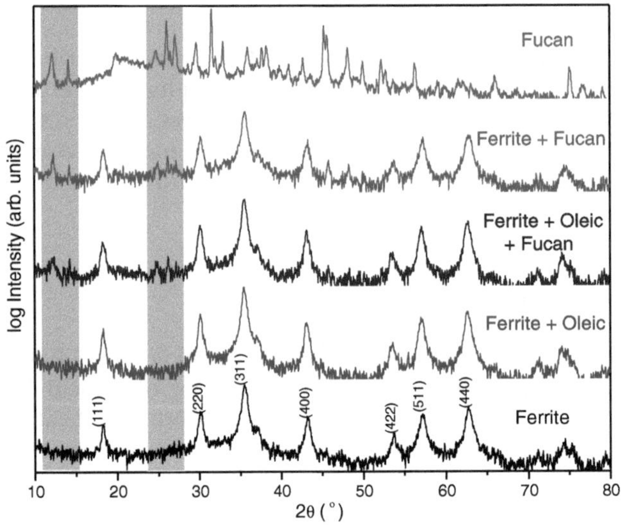

Fig. 1 X-ray diffraction patterns (in logarithmical scale) of uncoated and coated cobalt ferrite nanoparticles. The *grey columns* indicate two peak reflections around 14° and three peak reflections around 26° belonging to Fucan, and they are present in all fucan coated samples studied in this work

3 Results and discussion

Figure 1 shows the X-ray diffractogram (XRD), in logarithmical scale, of the cobalt ferrite nanoparticles before and after coating with Fucan and/or oleic acid. The diffractogram of the uncoated nanoparticles (see bottom part of the figure) shows the characteristic of a cubic spinel structure, in accordance with the standard JCPDS card N° 22-1086. The diffractogram does not vary after coating with fucan, oleic acid or combination of them, indicating that the functionalization does not degrade the core cobalt ferrite nanoparticles. However, note that the XRD of the cobalt ferrite nanoparticles coated with fucan shows peaks characteristic of the fucan around 14° and 26°, corroborating the functionalization. No peaks of contaminants were detected in samples.

In order to study the functional groups formed in the coated cobalt ferrite sample, FTIR measurements were performed (Fig. 2). In the figure, the presence of the OH band (3500 cm^{-1}) in all the spectra in indicates the presence of water, which is reasonable since the functionalization was performed in aqueous solutions as mentioned in the experimental section. However, in the case of the presence of water in the cobalt ferrite case, it should indicate that the powder were not dry enough. Moreover, the characteristic absorption band of Co (Fe)-O can be seen at 583 cm^{-1}. In the case of the cobalt ferrite coated with fucan, the bands at 2921 and 2851 cm^{-1} are related to the vibration of C-H. The band Co-OH is indicated by the vibration at 430 cm^{-1}. The band at 1626 cm^{-1} corresponds to H-O-H absorption, and that at 904 cm^{-1} is characteristic of the FeOOH (goethite) [19]. These results clearly indicate the functionalization of the cobalt ferrite nanoparticles with fucan and/or oleic acid.

Figure 3 shows the magnetic hysteresis loops of the cobalt ferrite nanoparticles without coating at three different temperatures (173, 243 and 300 K). No saturation is observed under the maximum applied magnetic field (75 kOe). In addition, from the inset in the figure, it is observed that the remanence and coercivity slightly increase when the temperature is decreased. Despite the ratio of increment is not linear

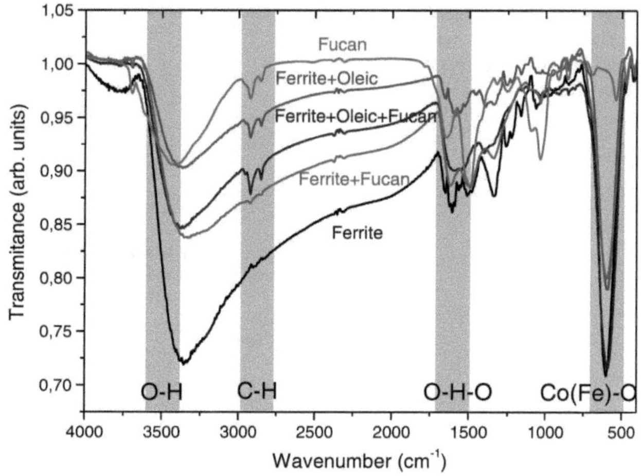

Fig. 2 FTIR of uncoated and coated cobalt ferrite nanoparticles

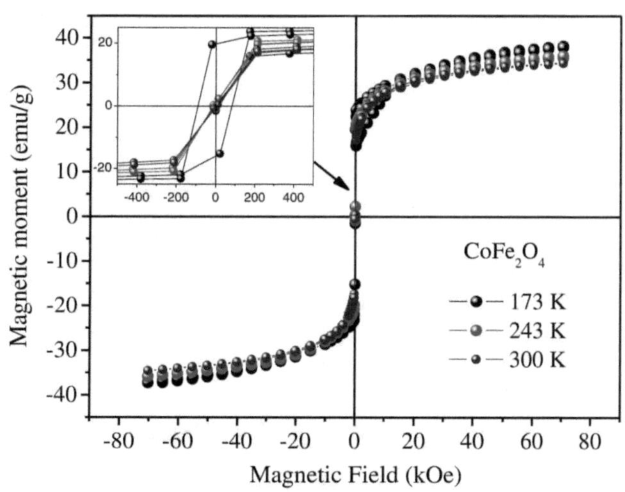

Fig. 3 Hysteresis loops of the cobalt ferrite nanoparticles at different temperatures

(especially in the case of the coercivity), we can infer from the last observations that the sample is superparamagnetic, which is typical for iron oxide based nanoparticles. The hysteresis shows remanence and coercive field at 173 K and not at 243 K and 300 K, indicating that the blocking temperature (T_B) should fall between 173 and 243 K. At temperatures above the T_B, the thermal effects allow rotation of the single domains to the easiest magnetization directions by getting over the energy barriers in zero field and consequently there is no coercivity. Recent studies also demonstrate the lack of coercivity in ferromagnetic microspheres in liquid suspension [20]. In that case the effect is originated by physical rotation of the magnetic spheres under applying a magnetic field.

It is unlikely that for applied magnetic fields above 75 kOe, the magnetization turns higher than 40 emu/g (3.25 μ_B/ formula unit). Moreover, considering that cobalt ferrites generally form an inverse spinel with the ion Co^{2+} replacing the Fe^{2+} in

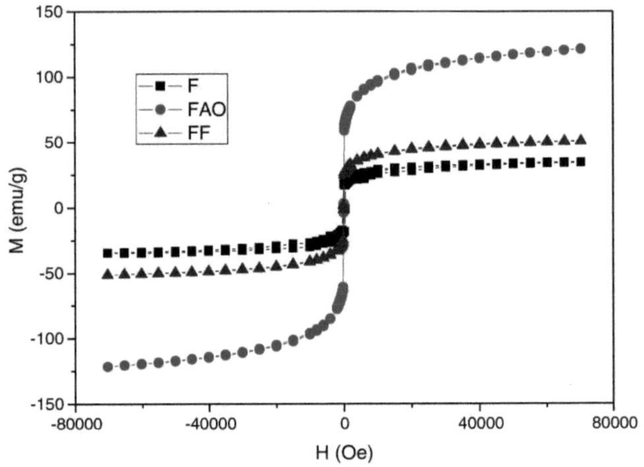

Fig. 4 M(H) loops of the ferrite (F), ferrite with oleic acid (FAO) and ferrite with fucan (FF)

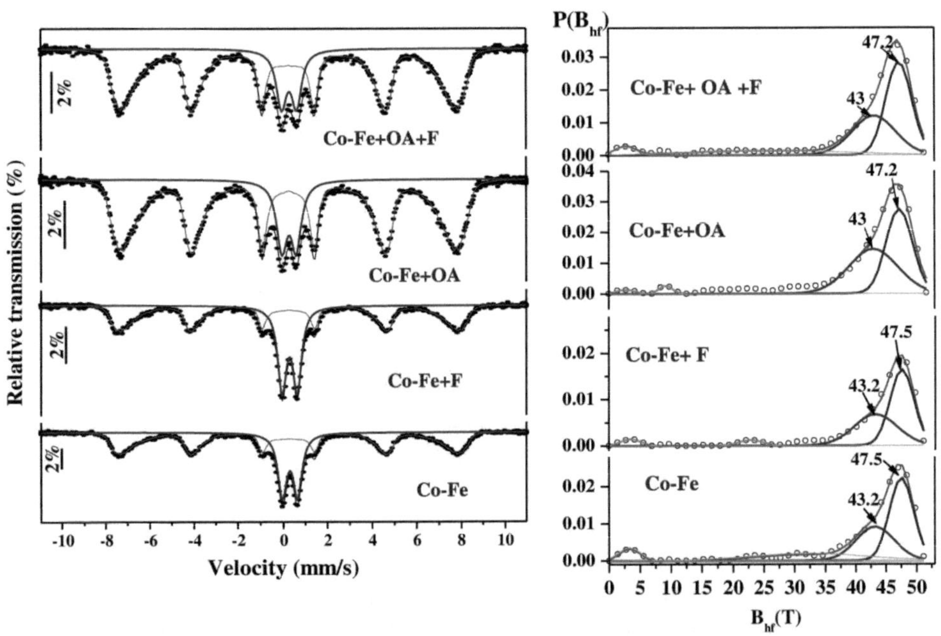

Fig. 5 MS of the cobalt ferrite with oleic acid and fucan (Co–Fe+OA+F), cobalt ferrite with oleic acid (Co–Fe+OA), ferrite with fucan (Co–Fe+F) and cobalt ferrite (Co–Fe). The corresponding hyperfine field distributions $P(B_{hf})$ and two average magnetic fields are shown on the right hand size

the site B of an octahedral structure and that this structure produces a theoretical magnetic moment of 3.0 μ_B/f.u. (at 0K), then our result (3.25 μ_B) is consistent with the theoretical value of 3.6 μ_B/f.u. calculated by Slonczewski [21] and Tachiki [22] in which they assumed that the structure of cobalt ferrite is not completely inverse spinel and the degree of the reversal depends on the heat treatment. Figure 4 shows the magnetic hysteresis loops of the coated cobalt ferrite nanoparticles measured at

Table 1 Mössbauer hyperfine parameters obtained at RT

Samples	δ (mm/s) $\pm$ 0.03	Δ (mm/s) $\pm$ 0.02[a]	B_{hf} (T)[b]	Area (%) $\pm$ 2
Cobalt ferrite	0.413	0.68	47.5–43.2	36.1
	0.370			63.9
Cobalt ferrite with fucan	0.408	0.68	47.5–43.2	40.3
	0.370			59.7
Cobalt ferrite with oleic acid	0.388	0.64	47.23–43.0	17.5
	0.370			82.6
Cobalt ferrite with oleic acid and fucan	0.405	0.67	47.2–43.0	17.1
	0.369			82.9

[a] Representing the superparamagnetic doublet

[b] Average magnetic fields are show in the right side of Fig. 5

δ is the isomer shift relative to α iron, Δ is the quadrupole splitting and A is the spectral fraction as obtained from the fit. B_{hf} is the hyperfine field in Tesla

room temperature. It is observed that the saturation magnetization of cobalt ferrite nanoparticles coated with oleic acid is far higher than the uncoated cobalt ferrite. This also occurs with the nanoparticles coated with fucan. This effect should be caused by magnetic disorder on the surface due to the effect of external media interactions (i.e interaction between particles). It is almost sure that fucan and oleic acid coatings weaknesses the interaction between particles, which in turns reduces the disorder of the spins on the surface, thus resulting in the increase of the magnetization.

The uncoated cobalt ferrite and fucan cobalt ferrite nanoparticles (with average size of 20 nm) have similar Mössbauer spectra at RT (Fig. 5). In the Mössbauer spectra of the coated with fucan polysacharides, the coating minimizes the interactions between particles which are reflected in a diminution of the hyperfine field and showing one sextet with broadened linewidth and one doublet. The fitting was made using an hyperfine field distribution $P(B_{hf})$ and one doublet. The spectrum is typical for magnetic relaxation of ferrites nanoparticles. In this case, the magnetic moment of an individual particle fluctuates through the easiest axis of magnetization during a time $\tau = \tau_0 \exp(KV/k_B T)$ where K is the magnetic anisotropy constant, V is the volume of the particle, k_B is the Boltzmann constant, T is the temperature and τ_0 is a constant characteristic of material. Therefore, for a fluctuation time longer than the characteristic Mössbauer time (10^{-8} s). A magnetically splitted spectra can be measured due to slow relaxation. Their Mössbauer hyperfine parameters are similar and they are listed in Table 1. The presence of the cobalt ferrite cores in each sample are confirmed by the sextets with hyperfine fields <47.7 T (two representative values are 47.5 and 43.2 T) with areas 63.9 % and 59.7 % for the uncoated and the coated particles, respectively. The doublets, which denote the superparamagnetic effect, have slightly different peak areas, 36.1 % for the uncoated and 40.3 % for the coated.

The Mössbauer spectra of the cobalt ferrite nanoparticles coated with oleic acid and oleic plus fucan, are very similar and the effect of oleic acid to produce a decrease in particle size, which on average are 16 nm and a decrease of the dipole-dipole interaction that is reflected in a slight decrease in magnetic field hyperfine giving a relaxation time more slow, manifested through a superparamagnetic doublet of lower intensity. What is striking is that the peak area or the superparamagnetism of the samples are reduced to half, with areas of 17.5 % ferrite with oleic acid and 17.1 % ferrite with oleic acid and fucan compared to the previous two samples.

4 Conclusions

Oleic acid prevents agglomeration of the cobalt ferrite nanoparticles with and without fucan coating. The presence of polysaccharides in the fucan coated cobalt ferrite nanoparticles was confirmed by X-ray diffractogram. The crystallinity of the cobalt ferrite nanoparticles are not affected by any of the types of coatings used in this work. Magnetization measurements of the cobalt ferrite nanoparticles with and without coating reveal superparamagnetism and this behavior is produced by the core nanoparticles rather than the coatings. The cobalt ferrite nanoparticles coated with oleic acid presented the highest magnetization than when coating with fucan. This might cause by the reordering of the spins located in the surface of the nanoparticles caused by their functionalization with oleic acid. Mössbauer spectroscopy confirms that the cobalt ferrite nanoparticles with and without coating have superparamagnetic relaxation. The nanoparticles covered with fucan has a faster relaxation time than those covered with oleic acid.

Acknowledgements This work was supported by the Brazilian science agencies Coordenação de Aperfeiçoamento de Pessoal de Nível Superior (CAPES), The Conselho Nacional de Desenvolvimento Cientifico e Tecnológico (CNPq) and the Fundação de Amparo à Ciência e tecnologia do estado de Pernambuco (FACEPE) (APQ-0589–1.05/08).

References

1. Praveena, K., Sadhana, K., Bharawaj, S., Murthy, S.R.: J. Magn. Magn. Mater. **321**, 2433 (2009)
2. Rajath, P.C., Manna, R.S., Banerjee, D., Varma, M.R., Suresh, K.G., Nigam, A.K.: J. Alloy. Compd. **453**, 155 (2006)
3. Tanaka, T., Shimazu, R., Nagai, H., Tada, M., Nakagawa, T., Shandhu, A., Handa, H., Abe, M.: J. Magn. Magn. Mater. **321**, 1417 (2009)
4. Lee, S.W., Bae, S., Takemura, Y., Shim, I.B., Kim, T.M., Kim, J., et al.: J. Magn. Magn. Mater. **310**, 2868 (2007)
5. Pita, M., Abad, J.M., Dominguez, C.V., Briones, C., Martí, E.M., Gago, J.A.M., Morales, M.P., Fernández, V.M.: J. Colloid Interface Sci. **321**, 484 (2008)
6. Boisson-Vidal, C., Haroun-Bouhedja, F., Ellouali, M., Blondin, C., Fischer, A.M., Agostini, A., et al.: Biological activities of polysaccharides from marine algae. Drugs Future **20**, 1237–1249 (1995)
7. Albuquerque, I.R.L., Queiroz, K.C.S., Alves, L.G., Santos, E.A., Leite, E.L., Rocha, H.A.O.: Heterofucans from Dictyota menstrualis have anticoagulant activity. Braz. J. Med. Biol. Res. **37**, 167–171 (2004)
8. Dos, J.G., Duque, S., Macedo, M.A., Moreno, N.O., Lopez, J.L., Pfanes, H.D.: J. Magn. Magn. Mater. **1424**, 226–230 (2001)
9. Panda, R.N., Shih, J.C., Chin, T.S.: J. Magn. Magn. Mater. **257**, 79 (2003)
10. Sartale, S.D., Lokhande, C.D.: Ceram. Int. **28**, 467 (2002)
11. Yan, C.H., Xu, Z.G., Cheng, F.X., Wang, Z.M., Sun, L.D., Liao, C.S., Jia, J.T.: Solid State Commun. **111**, 287 (1999)
12. Godinho, M.I., Catarino, M.A., da Silva Pereira, M.I., Mendonc, M.H., Costa, F.M.: Electrochim. Acta **47**, 4307 (2002)
13. Ding, J., Miao, W.M., McCormick, P.G., Street, R.: Appl. Phys. Lett. **65**, 3135 (1994)
14. Blaskov, V., Petkov, V.: J. Magn. Magn. Mater. **162**, 331 (1996)
15. El-Shobaky, G.A., Turky, A.M., Mostafa, N.Y., Mohamed, S.K.: J. Alloy. Compd. **493**, 415–422 (2010)
16. Carneiro Leao, A.M.A., Oliveira, E.A., Carvalho, L.B. Jr.: Appl. Biochem. Biotechnol. **32**, 53–58 (1991)
17. Cullity, B.D.: Elements of X-ray Diffraction, 3rd edn. Prentice-Hall, Prentice-Hall International, Upper Saddle River (2000)

18. Brand, R.A.: Normos Programs. Duisburg University (1989)
19. Luqueze, R. C.: Síntese e caracterização de nanopartículas magnéticas de ferrita de cobalto recobertas por 3-aminopropiltrietoxissilano para uso como material híbrido em nanotecnologia. Thesis (in Portuguese), Doutorado em Ciências na Área de Tecnologia Nuclear-Materiais, Instituto de Pesquisas Energéticas e Nucleares/USP, SP (2006)
20. de los Santos Valladares, L., Llandro, J., Lee, D., Mitrelias, T., Palfreyman, J., Hayward, T., Cooper, J., Bland, J.A.C., Barnes, C.H.W., Arroyo, J., Lees, M.: J. Magn. Magn. Mater. **321**, 2129–2134 (2009)
21. Slonczewski, J.L.: Phys. Rev. **110**, 1341 (1958)
22. Tachiki, M.: Prog. Theor. Phys. (Kyoto) **23**, 1055 (1960)

DOI 10.1007/978-94-007-6482-8_26
Reprinted from *Hyperfine Interactions*, DOI 10.1007/s10751-013-0875-9

Magnetic and Mössbauer studies of fucan-coated magnetite nanoparticles for application on antitumoral activity

V. A. J. Silva · P. L. Andrade · Angel Bustamante · L. de los Santos Valladares · M. Mejia · I. A. Souza · K. P. S. Cavalcanti · M. P. C. Silva · J. Albino Aguiar

Published online: 23 July 2013

Abstract Fucan-coated magnetite (Fe_3O_4) nanoparticles were synthesized by the co-precipitation method and studied by Mössbauer spectroscopy and magnetic measurements. The sizes of the nanoparticles were 8–9 nm. Magnetization measurements

Proceedings of the Thirteenth Latin American Conference on the Applications of the Mössbauer Effect, (LACAME 2012), Medellín, Columbia, 11–16 November 2012.

V. A. J. Silva · P. L. Andrade · J. Albino Aguiar
Programa de Pós-Graduação em Ciências de Materiais, CCEN, Universidade Federal de Pernambuco, Av. Prof. Moraes Rego s/n, Cidade Universitária, Recife, PE, CEP 50670-901, Brazil

V. A. J. Silva (✉) · P. L. Andrade · M. P. C. Silva
Laboratório de Imunopatologia Keizo Asami (LIKA), Departamento de Bioquímica, Universidade Federal de Pernambuco, Av. Prof. Moraes Rego s/n, Cidade Universitária, Recife, PE CEP 50670-901, Brazil
e-mail: valdeene@hotmail.com

A. Bustamante · M. Mejia
Laboratorio de Cerámicos y Nanomateriales, Facultad de Ciencias Físicas, Universidad Nacional Mayor de San Marcos, Ap. Postal 14-0149, Lima, Peru

L. de los Santos Valladares · J. Albino Aguiar
Departamento de Física, Universidade Federal de Pernambuco, Av. Prof. Moraes Rego s/n, Cidade Universitária, Recife, PE CEP 50670-901, Brazil

L. de los Santos Valladares
Cavendish Laboratory, Department of Physics, University of Cambridge, J.J. Thomson Avenue, Cambridge, CB3 0HE, UK

I. A. Souza
Departamento de Antibióticos, Universidade Federal de Pernambuco, Av. Prof. Moraes Rego s/n, Cidade Universitária, Recife, PE CEP 50670-901, Brazil

K. P. S. Cavalcanti
Hospital das Clínicas, Universidade Federal de Pernambuco, Av. Prof. Moraes Rego s/n, Cidade Universitária, Recife, PE CEP 50670-901, Brazil

and Mössbauer spectroscopy at 300 K revealed superparamagnetic behavior. The magnetic moment of the Fe_3O_4 is partly screened by the Fucan coating aggregation. When the magnetite nanoparticles are capped with oleic acid or fucan, reduced particle-particle interaction is observed by Mössbauer and TEM studies. The antitumoral activity of the fucan-coated nanoparticles were tested in Sarcoma 180, showing an effective reduction of the tumor size.

Keywords Magnetic nanoparticles · Magnetite · Fucan · Antitumoral activity

1 Introduction

Currently, magnetic nanoparticles (MNPs) are widely used in the fields of biology and medicine, such as protein and enzyme immobilization, bioseparation, immunoassay, hyperthermia, drug delivery, magnetically enhanced transfection, tissue engineering and magnetic resonance imaging (MRI) [1–5]. Magnetic nanoparticles are usually composed of magnetic elements, such as iron, nickel, cobalt and their respective oxides [6–8]. Iron oxides, such as magnetite (Fe_3O_4), are biodegradable, biocompatible, present stable magnetic response and superparamagnetic effects [9, 10]. Magnetite has recently attracted attention because it has a high Curie temperature (T_C ~850 K) and nearly full spin polarization at room temperature. Both properties are of great potential for applications in giant magnetoelectronic and spin-valve devices based on magnetite films [11]. Mössbauer and magnetic studies performed in magnetite confirms that the iron cations in Fe_3O_4 present two valence states, $Fe^{2.5+}$ and Fe^{3+} distributed in an inverse spinel structure (space group Fd_3m). Fe^{+3} cations locate on the tetrahedral co-ordination (FeO_4) site (A); and both Fe^{+3} and $Fe^{+2.5}$ locate on the octahedral co-ordination (FeO_6) site (B) in antiparallel arrangement, yielding ferrimagnetic order below T_C [2, 11].

Magnetite nanoparticles have reduced magnetic dipole-dipole interaction; hence, no magnetization is retained in these particles after removal of the external field. They are thus considered as superparamagnetic, and their magnetic properties are affected by coating them with different capping agents [12]. Several coating materials have been used to modify the surface chemistry of the magnetite nanoparticles, including organic polymers, organic surfactants, metals [13], oxides and bioactive molecules and structures [14]. One of the main problems in producing stable magnetic fluid consisting on individual coated magnetite nanoparticles is to prevent their agglomeration during the synthesis process [15]. In this sense, oleic acid is commonly used as surfactant to modify the surface of the magnetite particles since it has higher affinity to the surface compared to other surfactants [16].

Fucoidans are water soluble sulfated polysaccharides with high molecular weight. They are the main constituents of brown algae and can be found also in some marine invertebrates [17–20]. These complex polysaccharides show a wide variety of biological activities, such as anti-adhesive [21], anti-coagulant [22, 23], anti-complimentary [24], anti-oxidant [25], anti-proliferative [26], anti-thrombotic [27], anti-platelet aggregation [28], anti-tumor [29] and anti-viral properties [30, 31]. Thus, obtaining magnetite nanoparticles coated with polysaccharides is attractive for medical applications.

In the present work we study the influence of the Fucan coating on the magnetic properties of the magnetite nanoparticles by Mössbauer spectroscopy and magnetic measurements. The results are compared to similar works reported by other authors for magnetite coated with silica [32], oleic acid and polyethylene glycol [12, 33]. In addition, preliminary study of the antitumoral activity of coated magnetite nanoparticles is reported.

2 Methodology

2.1 Preparation of the fucan-magnetite nanoparticles

Algae *Sargassum cymosum* were dried, pulverized and immersed in 300 mL of acetone and stirred during 12 h to remove pigments and lipids. This process was repeated twice. Then, it was decanted and the residue was dried at 45 °C under aeration to obtain a Ketonic powder. Next, a solution of 0.15 M NaCl (250 mL) was added to the ketonic powder and the pH was adjusted to 8.0 with NaOH. The proteolysis of this ketonic powder was carried out by papain enzyme (15 mg/g of ketonic powder) and the mixture was incubated at 45 °C overnight. The supernatant was collected by centrifugation at 10,000 × g for 10 min and dried in a lyophilizer Multi-Tainer (FTS Systems, INC).

In parallel to the fucan extraction, an aqueous suspension of magnetite nanoparticles was prepared by co-precipitation of Fe(III) and Fe(II) in the presence of NH_4OH and oleic acid following standard procedure [34]. 10 mL of ferrous chloride, 10 mL of ferric chloride and 500 μL of oleic acid were added to 100 mL of distilled water under stirring. Then a NH_4OH solution was dropped to the mixture to raise the pH to 11.0 while maintaining a vigorous stirring. After this step, the mixture was heated at 85 ± 3 °C for 30 min while stirring it (7,000 rpm). The co-precipitated magnetite nanoparticles were then thoroughly washed with distilled water. The material was dried and kept at room temperature (25 °C). The coating of magnetite nanoparticles with the polysaccharide fucan was achieved by the adsorption method. The fucan solution in distilled water (50 mg/mL) was added to the obtained magnetite nanoparticles (100 mg) and maintained at 25 °C for 16 h under stirring. Afterward, the coated nanoparticles were thoroughly washed with distilled water. The material was dried and kept at room temperature (25 °C).

2.2 Characterization of the fucan-magnetite nanoparticles

The sizes of the MNPs were observed through a transmission electron microscopy (TEM, TECNAI G2 Spirit—FEI Company) with an acceleration voltage of 80 kV. For TEM analysis, the samples were deposited on copper grids of 400 mesh coated with carbon. The crystalline properties and phase identification were characterized by X-rays diffraction (XRD), using a Siemens D5000 diffractometer. The diffractogram was obtained by using Cu-Kα radiation ($\lambda = 1.5406$ Å) in the range $10° < 2\theta° < 80°$ with a step of 0.02° and acquisition time of 1.0 s/step.

The Mössbauer characterization was performed with a conventional transmission Mössbauer spectrometer, operating with 1,024 channels (after folding is 512 channels) and a Wissel INC. velocity module with a sinusoidal signal. The measurements were taken at room temperature (RT) and the obtained data were adjusted with the help of the program NORMOS, generating the data file with the extension PLT and determining the difference between the experimental data and the calculated data. In this program, the good fitting is controlled by the value of the χ^2. The source employed was a ^{57}Co in rhodium matrix with a of 25 mCi. The isomer shift and the velocity scale were calibrated with respect to a α-Fe film at RT. The sample holder used has a diameter of 1 cm (0.7854 cm^2) which permitted to ascertain and quantify the small systematic effects of cosine smearing which usually occur in the folded Mössbauer spectra when relatively large collection solid angles are used. These conditions were appropriate to obtain a rating of 8,500 counts per second.

The magnetic properties of the magnetite nanoparticles (MNPs) were measured on a MPMS-5S (Magnetic Property Measurement System) magnetometer from Quantum Design, with sensor SQUID (Superconducting Quantum Interference Device). About 5 mg of each sample were encapsulated and mounted in the equipment rod. The measurements were taken at room temperature (25 °C) and under different applied magnetic fields from −7 to +7 kOe, and the magnetization (in emu/g) was obtained by dividing the raw magnetization by the sample mass.

2.3 Antitumoral activity in vivo

Sarcoma 180 tumor cells were subcutaneously inoculated (3×10^6 cells/mouse) into 3 months old male Swiss mice. Subsequently, the fucan-coated and uncoated magnetite nanoparticles were dissolved in saline and injected intraperitoneally (ip) once a day for 7 days, starting at 48 h after tumor inoculation. The same volume of saline was injected into the control mice. Despite the fact that magnetite was reported to have high biocompatibility without toxicity [35] the uncoated magnetite nanoparticles were injected to a mouse to confirm it and to observe possible side effects. The mice were sacrificed on the next day after the last injection, and the tumors were excised. The tumor weights were compared with those in the control mice.

3 Results and discussion

In Fig. 1a, the uncoated MNPs showed the aggregate of particles occurred since the specific surface area (surface-to-volume ratio) is large resulting in a high surface energy. After fucan coating, the aggregation was reduced and the nanoparticle dispersion was improved (see Fig. 1b). The lack of aggregation is probably caused by the fucan coatings which weaknesses the magnetic interactions between particles.

From many TEM images, 261 and 728 uncoated and coated particles respectively, with nearly spherical shape, were counted to estimate the mean diameter using the ImagJ software. Subsequently a size histogram was mounted using the Sturges method. The bin-width (W) is obtained from the relation: $W = (D_{\max} - D_{\min})/k$, where $k = 1 + 3.322 \log(N)$. The particle size distribution was then modeled with a log-normal distribution: $f(D) = \left(A/\sqrt{2\pi\sigma \ D}\right) \exp\left[-\ln^2(D/D_0)/2\sigma^2\right]$. Here the D_0 is the statistical mean diameter, A is a constant to be determined by fitting the statistical distribution and σ represents the degree of polydisperion. The mean value and its standar deviation are given by $\langle D\rangle = D_0 e^{\sigma^2}/2$ and $\sigma_D = \langle D\rangle \sqrt{e^{\sigma^2} - 1}$, respectively. The histogram distribution of the size for the uncoated and coated

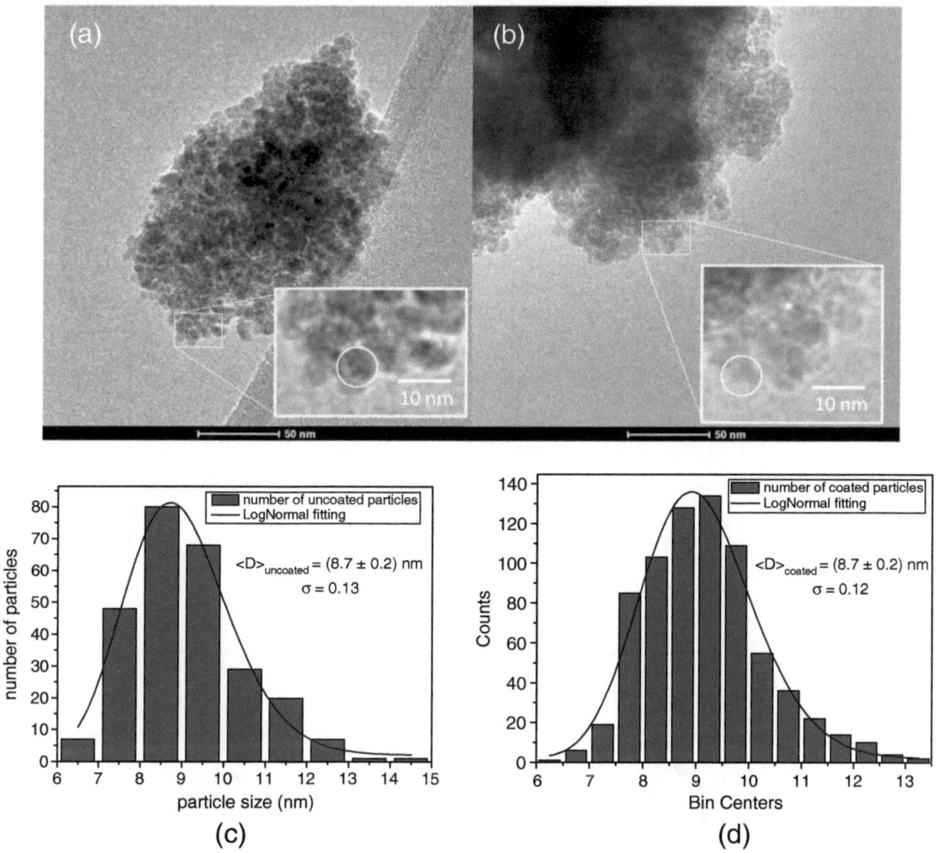

Fig. 1 **a** TEM micrograph of the uncoated MNPs, **b** TEM micrograph of fucan- coated MNPs and histogram of the size distributions of the uncoated **c** and Fucan-coated **d** Nanoparticles

nanoparticles are shown in Fig. 1c and d respectively. Table 1 list the statistical values obtained during fitting. Note that in the table, the coated nanoparticles are slightly bigger than the uncoated ones due to the presence of the Fucan.

Figure 2 shows the XRD of the uncoated magnetite nanoparticles, magnetite-oleic acid nanoparticles, magnetite-fucan nanoparticles and magnetite-oleic acid-fucan nanoparticles. The principal reflections (220), (311), (400), (422), (511) and (440) were indexed with the PDF card No 85-1436. From the line broadening of the diffraction peaks the particle size were determined using the Scherrer formula $D = \frac{0.916\lambda}{\beta_{\rm hkl}\cos\theta_{\rm hkl}}$ where $\lambda = 1.5406$ Å is the wavelength of the applied X-ray radiation ($K_{\alpha 1}$ radiation of copper), $\beta_{\rm hkl}$ is the pure diffraction line broadening (in radians), which can be easily found by measuring the full width at half maximum (FWHM) of the principal hkl reflections and $\theta_{\rm hkl}$ is the Bragg angle. In order to obtain more accurate estimations on the Scherrer formula, each peak was fitted with Gaussian functions and levelling the background in each diffractogram to zero. For example, Fig. 2b and c show the (311) reflections of the uncoated and Fucan-coated samples. Note that the diffractogram for the coated sample is more dispersed due to the

Table 1 Statistical values obtained after fitting the histogram distribution of the N = 261 and 728 uncoated and coated nanoparticles respectively taken from the TEM micrographs

	N	<*D*> (nm)	D_0	Σ	*A*
Uncoated	261	8.7 ± 0.2	8.8	0.13	242.21
Coated	728	8.9 ± 0.2	9.0	0.11	348.98

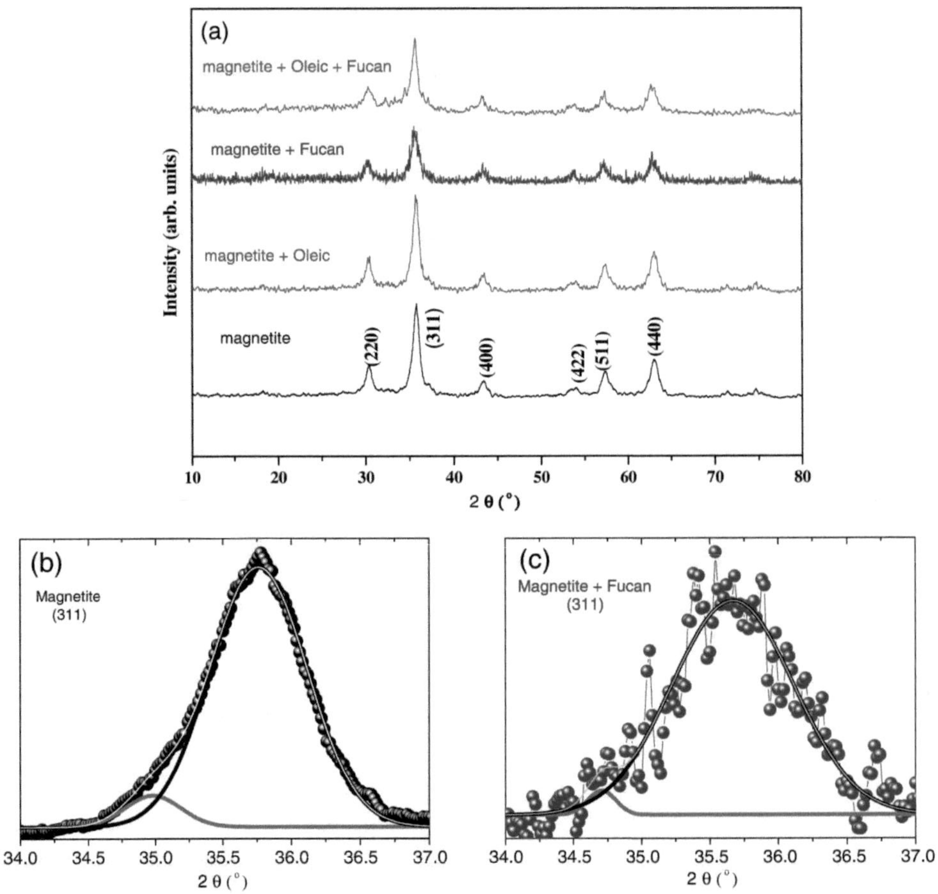

Fig. 2 **a** X-ray diffraction patterns of the Fe_3O_4 samples with different shells, **b** (311) peak for magnetite **c** (311) peak for magnetite-fucan

presence of the organic shells. Table 2 show the calculated average sizes for the four samples. The average sizes for the uncoated and Fucan-coated particles are 8.5 and 8.7 nm respectively, which are very close to those values estimated in the TEM images above.

The measurements of the magnetization with the applied magnetic field, M(H), of the uncoated and fucan-coated magnetite nanoparticles at 300 K are presented in Fig. 3. The magnetization values were obtained by dividing the raw data provided by the DC-MPMS magnetometer by the sample mass, thus in the case of the fucan-coated nanoparticles, these values correspond to the raw data divided by the total

Table 2 Average size of the crystallites obtained from the principal peaks on the XRD of the uncoated and coated magnetite particles

Hkl	Magnetite grain size (nm)	Magnetite + oleic acid grain size (nm)	Mag + fucan grain size (nm)	Mg + F+ oleic acid grain size (nm)
220	8.4	8.9	8.6	9.1
311	8.7	9.9	9.0	13.4
400	8.3	8.0	8.6	9.6
511	8.4	7.8	8.6	10.5
440	8.5	10.2	8.5	9.3
Mean size (nm)	8.5	8.9	8.7	10.4

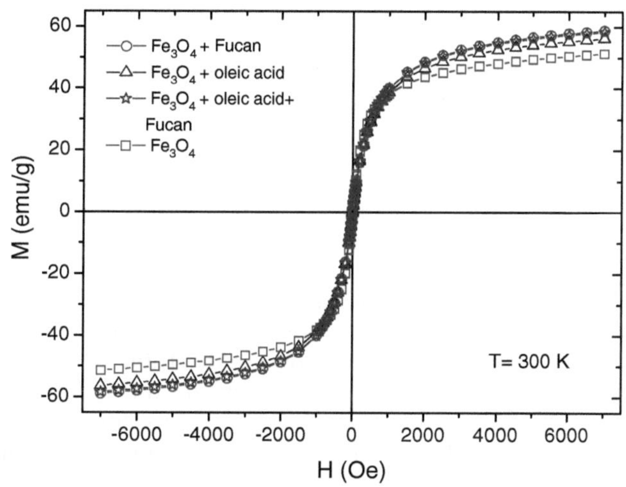

Fig. 3 Magnetization (M) curves as the function of the applied magnetic field (H) of the Fe_3O_4 nanoparticles

mass (about 5 mg). In this way, the mass of the core magnetite in each sample are slightly different. Since we can not estimate the mass of the organic material, the data provided for the coated particles in the figure should be considered only as a first approximation.

The plots are typical from superparamagnetic iron nanoparticles with almost zero remanence and coercivity [11]. The superparamagnetic behavior is mainly generated by the magnetite cores in each sample. At 300 K, the saturation magnetization (M_S), determined by using thc law of approach to saturation (a linear term is included in order to account for the linear increase of M at high fields), is around 52 emu/g for the uncoated Fe_3O_4 NPs. This saturation magnetization value is consistent with the value reported in the literature for uncoated magnetite NPs with sizes smaller than 10 nm [11]. The saturation value for the uncoated sample should be considered as a first approximation (see above).

Figure 4a shows the Mössbauer spectra at RT of the uncoated magnetite nanoparticles and its correspondent hyperfine magnetic field distribution. The spectrum of the magnetite nanoparticles show one sextet with broadened line width and it was fitted with one hyperfine magnetic field distributions with $\chi^2 \sim 1.47$, showing a principal most probably magnetic field of 46.10 T due to magnetic relaxation of magnetite nanoparticles. In this case, the magnetic moment of an individual particle fluctuates through the most easy axis of magnetization during a time $\tau = \tau_0 \exp(KV/k_BT)$,

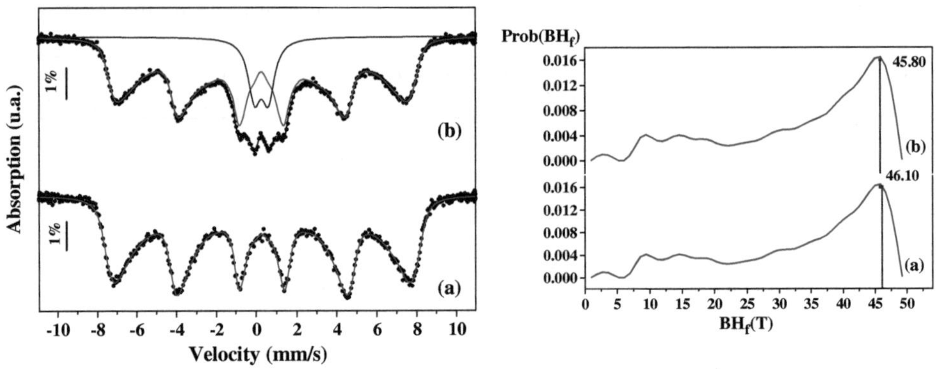

Fig. 4 Mossbauer spectra at RT. **a** Uncoated nano-magnetite particles and **b** Fucan coated nano-magnetite. The hyperfine magnetic field distributions are on the *right* panel

where K is the magnetic anisotropy constant, V is the volume of the particle, k_B is the Boltzmann constant, T is the temperature and τ_0 is a constant characteristic of the material. Therefore, for a fluctuation time longer than the characteristic Mössbauer time (10^{-8} s), magnetically splitted spectra can be measured due slow relaxation. The hyperfine parameters are shown in the Table 3.

Figure 4b show the Mössbauer spectra at RT of the magnetite nanoparticles coated with fucan polysacharides and its corresponding hyperfine magnetic field distribution. The spectrum was fitted in first approximation using one distribution and one doublet with $\chi^2 \sim 1.38$, showing an asymmetrically broadened spectra. This is originated by a complex mixture of static and dynamic broadenings due to the broad size distribution and the strength of interparticle interactions with slow particles (sextet with area of 87.1 %) and rapid particles (the doublet SP with area of 12.9 %). In this way, a reduced magnetic dipole-dipole interaction is produced by the coated fucan as observed in magnetization measurements (Fig. 3) and also in other systems [34]. The hyperfine magnetic field distribution shows a principal magnetic field most probably of 45.80 T.

Figure 5 shows the Mössbauer spectra at RT of the oleic acid-magnetite nanoparticles and fucan-oleic acid- magnetite nanoparticles. In the case of the Mössbauer spectra of the magnetite coated with oleic acid + fucan polysacharides, the coating minimizes the interactions between particles which are reflected in a diminution of the hyperfine field, showing a superposition of one hyperfine field distribution and one doublet, with a principal magnetic field most probably of 13.3 T and area of 71.3 % due to magnetic relaxation of magnetite nanoparticles. The doublet would be magnetic collective excitations or superparamagnetic signal with $QS = 0.68$ mm/s, $\delta = 0.389$ mm/s and 28.7 %.

While the Mössbauer spectra of the magnetite coated with oleic acid, the spectrum is typical of magnetic nanoparticles with weak dipole-dipole interaction. The presence of a doublet of 17.7 % area in contrast to the hyperfine field distribution $P(B_{hf})$ of 82.3 % area indicates the superparamagnetic effect has intensified. The Mössbauer spectrum of the magnetite nanoparticles coated with fucan and oleic acid is similar to the previous one. However in this case it seems that the magnetic distribution of the

Table 3 Mössbauer hyperfine parameters obtained at RT

Samples - sites	δ (mm/s)	ε, Δ (mm/s)[a]	B_{hf} (Tesla)	Area (%)
Magnetite Dist. Fe^{3+}-$Fe^{2.5+}$	0.391	−0.01	46.10	100.0
Magnetite with fucan Dist				
Fe^{3+}-$Fe^{2.5+}$	0.380	0.00	45.80	87.1
Doublet SP	0.390	0.67	–	12.9
Magnetite with oleic acid				
Dist	0.403	−0.01	44.8	82.3
Doublet SP	0.395	0.63	–	17.7
Magnetite with oleic acid				
and fucan Dist	0.386	−0.02	13.3	71.3
Doublet SP	0.389	0.68	–	28.7

[a]Representing the superparamagnetic doublet
δ is the isomer shift relative to α-iron, ε and Δ is the quadrupole splitting and A is the spectral fraction as obtained from the fit. B_{hf} is the hyperfine field. *SP* Superparamagnetic

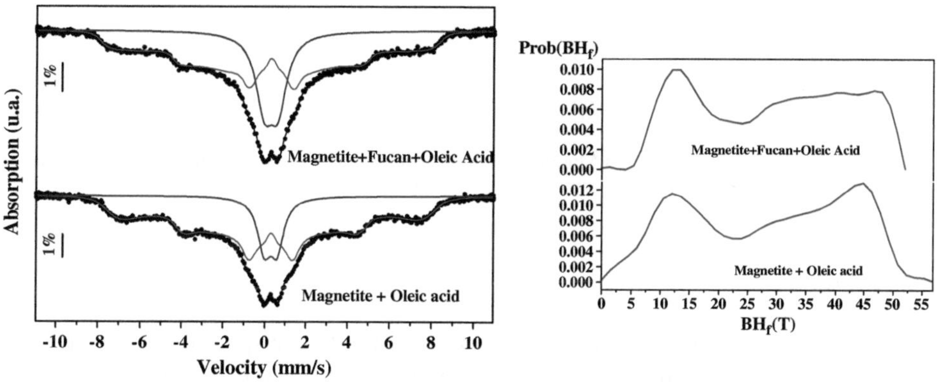

Fig. 5 Mössbauer spectra at RT for oleic acid-magnetite nanoparticles and fucan-oleic acid- magnetite nanoparticles. *Solid lines* are the best fits using hyperfine field distributions $P(B_{hf})$ shown on the *right panel*

surface affects the anisotropy energy and the doublet increases in area (to 28.7 %) whereas the hyperfinc field distribution $P(B_{hf})$ area is 71.3 %

The magnetite-fucan nanoparticles were applied in vivo to analyse the antitumoral activity of Sarcoma 180. The samples of pure fucan and a control group with only saline solution were also applied. In Fig. 6, we can observed the Sarcoma 180 tumor after treated by saline as control (A), by magnetite nanoparticles (B) and by fucan-magnetite nanoparticles (C). It is observed that the magnetite nanoparticles not presented antitumoral effect (Fig. 6b), but in Fig. 6c he tumor size is reduced, demonstrating that fucan-magnetite nanoparticles are effective in the antitumoral treatment for Sarcoma 180. The animals that were treated with pure fucan injection, died within the 24 h after the application. Our research group is currently performing more tests in vivo and in vitro in order to understand better what happened. However, up to date is is assumed that the pure fucan killed the animals due a hyperestimulation immunology.

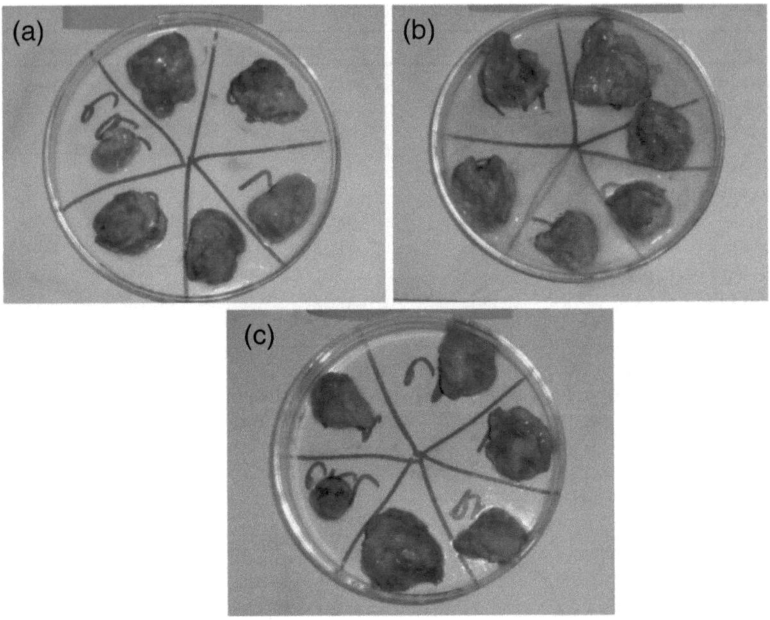

Fig. 6 In vivo effect of **a** Control, **b** Magnetite nanoparticles and **c** Fucan-magnetite nanoparticles in Sarcoma 180

4 Conclusions

In summary, the magnetic measurements and Mössbauer spectroscopy at RT revealed that the superparamagnetic property of the Fe_3O_4 nanoparticles is affected by the fucan polysaccharide and oleic acid coating, thus preventing aggregation. The fucan-magnetite nanoparticles are promising for the treatment of the Sarcoma 180 tumor size.

Acknowledgements This work was supported by the Brazilian science agencies Coordenação de Aperfeiçoamento de Pessoal de Nível Superior (CAPES), Conselho Nacional de Desenvolvimento Científico e Tecnólogico (CNPq) and the Fundação de Amparo à Ciência e Tecnologia do Estado de Pernambuco (FACEPE) (APQ-0589-1.05/08).

References

1. Zhang, L.Y., Zhu, X.J., Sun, H.W., Chi, G.R., Xu, J.X., Sun, Y.L.: Control synthesis of magnetic Fe_3O_4–chitosan nanoparticles under UV irradiation in aqueous system. Curr. Appl. Phys. **10**, 828–833 (2010)
2. Zhan, Y., Meng, F., Yang, X., Zhao, R., Liu, X.: Solvothermal synthesis and characterization of functionalized graphene sheets (FGSs)/magnetite hybrids. Mater. Sci. Eng. B. **176**, 333–1339 (2011)
3. Corchero, J., Villaverde, A.: Biomedical applications of distally controlled magnetic nanoparticles. Trends Biotechnol. **27**, 468–76 (2009)
4. Rahimi, M., Wadajkar, A., Subramanian, K., Yousef, M., Cui, W., Hsieh, J., Nguyen, K.: In vitro evaluation of novel polymer-coated magnetic nanoparticles Q13 for controlled drug delivery. Nanomedicine **6**, 672–680 (2010)

5. de los Santos Valladares, L., Llandro, J., Lee, D., Mitrelias, T., Palfreyman, J.J., Hayward, T.J., Cooper, J., Bland, J.A.C., Barnes, C.H.W., Arroyo, J.L., Lees, M.: Magnetic measurements of suspended functionalised ferromagnetic beads under DC applied fields. J. Magn. Magn. Mater. **321**, 2129–2134 (2009)
6. Shubayev, V.I., Pisanic, T.R., Jin, S.: Magnetic nanoparticles for theragnostics. Adv. Drug Deliv. Rev. **61**, 467–477 (2009)
7. Andrade, P.L., Silva, V.A.J., Maciel, J.C., Santillan, M.M., Moreno, N.O., De Los Santos Valladares, L., Bustamante Domínguez, A., Pereira, S.M.B., Silva, M.P.C., Albino Aguiar, J.: Preparation and characterization of cobalt ferrite nanoparticles coated with fucan and oleic acid. Hyperfine Interact. doi:10.1007/s10751-013-0835-4
8. Ramos, J.A., Bustamante Domínguez, A., Flores Santibañez, J., Mejia, M., Osorio Anaya, A., Martínez, A.I., De Los Santos Valladares, L., Barnes, C.H.W.: Mössbauer study of intermédiate superparamagnetic relaxation of maghemite (γ-Fe2O3) nanoparticles. Hyperfine Interact. doi:10.1007/s10751-013-0864-z
9. Jain, T.K., Richey, J., Strand, M., Leslie-Pelecky, D.L., Flask, C.A., Labhasetwar, V.: Magnetic nanoparticles with dual functional properties: drug delivery and magnetic resonance imaging. Biomaterials **29**, 4012–4021 (2008)
10. Kumar, A., Prasanna, K.J., Behera, S., Lockey, R.F., Monapatra, S.M., Shy, A.M.: Multifunctional magnetic nanoparticles for targeted delivery. Nanomedicine **6**, 64–69 (2010)
11. Goya, G.F., Berquó, T.S., Fonseca, F.C., Morales, M.P.: Static and dynamic magnetic properties of spherical magnetite nanoparticles. J. Appl. Phys. **94**, 3520–3528 (2003)
12. Mikhaylova, M., Kim, D.K., Bobrysheva, N., Osmolowsky, M., Semenov, V., Tsakalakos, T., Muhammed, M.: Superparamagnetism of magnetite nanoparticles: dependence on surface modification. Langmuir **20**, 2472–2477 (2004)
13. León Félix, L., Chaker, J., Parise, M., Coaquira, J.A.H., de los Santos Valladares, L., Bustamante, A., Garg, V.K., Oliveira, A.C., Morais, P.C.: Synthesis and characterization of uncoated and gold-coated magnetite nanoparticles. Hyperfirne Interact. doi:10.1007/s10751-013-0857-y
14. Cabrera, L., Gutierrez, S., Menendez, N., Morales, M.P., Herrasti, P.: Magnetite nanoparticles: electrochemical synthesis and characterization. Electrochim. Acta **53**, 3436–3441 (2008)
15. Kim, D.K., Zhang, Y., Voit, W., Rao, K.V., Muhammed, M.: Synthesis and characterization of surfactant-coated superparamagnetic monodispersed iron oxide nanoparticles. J. Magn. Magn. Mater. **225**, 30–36 (2001)
16. Yang, K., Peng, H., Wen, Y., Li, N.: Re-examination of characteristic FTIR spectrum of secondary layer in bilayer oleic acid-coated Fe3O4 nanoparticles. Appl. Surf. Sci. **256**, 3093–3097 (2010)
17. Araujo, P.M., Oliveira, G.B., Cordula, C.R., Leite, E.L., Carvalho, L.B., Silva, M.P.C.: Sulfated fucan as support for antibiotic immobilization. Braz. J. Med. Biol. Res. **37**, 301–305 (2004)
18. Yoon, S.J., Pyun, Y.R., Wang, J.K.H., Moura, P.A.S.: A sulfated fucan from the brown alga Laminaria cichorioides has mainly heparin cofactor II dependent anticoagulant activity. Carbohydr. Res. **342**, 2326–2330 (2007)
19. Queiroz, K.C.S., Medeiros, V.P., Queiroz, L.S., Abreu, L.R.D., Rocha, H.A.O., Ferreira, C.V., Juca, M.B., Aoyama, H., Leite, E.L.: Inhibition of reverse transcriptase activity of HIV by polysaccharides of brown algae. Biomed. Pharmacother. **62**, 303–30 (2008)
20. Karmakar, P., Ghosh, T., Sinha, S., Saha, S., Mandal, P., Ghosal, P.K., Ray, B.: Polysaccharides from the brown seaweed Padina tetrastromatica: characterization of a sulfated fucan. Carbohyd. Polym. **78**, 416–421 (2009)
21. McCormick, C.J., Newbold, C.I., Berendt, A.R.: Sulfated glycoconjugatesenhance CD36-dependent adhesion of Plasmodium falciparum-infectederythrocytes to human microvascular endothelial cell. Blood **96**, 327–333 (2000)
22. Cumashi, A., Ushakova, N.A., Preobrazhenskaya, M.E., D'Incecco, A., Piccoli, A., Totani, L.: A comparative study of the anti-inflammatory, anticoagulant, anti-angiogenic and anti-adhesive activities of nine different fucoidans from brown seaweeds. Glycobiology **17**, 541–552 (2007)
23. Mourão, P.A.S.: Use of sulfated fucans as anticoagulant and antithrombicagents: future perspective. Curr. Pharm. Des. **10**, 967–981 (2004)
24. Tissot, B., Daniel, R.: Biological properties of sulfated fucans: the potent inhibiting activity of algal fucoidan against the human complement system. Glycobiology **13**, 29G–31G (2003)
25. Chew, Y.L., Lim, Y.Y., Omar, M., Khoo, K.S.: Antioxidant activity of three edible seaweeds from two areas in South East Asia. LWT-Food Sci. Technol. **41**, 1067–1072 (2008)

26. Patel, M.K., Mulloy, B., Gallagher, K.L., O'Brien, L., Hughes, A.D.: The antimitogenic action of the sulfated polysaccharide sulfated fucan differs from heparin in human vascular smooth muscle cells. Thromb. Haemost. **87**, 149–154 (2002)
27. Nishino, T., Fukuda, A., Nagumo, T., Fujihara, M., Kaji, E.: Inhibition of the generation of thrombin and factor Xa by a sulfated fucan from the brown seaweed Ecklonia kurome. Thromb. Res. **96**, 37–49 (1999)
28. Alwayn, I.P., Appel, J.Z., Goepfert, C., Buhler, L., Cooper, D.K., Robson, S.C.: Inhibition of platelet aggregation in baboons: therapeutic implications for xenotransplantation. Xenotransplantation **7**, 247–257 (2000)
29. Alekseyenko, T.V., Zhanayeva, S.Y., Venediktova, A.A., Zvyagintseva, T.N., Kuznetsova, T.A., Besednova, N.N.: Antitumor and antimetastatic activity of fucoidan, a sulfated polysaccharide isolated from the Okhotsk sea Fucus evanescens brown alga. Bull. Exp. Biol. Med. **147**, 730–732 (2007)
30. Ghosh, T., Chattopadhyay, K., Marschall, M., Karmakar, P., Mandal, P., Ray, B.: Focus on antivirally active sulfated polysaccharides: from structure activity analysis to clinical evaluation. Glycobiology **19**, 2–15 (2009)
31. Chandia, N.P., Matsuhiro, B.: Characterization of a fucoidan from Lessonia vadosa (Phaeophyta) and its anticoagulant and elicitor properties. Int. J. Biol. Macromol. **42**, 235–240 (2008)
32. Ferreira, R.V., Pereira, I.L.S., Cavalcante, L.C.D., Gamarra, L.F., Carneiro, S.M., Amaro Jr., E., Fabris, J.D., Domingues, R.Z., Andrade, A.L.: Synthesis and characterization of silica-coated nanoparticles of magnetite. Hypefine Interact. **195**, 265–274 (2010)
33. Ghosh, R., Pradhan, L., Devi, Y.P., Meena, S.S., Tewari, R., Amit, A.K., Sharma, S., Gajbhiye, N.S., Vatsa, R.K., Pandey, B.N., Ningthoujam, R.S.: Induction heating studies of Fe_3O_4 magnetic nanoparticles capped with oleic acid and polyethylene glycol for hyperthermia. J. Mater. Chem. **21**, 13388 (2011)
34. Carneiro Leão, A.M.A., Oliveira, E.A., Carvalho Jr., L.B.: Immobilization of protein on ferromagnetic dacron. Appl. Biochem. Biotechnol. **32**, 53–58 (1991)
35. Theerdhala, S., Bahadur, D., Vita, S., Perkas, N., Zhong, Z., Gedanken, A.: Sonochemical stabilization of ultrafine colloidal biocompatible magnetite nanoparticles using amino acid, L-arginine, for possible bio applications. Ultrason. Sonochem. **17**, 730–737 (2010)

DOI 10.1007/978-94-007-6482-8_27
Reprinted from *Hyperfine Interactions*, DOI 10.1007/s10751-013-0830-9

Structural and hyperfine properties of Mn and Co-incorporated akaganeites

Ana E. Tufo · Karen E. García · Cesar A. Barrero · Elsa E. Sileo

Published online: 7 March 2013

Abstract The structural and hyperfine properties of pure and substituted akaganeites prepared in the presence of Mn, Co and urea are presented and discussed. In all samples, the chloride content increased with the increase in the urea concentration of the parent solution, and a small Mn-for-Fe or Co-for-Fe substitution occurred. In pure akaganeites, the increase of urea concentration provoked an enlargement of the unit cell volume and a decrease of the crystallinity of the synthesised oxides. The incorporation of Mn and Co provoked changes in cell parameters and an increase in the crystallinity of the samples. The hyperfine parameters for both iron sites of the akaganeites remained practically unchanged, and the spectral areas of the iron sites located close to the chlorides decreased for the doped samples. The recoilless f-factor increased for the substituted akaganeites, indicating an increase in the strength of the atomic bonding of the iron ions.

Keywords Akaganeite · Substitution · Mössbauer spectroscopy · X-ray diffraction · Recoilless f-factor

Proceedings of the Thirteenth Latin American Conference on the Applications of the Mössbauer Effect, (LACAME 2012), Medellín, Columbia, 11–16 November 2012.

A. E. Tufo · E. E. Sileo (✉)
INQUIMAE y Facultad de Ciencias Exactas y Naturales, Ciudad Universitaria, Pabellón II, Piso 3, CABA, Universidad de Buenos Aires, Buenos Aires, Argentina
e-mail: sileo@qi.fcen.uba.ar

K. E. García · C. A. Barrero
Grupo de Estado Sólido, Facultad de Ciencias Exactas y Naturales y Sede de Investigación Universitaria, Universidad de Antioquia, Medellín, Colombia

K. E. García
e-mail: kgarciat05@yahoo.com

1 Introduction

Akaganeite (β-FeOOH) is an uncommon mineral in nature, which is mainly found in soils and geothermal brine deposits as an oxidation product of pyrite [1], and as a corrosion product of natural, non-meteoritic minerals [2]. The oxyhydroxide is often detected on iron objects exposed to chloride-containing marine environments [3], and has also been detected as a corrosion product of iron meteorites [4].

Post and Buchwald [5], have studied the crystal structure of the akaganeite formed on an iron meteorite and confirmed that the structure is monoclinic (I2/m), and not tetragonal as previously reported by Keller [6]. The solid is formed by four double chains of edge-linked Fe(III)(O,OH)$_3$ octahedron that shares corners, forming a tunnel hollandite-like (Ba_8(Mn(IV), Mn(II)$_8O_{16}$) arrangement, that runs along [010]. The Fe-O octahedron in akaganeite is distorted with Fe-(O,OH) distances ranging from 1.94 to 2.13 Å and shows three longer and three shorter Fe-O distances; as expected the longer distances are associated with the OH^- anions. Depending on the nature of the precursor, the tunnel is partly occupied by chloride or similar anions such us OH^- or F^- [7, 8] that are believed to be essential to maintain the structure. The anions in the tunnel may be exchanged [9]. Post et al. [10] using neutron powder diffraction data have shown that weak hydrogen bonds are formed between the framework O atoms and Cl ions, and that the Cl^- fills approximately 2/3 of the tunnel sites, suggesting an ordering scheme in a given tunnel with every third tunnel site vacant.

When akaganeite is synthesized in acidic solutions, extra protonated oxide ligands are necessary to balance the Cl anions in the channels or are necessary to balance the charge of substituting lower-valence cations into the octahedral sites. Stahl et al. [11] have proposed that β-FeOOH can be represented by the formula ($FeO_{1-x}(OH)_{1+x}Cl_x$), and showed that the structure did not contain H_2O molecules.

Akaganeite is an antiferromagnetic material with a Néel temperature of 299 K, and its Mössbauer spectra have attracted a lot of attention because of their complexity in both the paramagnetic and magnetically-ordered states [12]. At the magnetic ordering temperature, the Mössbauer spectra changes from a broad doublet to a sextet, and a careful analysis shows that the broad doublet consists of two doublets rather than just one broadened doublet due to high site distortion. In the magnetically-ordered state, the spectra have an asymmetric appearance, and at least three sextets with differing magnetic hyperfine fields and quadrupole splittings, are needed to fit the spectra. Barrero et al. [13] have found that the two crystallographic sites required by the monoclinic symmetry are only discernible in the low temperature magnetic region. At room temperature, in the paramagnetic state, the spectrum may be fitted with two doublets whose origin is related to the chlorine content, i.e. one doublet is assigned to the Fe(III) ions located close to chloride ions, and the other doublet is assigned to the iron ions located close to chloride vacancy sites. The low temperature magnetic spectra can be adequately fitted with four sextets, whose hyperfine parameters must be subjected to some constraints. The origin of these components is related to the two different crystallographic sites and to the chlorine content. One important parameter which defines the usefulness of the Mössbauer transition is the fraction of gammas which are emitted and absorbed without recoil, i.e. the recoilless f-fraction [14]. The calculation of the f-fraction is very important for complete quantitative characterization of a given sample by Mössbauer spectrometry. Among other things, this parameter can be related with the strength of the atomic bonding, in such a way that at a given temperature a sample with a high f-fraction could be related with high iron bonding. The f-fraction also allows the determination of the relative atomic or weight fractions of each iron site in a single phase or of each iron phase present in mixed iron phase [15]. Only few articles have reported f-fractions of pure akaganeite [15–17] but to the best of our knowledge there are no reports of f values for doped akaganeites. This work is still to be done.

On the other hand, the substitution of foreign elements in the crystal structure of akaganeite can strongly affect the magnetic hyperfine fields. In natural environments Fe-for-Al substitution is often found because Al is more common in the earth's crust than Fe, and the best documented substitution is that of Al^{3+} for Fe^{3+}. Cations substitution have been reported for natural and synthetic oxyhydroxides, particularly, goethite [18]. The substitution of Fe(III) for a foreign metal cation usually affects, among others, the unit cell parameters, the chemical reactivity, the crystallinity and the superficial properties of the substituted oxides. If the foreign cation is diamagnetic, the substitution constitutes a strong perturbation of the magnetic properties of Fe oxides, lowering the ordering temperatures and reducing the hyperfine fields at all temperatures. However the magnetic hyperfine fields react similarly to other perturbations: thus small particle size has a similar effect to diamagnetic substitution on the field reduction. Ishikawa et al. [19] have synthesized akaganeite in the presence of Cu(II), Ti(IV), Ni(II), and Cr(III) ions and found that Ti drastically affects the particle size and the crystallite size of akaganeite in comparison to the other cations. A similar result for Ti(IV) was later confirmed [20]. Also the synthesis of β-FeOOH in solutions containing urea and Al(III), Cr(III), and Cu(II) ions (concentration 10 mol%) have been studied by García et al. [21] and the authors reported that the formed akaganeite presented only subtle differences in the crystallographic and hyperfine properties. However when Al(III) (30 mol%) was used, the average crystallite was significantly reduced. The incorporation of Si in akaganeite was studied by Cornell [22] and it was found that the oxyhydroxide incorporated up to 4 mole% Si, and that the presence of Si strongly promoted twinning of akaganeite. The dissolution data indicated that 20 % of the incorporated Si is located on the surfaces of the crystals and the remainder is distributed uniformly in the crystal structure. Akaganeite has been prepared in the presence of manganese with different valence states [9]; the addition of manganese appears to impede crystallization at low pH but improves it at high pH. Holm [23] has synthesized akaganeite in the presence of Cr, Mn, Co, Ni and Zn (50 mol%), and also reported small incorporations of foreign cations.

As a part of a general study about the feasibility of substitution on the structure of akaganeite at different pH values, and in order to determine if the substitution can be derived from Mössbauer spectroscopy measures, we now report the structural and hyperfine characteristics of a series of pure and incorporated akaganeites formed in the presence of Co(II) and Mn(II) ions, and different urea concentrations.

2 Materials and methods

Two samples of cobalt doped β-FeOOH were prepared by dissolving in bidistilled water (2000 mL), $FeCl_3.6H_2O$ (53.80 g, 0.198 mol), $CoCl_2.6H_2O$ (5.96 g, 0.025 mol), and urea $(NH_2)_2CO$ (12.00 or 48.00 g, 0.20 or 0.80 mol). Similarly, two Mn-doped samples were prepared using $MnCl_2.4H_2O$ (3.96 g, 0.016 mol). The four solutions obtained were aged for 48 h at 70 °C in closed polyethylene flasks. The solids were recovered from the suspensions by dialysis through high-purity membranes. The dialysis was carried until the conductivity of the solutions was similar to that of bidistilled water. The suspensions were dried at 40 °C for 48 h. The solids were named as Aka-Co-0.10, Aka-Co-0.40, Aka-Mn-0.10, and Aka-Mn-0.40. Two samples of pure akaganeites were also prepared following the same procedure by dissolving $FeCl_3.6H_2O$ (53.80 g, 0.198 mol), and urea (12.00 or 48.00 g, 0.20 or 0.80 mol) in bidistillated water (final volume 2000 mL). A third sample of pure akaganeite was also prepared, following Cornell and Schwertmann [18], in absence of urea. Samples were named Aka-0.10, Aka-0.40, and Aka-0.

The Fe content in the samples was determined spectrophotometrically using the thioglycolic acid method [24], Mn and Co content was determined by absorption spectrometry, and Cl content was determined by semi-quantitative energy dispersive X-ray analysis (EDS) which was conducted with an Oxford Instruments INCA system. X-ray diffraction patterns were recorded in a Siemens D5000 diffractometer using graphite-monocromated Cu $K\alpha$ radiation. The data were analysed using the GSAS [25] system with the EXPGUI interface [26]. The whole X-ray diffraction patterns were used to refine the crystal structures. Starting unit-cell parameters and atomic coordinates for akaganeite were taken from the literature [11], and monoclinic akaganeite was refined using the I2/m group. Peak profiles were fitted using the Thompson-Cox-Hastings pseudo-Voigt function [27]. As powder diffraction patterns show an anisotropic line-shape broadening that is not a smooth function of d-spacing, the mean coherence path dimensions (MCP) or crystallite sizes, were determined in the parallel direction (P_{paral}) and perpendicular direction (P_{perp}) to the anisotropic broadening (002) axis. Crystallite dimensions were calculated making allowances for the instrument broadening function that was previously modelled using NIST SRM 660 lanthanum hexaboride (LaB_6) standard. Preferential orientation was refined using spherical harmonics. ^{57}Fe Mössbauer spectra were recorded in a conventional Mössbauer spectrometer working in constant acceleration at room temperature. The spectra were analyzed using the MOSF program [28], which is based on non-linear least squares fitting procedures assuming lorentzian Mössbauer lines while the isomer shift value is quoted relative to that of α-Fe at 300 K. In order to determine the relative Mössbauer recoilless f-fractions at room temperature for all akaganeites we prepare two types of absorbers, one of about 15 mg (named sample A) and other of about 20 mg (named sample B), each consisting of homogenous mixtures of the akaganeites and standard natural iron powder. For example, the absorber named as Aka-Co-0.10-Fe-A means that about 15 mg of Aka-Co-0.10 and iron powder were mixed. Similar notation was used for the other samples. The mass of the absorbers was obtained using a microbalance. The absorbers were then carefully placed in a sample holder of 1.26 cm in diameter.

In this work we obtained the recoilless f-fraction of akaganeite, f_{Ak}, relative to that of iron powder, f_{Fe}, proceeding similarly as in the work by Oh and Cook [15]. For that purpose we have used the following equation:

$$\frac{f_{\mathrm{Ak}}}{F_{\mathrm{Fe}}} = \left(\frac{m_{\mathrm{Fe}}}{m_{\mathrm{Ak}}}\right)\left(\frac{MW_{\mathrm{Ak}}}{MW_{\mathrm{Fe}}}\right)\left(\frac{A_{\mathrm{Ak}}}{A_{\mathrm{Fe}}}\right) \tag{1}$$

where m_{Ak} and m_{Fe} stand for the sample masses, MW_{Ak} and MW_{Fe} are the molar weights, and A_{Ak} and A_{Fe} are the relative areas of akaganeite and iron powder, respectively.

3 Results and discussion

3.1 Chemical analysis

Table 1 shows the nominal preparative and the final concentrations of Co and Mn, expressed as μ_{Me} ratio (μ_{Me} [mol mol^{-1} %] = [Me] × 100 / [Me] + [Fe]; Me: Co or Mn; [Me]: mol L^{-1}), in the prepared samples, also the final Cl content, and the initial and final pH values are also shown for all preparations.

From the data reported in Table 1, it is noticed that a major metal incorporation is observed in those samples showing the lower pH values, however in all cases the incorporation is low and in the range 0.05–0.11 mol mol^{-1} % for Co, and 0.08–0.16 mol mol^{-1} % for Mn. The low Mn (or Co) for iron substitution in akaganeite may be ascribed to the difference in coprecipitation abilities of the ions with Fe(III) via hydrolysis during the aging. The degree of hydrolysis of the ions can be roughly compared by their first hydrolysis constants, being in the order Fe(III) (pKa: 2.73) >> Mn(II) (pKa: 10.46) ~ Co(II) (pKa: 10.20) [19, 29].

The final Cl content also varied with the urea concentration of the parent solution, and the incorporation increases at higher preparative pH values. The same trend is found in the Co and Mn doped samples. Both the Mn and the Co incorporations are low with values in the range 0.05–0.16, being the incorporation higher in the Mn-doped akaganeites. The foreign metal incorporation decreases with the increase in urea content.

3.2 Structural properties

The results of the Rietveld simulation are presented in Table 2. The XRD diagrams of all samples corresponded to akaganeite. The quality of the refinements was satisfactory and similar for all samples. The profiles of the observed and calculated diffraction patterns and the difference plot for sample Aka-Co-0.40 are shown in Fig. 1. As can be observed, the experimental diagram is well described by the akaganeite structural model that has been employed.

The unit cell volume of pure akaganeites increased from 336.047(23) Å^3 to 337.339(34) Å^3 with the increase of urea concentration of the preparative solution, and with the increase in Cl content. The calculated crystallite sizes parallel (L_{paral}), and perpendicular (L_{perp}) to the (002) axis of pure akaganeites decrease, from 57 nm to 44 nm for L_{paral} and from 140 nm to 117 nm for L_{perp}, with the increase of urea content of the parent solution, indicating that the presence of urea decreases the crystallinity of the synthesised oxide.

The cell volumes of undoped and Co-doped akaganeites markedly increase with the increment in urea content. However, the opposite trend occurs in the Mn-doped samples, for which the cell volume slightly decreases with increasing urea. Now, for those akaganeites with low urea content, the cell volume slightly increases with Mn (or Co) doping. But, for the samples with high urea content, the opposite behavior

Table 1 Chemical analysis of the prepared akaganeites

Sample	Aka-0	Aka-0.10	Aka-0.40	Aka-Co0.10	Aka-Co0.40	Aka-Mn0.10	Aka-Mn0.40
μ_{Fe} [mol mol^{-1}%] (nominal)	1	–	–	–	–	–	–
μ_{Co} [mol mol^{-1}%] (nominal)	–	–	–	11.21	11.21	–	–
μ_{Mn} [mol mol^{-1}%] (nominal)	–	–	–	–	–	7.47	7.47
Atomic % (EDS)	Fe: 25.14 Cl: 3.39	Fe:24.88 Cl: 3.48	Fe: 24.49 Cl: 4.25	Fe: 22.33 Cl: 3.74	Fe: 27.13 Cl: 5.22	Fe: 23.73 Cl: 3.46	Fe: 26.58 Cl: 4.00
ratio Fe/Cl (Atomic %)	7.42	7.15	5.76	5.97	5.20	6.86	6.67
μ_{Co} [mol mol^{-1}%] (incorporated)	–	–	–	0.11	0.05	–	–
μ_{Mn} [mol mol^{-1}%] (incorporated)	–	–	–	–	–	0.16	0.08
Initial and final pH	1.73-1.20	1.74-1.03	1.86-1.42	2.07-1.57	2.17-1.82	2.08-1.54	2.23-1.76

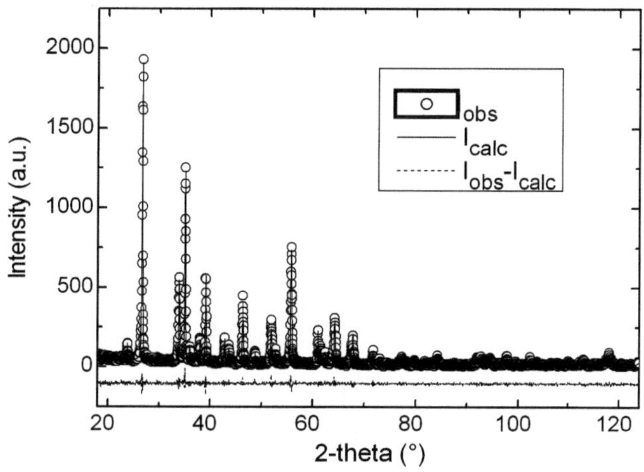

Fig. 1 Observed and calculated DRX profiles for sample Aka-Co-0.40

is noticed. Therefore, the observed variation in the unit cell parameters can be attributed to concomitant effects coming from the very small Mn (or Co) for Fe substitution, and principally due the variation in the Fe and Cl contents which can be accompanied by the presence of other defects, like more OH- groups, vacancies, etc. Moreover, in this analysis we have to take into consideration the coordination number, the oxidation state and the spin state of the cations, which affects their ionic radii. For example if we assume that Mn and Co are octahedrally coordinated, and that they have 2+ oxidation and high spin states, then the. ionic radii for Fe^{3+}, Co^{2+}, Mn^{2+} and Cl^{-1} should be of 0.645 Å, 0.745 Å, 0.83 Å, and 1.81 Å, respectively [30]. But other radii may occur if other assumptions are considered, which makes the analysis more complex.

The solids obtained in 0.10 M urea showed L_{paral} and L_{perp} values that changed from 66/118, in pure akaganeite, to 97/108, and to 52/215 in the Co and Mn added samples, indicating that the foreign cation incorporations provoked an increasing elongated form in the domains. In the solids obtained at higher urea concentration (0.40 M) the values changed from 44/117 to 146/40, with the Co incorporation, and to 84/86 with Mn substitution, indicating a marked variation in the form of the domains. These results confirm that crystallinity depend not only on the Me-for-Fe substitution but also on the preparative pH conditions.

3.3 Hyperfine properties

Figure 2a shows the room temperature Mössbauer spectrum of Aka-Mn-0.10. The spectrum was fitted with two components, doublet 1 (D1) and doublet 2 (D2), whose hyperfine parameters are reported in Table 3. Component D1 originates from iron atoms located close to the chloride sites, while component D2 is attributed to iron atoms located close to the chloride vacancy sites. Similar spectra (not shown) were obtained for the other samples. These values are in good agreement with those reported in the literature for pure akaganeites [20]. No signals from other iron phases were detected in good agreement with the XRD findings. Table 3 presents the values of the center shifts, δ, the line widths, Γ, and the quadrupole splittings, Δ, for both

Table 2 Structural analysis of the prepared akaganeites

Sample	Aka-0	Aka-0.10	Aka0.40	Aka-Co0.10	Aka-Co0.40	Aka-Mn0.10	Aka-Mn-0.40
a[Å]	10.531(8)	10.534(9)	10.558(9)	10.539(4)	10.544(5)	10.539(8)	10.543(11)
b[Å]	3.031(1)	3.031(1)	3.0328(1)	3.032(1)	3.032(1)	3.0317(1)	3.0309(1)
c[Å]	10.528(5)	10.530(1)	10.534(1)	10.527(1)	10.544(5)	10.544(1)	10.532(8)
β [°]	89.99(1)	90.00(1)	89.92(1)	90.03(1)	89.97(1)	89.87(1)	90.00(2)
Volume [Å^3]	336.047(23)	336.322(46)	337.339(34)	336.432(22)	337.148(25)	336.797(57)	336.605(35)
L_{paral} [nm]	57	66	44	97	146	52	84
L_{perp} [nm]	140	118	117	108	40	215	86
Rwp	8.12	11.88	12.92	14.65	13.02	13.52	13.09
Rp	6.67	9.58	9.94	10.14	10.25	10.47	7.69
GofF	1.11	1.06	1.09	1.08	1.16	1.16	1.21
R_{Bragg}	4.83	10.70	7.62	12.13	9.00	9.80	9.55

Atomic parameter: z = 0.250; Rp: 100 Σ|Io-Ic|/Σ Io; Rwp: 100 [Σ wi (Io-Ic)2/Σ wi Io2]0.5; RBragg: 100 Σ|Iko-Ikc| /Σ Iko; GofF: Σ wi (Io-Ic) 2/(N-P); I_o and I_c: observed and calculated intensities; *wi* weight assigned to each step intensity; *I*ko and Ikc observed and calculated intensities for Bragg k-reflection; *N* and P number of data points in the pattern and number of parameters refined. Values in parentheses are esd for the least significant figures of the data shown, the esd values are taken from the final cycle of the Rietveld refinement

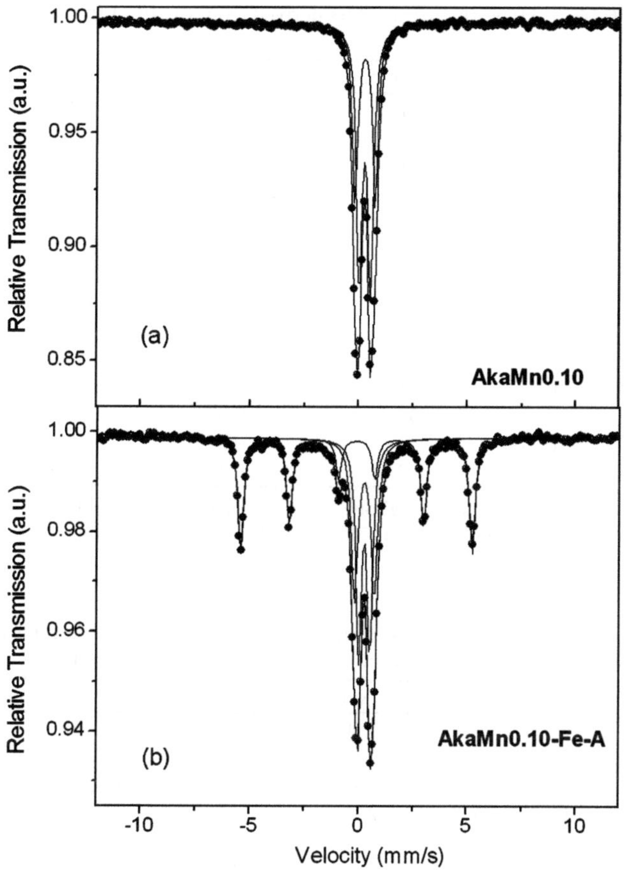

Fig. 2 Room temperature Mössbauer spectra of: **a** Aka-Mn-0.10, and **b** mixture of 9.1 mg of Aka-Mn-0.10 and 5.9 mg of standard iron powder

iron sites in the akaganeites. A comparison between pure and doped samples shows that the values of the parameters remain practically unchanged independently of Co, Mn and urea concentrations. These results suggest that only a very small amount of these cations may have entered into the structure and that the incorporation does not provoke appreciable changes in these hyperfine parameters.

A comparison between the relative area of doublet D1 with the Cl content and also with the Fe/Cl ratio shows an interesting trend for the pure akaganeites (see Tables 1, 2 and 3). The area of D1 behaves similarly as the Fe/Cl ratio, but not similarly with respect to the Cl content, i.e. the area of D1 increase when going from Aka-0, to Aka-0.10, and to Aka-0.40, with the increase of Cl/Fe ratio.

On the other hand, the relative areas of the D1 sites slightly decrease for the doped samples in comparison to those of the pure akaganeites, this effect being more pronounced for Mn-doped than for Co-doped akaganeites.

The small differences between the Fe/Cl ratio of pairs Aka-Co-0.10/Aka-Co-0.40 and Aka-Mn-0.10/Aka-Mn-0.40 are not detected through changes in D1, and the value remains unaltered within each series, presenting values of 61 and 59 for the Co- and Mn-akaganeites, respectively. However, a decrease in the D1 value is observed when samples Aka-0.10, Aka-Co-0.10 and Aka-Mn-0.10 are compared. This change

Table 3 Hyperfine parameters derived from the fit of the room temperature Mössbauer spectra of the prepared samples

Sample	Component	δ (mm/s)	Γ (mm/s)	Δ (mm/s)	Area (%)
Aka-0	D1	0.36	0.30	0.54	65
	D2	0.37	0.30	0.97	35
Aka-0.10	D1	0.36	0.28	0.54	63
	D2	0.37	0.30	0.95	37
Aka-0.40	D1	0.36	0.28	0.54	61
	D2	0.37	0.31	0.95	39
Aka-Co-0.10	D1	0.37	0.32	0.53	61
	D2	0.37	0.31	0.95	39
Aka-Co-0.40	D1	0.37	0.31	0.53	61
	D2	0.37	0.30	0.95	39
Aka-Mn-0.10	D1	0.37	0.31	0.52	59
	D2	0.37	0.31	0.95	41
Aka-Mn-0.40	D1	0.37	0.31	0.52	59
	D2	0.37	0.30	0.95	41

Estimated errors are of about $\pm$ 0.01 mm/s for the centre shift, δ, line broadening, Γ, and quadrupolar splitting, Δ, and of about $\pm$ 2 % for the areas. D1 and D2 stand for doublet 1 and doublet 2, respectively

Table 4 Masses (m_{Ak} and m_{Fe}) of akaganeite and iron powder used to prepare the absorbers and the relative Mössbauer areas (A_{Ak} and A_{Fe}), for all samples

Sample	m_{Ak}(mg)	m_{Fe}(mg)	MW_{Ak}(g/mol)	A_{Ak}	A_{Fe}	f_{Ak}/f_{Fe}
Aka-0-Fe-A	9.30	5.9	93.59	47.92	52.08	0.98
Aka-0-Fe-B	12.3	7.9	93.59	45.19	54.81	0.89
Aka-Co-0.10-Fe-A	9.1	5.9	95.39	60.27	39.73	1.68
Aka-Co-0.10-Fe-B	12.2	7.8	95.39	58.69	41.31	1.55
Aka-Co-0.40-Fe-A	9.1	5.9	95.93	52.41	47.59	1.23
Aka-Co-0.40-Fe-B	12.2	7.8	95.93	57.77	42.23	1.50
Aka-Mn-0.10-Fe-A	9.1	5.9	93.81	61.71	38.29	1.75
Aka-Mn-0.10-Fe-B	12.2	7.8	93.81	55.98	44.02	1.37
Aka-Mn-0.40-Fe-A	9.1	5.9	94.25	59.34	40.66	1.60
Aka-Mn-0.40-Fe-B	12.2	7.8	94.25	56.67	43.33	1.41

MW_{Ak} molar weight of akaganeite calculated considering the formula of β-$Fe_{1-y}Me_yO_{(1-x)}(OH)_{1+x}Cl_x$ and the values reported in Table 1, where Me accounts for Co or Mn
f_{Ak}/f_{Fe} ratio of the recoilless f-factor of akaganeite to that of iron powder, calculated using (1)
MW_{Fe} molar weight of the iron powder (55.845 g/mol)

Table 5 Average relative recoilless f-fractions for the akaganeites

Sample	f_{Ak}/f_{Fe}
Aka-0	0.93 $\pm$ 0.06
Aka-Co-0.10	1.62 $\pm$ 0.09
Aka-Co-0.40	1.36 $\pm$ 0.19
Aka-Mn-0.10	1.56 $\pm$ 0.27
Aka-Mn-0.40	1.54 $\pm$ 0.13

may be attributed to the fact that the relative number of the iron ions in the D1 site decreases, because the dopants (Co or Mn ions) prefer to be located at the D1 sites than at the D2 sites, i.e. the dopants prefer to be located at the least distorted sites.

This decrease in the number of Fe atoms in the D1 sites is observed along with an increase in the population of D2 sites, the effect is contrary to that observed with the increase in the Fe/Cl ratio that should indicate a decrease in D2 sites as observed for the pure akaganeites. The decrease in D1 values also may be attributed to variations in the Fe/Cl ratio, however the observed ratio (7.15, 5.97, and 6.86) do not follow the same trend of D1 (63, 61, and 59).

Figure 2b shows the room temperature Mössbauer spectrum of Aka-Mn-0.10-Fe-15. The spectrum was fitted with three components, two doublets (D1 and D2) related to the akaganeite phase and one sextet related to the iron powder. Table 4 list the data related with the calculation of the relative recoilless f-fractions for all samples.

The averaged relative recoilless f-fractions (f_{Ak}/f_{Fe}) calculated using (1), and the errors are reported in Table 5. It is found that within the error bars the f_{Ak}/f_{Fe} values for the the Co- and Mn-akaganeites are relatively similar to each other but markedly increased in comparison to the undoped sample. It is difficult to find a quantitative interpretation for these results because there is no simple relation between the structural, morphological and compositional properties of the samples on the one hand and the observed f_{Ak}/f_{Fe} values on the other hand. We could speculate that the increment in the relative f-fractions can be related to an increment in the stiffening of the lattice, perhaps due to a strengthening of the atomic bonding of the iron ions in the doped samples. Of course additional measurements are required to support these ideas.

4 Conclusions

The hydrolysis of $FeCl_3$ solutions at low pH values and in the presence of urea, Co and Mn ions produces impurity-free akaganeites with only small metal for iron substitution in the structure, this incorporation being more effective at the lower urea concentrations.

The urea and the small metal incorporation in the structure induced changes in the Fe to Cl ratio, the form of the crystallographic domains, the relative Mössbauer spectral areas and the relative recoilless f-fractions. However, within the error bars no changes in the center shift and the line width were observed in the samples. On the other hand, because the quadrupole splittings of the akaganeites were not significantly affected by the presence of urea and Mn or Co, the symmetry of the D1 and D2 sites did not change appreciably.

The present results show that, the properties of akaganeite are slightly affected by Co or Mn at low concentrations. When formed as a corrosion product of weathering, Co- and Mn-doped akaganeite could exhibit very similar properties to that of synthetically pure akaganeite.

Acknowledgements This work was supported by the Universidad de Antioquia (Sustainability project of the Solid State Group 2011–2012), COLCIENCIAS (project CO/08/15), and Ministerio de Ciencia y Técnica de Argentina (project PICT 2008–0780)

References

1. Holm, N.G., Dowler, M.J., Wadsten, T., Arrhenius, G.: β-FeOOH.Cln (akaganéite) and Fe1-xO (wüstite) in hot brine from the Atlantis II Deep (Red Sea) and the uptake of amino acids by synthetic β-FeOOH.Cln. Geochim. Cosmochim. Acta **47**, 1465–1470 (1983)
2. Holtstam, D.: Akaganéite as a corrosion product of natural, non-meteoritic iron from Qeqertarsuaq. West Greenland GFF **128**, 69–71 (2006)
3. Rémazeilles, C., Refait, P.: On the formation of β-FeOOH (akaganéite) in chloride-containing environments. Corros. Sci. **49**, 844–857 (2007)
4. Buchwald, V.F., Clarke Jr., R.S.: Corrosion of Fe-Ni alloys by Cl-containing akageneite (beta-FeOOH): the Antarctic meteorite case. Am. Mineral. **74**, 656–67 (1989)
5. Post, J.E., Buchwald, V.F.: Crystal structure refinement of akaganéite. Am. Mineral. **76**, 272–277 (1991)
6. Keller, P.: Eigenschaften von (Cl,F,OH)<2$Fe_8(O,OH)_{16}$ und Akaganeite. Neues Jahrb. Mineral. Abh. **113**, 29–49 (1970)
7. Dasgupta, D.R., Mackay, A.L.: J. Phys. Soc. Jpn. **14**, 932–935 (1959)
8. Mackay, A.L.: Mineral. Mag. **32**, 545–557 (1960)
9. Cai, J., Liu, J., Gao, Z., Navrotsky, A., Suib, S.: Synthesis and anion exchange of tunnel structure akaganeite. Chem. Mater. **13**, 4595–4602 (2001)
10. Post, J.E., Heaney, P.J., Von Dreele, R.B., Hanson, J.C.: Am. Mineral. **88**, 782–788 (2003)
11. Stahl, K., Nielsen, K., Jiang, J., Lebech, B., Hanson, J.C., Norby, P., van Lanschot, J.: Neutron and temperature resolved synchrotron X-ray powder diffraction study of akaganeite. Corros. Sci. **45**, 2563–2575 (2003)
12. Murad, E.: Mössbauer spectroscopy of clays, soils and their mineral constituents. Clay Miner. **45**, 413–430 (2010)
13. Barrero, C.A., García, K.E., Morales, A.L., Kodjikian, S., Greneche J.M.: New analysis of the Mössbauer spectra of akaganeite. J. Phys.: Condens. Matter. **18**, 6827–6840 (2006)
14. Murad, E., Cashion, J.: Mössbauer spectroscopy of environmental materials and their industrial utilization. Kluwer, Academic Publishers, Boston (2004)
15. Oh, S.J., Cook, D.E.: Mössbauer effect determination of relative recoilless fractions for iron oxides. J. Appl. Phys. **85**, 329–332 (1999)
16. De Grave, E., Van Alboom, A.: Phys. Chem. Miner. **18**, 337–342 (1991)
17. Vertes, A., Czakó-Nagy, I.: Electrochim. Acta **34**, 721–758 (1989)
18. Cornell, R.M., Schwertmann, U.: The iron oxides: structure, properties, reactions, occurrence and uses. Wiley-VCH Verlag GmbH & Co. KGaA, Weinheim (2003)
19. Ishikawa, T., Katoh, R., Yasukawa, A., Kandori, K., Nakayama, T., Yuse, F.: Corros. Sci. **43**, 1727–1738 (2001)
20. García, K.E.,Morales, A.L., Greneche, J.M., Barrero, C.A.: ^{57}Fe Mössbauer study of β-FeOOH obtained in presence of Al^{3+} and Ti^{4+} íons. Physica B: Condens. Matter. **389**, 88–93 (2007)
21. García, K.E., Barrero, C.A., Morales, A.L., Greneche, J.M.: Characterization of akaganeite synthesized in presence of Al^{3+}, Cr^{3+}, and Cu^{2+}ions and urea. Mater. Chem. Phys. **112**, 120–126 (2008)
22. Cornell, R.M.: Preparation and Properties of Si Substituted Akaganeite (b-FeOOH). Zeitschrift für Pflanzenernährung und Bodenkunde. J. Plant. Nutr. Soil Sci. **155**, 449–453 (1992)
23. Holm, N.G.: Substitution selectivity of some transition elements (Cr, Mn, Co, Ni, Cu, Zn) during formation of β-FeOOH. GFF Stockholm **107**, 297–300 (1985)
24. Leussing, D.L., Newman, L.: Spectrophotometric study of the bleaching of ferric thioglycolate. J. Am. Chem. Soc. **78**, 552–556 (1956)
25. Larson, A.C., Von Dreele, R.B.: General structure analysis system (GSAS) Los Alamos National Laboratory Report LAUR, pp. 86–748 (1994)
26. Toby, B.H.: EXPGUI, a graphical user interface for GSAS. J. Appl. Crystallogr. **34**, 210–213 (2001)
27. Thompson, P., Cox, D.E., Hastings, J.B.: Rietveld refinement of Debye-Scherrer Synchrotron X-ray data from Al2O3. J. Appl. Crystallogr. **20**, 79–83 (1987)
28. Vandenberghe, R.E., De Grave, E., De Bakker, P.M.A.: On the methodology of the analysis of Mössbauer spectra. Hyperfine Interact. **83**, 29–49 (1994)
29. Baes, C.F.J., Mesmer, R.E.: The hydrolisis of cations. Wiley Interscience, New York (1976)
30. http://abulafia.mt.ic.ac.uk/shannon/ptable.php. Consulted on Jan. 21, 2013

DOI 10.1007/978-94-007-6482-8_28
Reprinted from *Hyperfine Interactions*, DOI 10.1007/s10751-013-0849-y

Fe^{2+}-Mg order–disorder study in orthopyroxenes from São João Nepomuceno (IVA) iron meteorite

E. dos Santos · R. B. Scorzelli · M. E. Varela · P. Munayco

Published online: 26 March 2013

Abstract São João Nepomuceno (SJN) is an iron meteorite belonging to IVA chemical group. It consists of Fe-Ni metal matrix and silicate inclusions made of a coarse-grained mixture of tridymite, orthopyroxene and clinopyroxene. In spite of the extensive work performed on the IVA group there is still no consensus about their origin and its thermal history is subject of ongoing debates. In this work, we report preliminary results on Fe^{2+} distribution in the non-equivalent octahedral sites in orthopyroxene crystals extracted from São João Nepomuceno in order to infer on the thermal history of this meteorite and therefore acquire more information related to the origin of the IVA iron meteorite group.

Keywords Meteorite · São João Nepomuceno · Orthopyroxene · Fe^{2+}-Mg order–disorder · ^{57}Fe Mössbauer spectroscopy

1 Introduction

São João Nepomuceno (SJN) is an iron meteorite belonging to the IVA chemical group. It consists of Fe-Ni metal matrix and silicate inclusions made of a coarse-grained mixture of tridymite, orthopyroxene and clinopyroxene. In spite of the extensive work performed on the IVA irons there is still no consensus about their origin and thermal history. Their particular chemistry and range in metallographic

Proceedings of the Thirteenth Latin American Conference on the Applications of the Mössbauer Effect (LACAME 2012), Medellín, Colombia, 11–16 November 2012.

E. dos Santos (✉) · R. B. Scorzelli · P. Munayco
Centro Brasileiro de Pesquisas Físicas, Rua Dr. Xavier Sigaud, 150, 22290-180 Rio de Janeiro, Brazil
e-mail: edisanfi@cbpf.br

M. E. Varela
CONICET – ICATE, Av España 1512 Sur, J5402DSP, San Juan, Argentina

cooling rate are difficult to explain under conventional models [1]. The IVA irons are extremely depleted in volatile elements (e.g., Ga, Ge) with respect to other meteorite groups. Their large spread of metallographic cooling rates—too large for an asteroidal core enclosed by silicate mantle—argues against an origin in a mantled asteroidal core. If this would be the case, uniform cooling rates should to be expected. Nonetheless, metal-silicate mixing of the IVA group remains a complex issue.

Depending on the temperature and pressure history of the host rock, Fe^{2+} and Mg in orthopyroxene crystals fractionate between two nonequivalent octahedral sites, M1 and M2. In slowly cooled crystals, the Fe^{2+} ions populate essentially the M2 position while Mg ions occupy, predominantly, the M1 position. However, in crystals that have been rapidly cooled, a more disorder Mg, Fe distribution over M1 and M2 sites is observed [2]. Hence, the intracrystalline Fe^{2+}-Mg distribution determination is very useful in tracing the thermal history of the host rock. To accomplish this task, ^{57}Fe Mössbauer spectroscopy has played a major role in obtaining site populations in orthopyroxenes through the relative areas of the Fe^{2+} doublets in M1 and M2 sites [2].

In this work, we report preliminary results on Fe^{2+} distribution in the non-equivalent octahedral sites in orthopyroxene crystals extracted from SJN in order to infer on the thermal history of the coarse-grained silicate fraction of the host meteorite and, as a consequence, acquire more information related to the origin of IVA iron meteorite group.

2 Materials and methods

All samples used in this study belong to the Museu Nacional/UFRJ in Rio de Janeiro.

The separation of silicate inclusions of the SJN meteorite from its metallic matrix was done in several thin slides. The cutting process was performed by means of a diamond saw using alcohol as lubrificant, to avoid oxidation and contamination. Under a binocular microscope, orthopyroxene crystals were subsequently separated from silicate inclusions. The separate grains were inspected under an electron microscope (Jeol JSM 6490 LV, CBPF, Brazil) to verify their chemical composition. This material was used for preparing the Mössbauer powder absorber with a thickness of 1.8 mg Fe/cm^2. The thickness effect, i.e. the saturation of absorption peaks, is a well-known feature in Mössbauer spectroscopy [3]. As was already shown, such effect in orthopyroxene results in an underestimate of the ordering degree. For instance, an absorber thickness of 5 mg Fe/cm^2 causes an overestimation of Fe at M1 site by 2 % [4]. Thus, a thin Mössbauer absorber is fundamental for an accurate site occupancy determination. In general, absorber thicknesses of 2–5 mg Fe/cm^2 are assumed as thin absorbers. It is worth mention that in our previous work involving clinopyroxene, an absorber thickness of 3 mg Fe/cm^2 yielded Mössbauer results compatible with single-crystal X-ray diffraction experiment, suggesting negligible thickness effect [5]. In this study, the Mössbauer powder absorber thickness (1.8 mg Fe/cm^2) is even thinner than those assumed as thin absorbers in previous works; therefore, we expect negligible thickness effect.

^{57}Fe Mössbauer spectroscopy, in standard transmission geometry using a 25 mCi ^{57}Co/Rh radioactive source in sinusoidal mode, was performed at different temperatures as follow: room temperature (RT), 80 K and 4.2 K, with uncertainties of ± 1 K.

Spectra were recorded for 24 h in a 512 channels spectrometer and the calibration was taken at RT with α-Fe foil. The error in source velocity is less than 1 %. The NORMOS code [6] was used for the analyses of the Mössbauer spectra. The isomer shifts are given relative to α-Fe.

Two polished thick sections from São João Nepomuceno were prepared for chemical analysis. Objects were selected using an optical microscope and the major element chemical compositions were obtained using an ARL-SEMQ (WDS) electron microprobe (Icate, Argentina). Electron microprobe analysis were performed at 15 kV acceleration potential, 20 nA sample current and peak and background count times of: 20 s; and 10 s, respectively. Estimated precision for major and minor elements is better than 3 % and for Na about 10 %. Natural and synthetic standards were used for calibration and an online ZAF correction was applied to the data.

3 Results

The orthopyroxene (opx) average chemical composition (mean of 39 individual analyses) is listed in Table 1. In terms of the end-members of pyroxene solid-solution (En: enstatite), (Fs: ferrosilite) and (Wo: wollastonite) opx corresponds to: $En_{85}Fs_{14}Wo_1$. The derived opx unit formula calculated on the basis of six oxygen atoms is given by: $(Fe^{2+}_{0.273}Mg_{1.678}Mn_{0.016}Ca_{0.021}Ti_{0.001}Cr_{0.018})(Si_{1.982},Al_{0.012})O_6$ (Table 1).

The RT ^{57}Fe Mössbauer spectrum of opx from SJN (Fig. 1a) is characterized by an asymmetry that arises from two overlapping doublets attributed to Fe^{2+} at two nonequivalent sites. The first doublet, identified as Fe^{2+}(M1), corresponds to Fe^{2+} ions in the less distorted site M1 (isomer shift (IS) = 1.24 mm/s, quadrupole splitting (QS) = 2.26 mm/s) while the second one, Fe^{2+}(M2), is attributed to Fe^{2+} in the more distorted site M2 (IS = 1.14 mm/s, QS = 2.10 mm/s) (hyperfine parameters are shown in Table 2). The absence of any Fe^{3+} doublets and magnetic components assigned to Fe oxides indicate that terrestrial weathering did not take place in these samples [7], thus Fe site occupancies determination will not be hampered by secondary phases. Previous works [8, 9] on natural opx show that RT splitting of Fe^{2+}(M1) and Fe^{2+}(M2) doublets approach a similar value for low iron content (<23.1 mol%) and the hyperfine parameters measured in these samples, especially the relative areas, become less precise. However, for samples with iron concentration close to 86 mol%, the doublets are well resolvable even at RT. Hence, our RT results suggest low iron content for opx samples from SJN meteorite, in accordance with chemical composition results for these crystals.

At a given iron concentration, as a result of the differential dependence of the quadrupole splitting on temperature, the resolution of the Fe^{2+}(M1) and Fe^{2+}(M2) doublets increase with decreasing temperature [10]. Aiming to improve the accuracy of the measured hyperfine parameters, especially the relative areas that play important role in Fe site occupancy determination, low velocity ^{57}Fe Mössbauer spectroscopy at 80 K was carried out in our opx samples (Fig. 1b). This spectrum was fitted with the same components as RT spectrum and indicates that Fe^{2+} mainly populate M2 site ($A = 86$ %) (see hyperfine parameters in Table 2).

The normalized Mössbauer relative areas at 80 K (Table 2) show that Fe^{2+} fractions at M1 and M2 sites are 0.14 and 0.86, respectively. The occupancies

Table 1 Electron microprobe average chemical composition (wt%) and unit formula (*apfu*) for orthopyroxenes from São João Nepomuceno (IVA) meteorite

SiO_2	56.45	Si	1.982
TiO_2	0.05	^{IV}Al	0.012
Al_2O_3	0.30	Σ	1.994
Cr_2O_3	0.64		
FeO	9.29	Ti	0.001
MnO	0.53	Cr	0.018
MgO	32.07	Fe^{2+}	0.273
CaO	0.57	Mn	0.016
Total	99.90	Mg	1.678
		Ca	0.021
		Σ	2.007

apfu atoms per formula unit

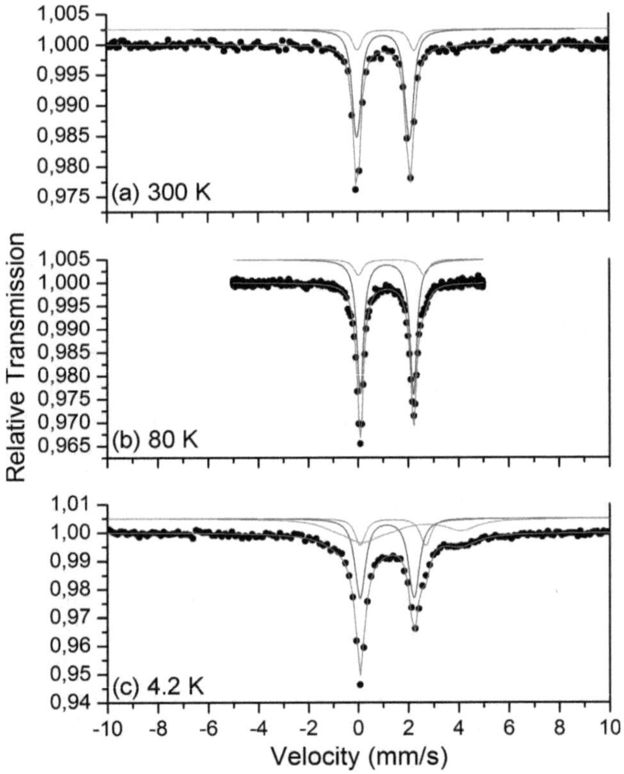

Fig. 1 ^{57}Fe Mössbauer spectra of the orthopyroxene samples from São João Nepomuceno (IVA) meteorite recorded at: **a** room temperature (300 K), **b** 80 K and **c** 4.2 K

of Fe^{2+} at M1 and M2 sites, $X_{Fe^{2+}(M1)}$ and $X_{Fe^{2+}(M2)}$, respectively, are obtained by multiplying the above mentioned Fe^{2+} fractions by the total amount of Fe^{2+} given by electron microprobe results. Thus, $X_{Fe^{2+}(M1)} = 0.14*0.273 = 0.038$ and $X_{Fe^{2+}(M2)} = 0.86*0.273 = 0.235$. Assuming complete order of Cr and Ti atoms at M1 site and Mn and Ca atoms at M2 site, Mg distribution over M1 and M2 sites can be calculated as follow: $X_{Mg(M1)} = 1 - (X_{Fe^{2+}(M1)} + N_{Cr} + N_{Ti}) = 0.943$ and $X_{Mg(M2)} = N_{Mg} - X_{Mg(M1)} = 0.735$, where N_{Cr}, N_{Ti}, N_{Mg} are the numbers of Cr, Ti and Mg atoms per formula unit (Table 1). The Fe^{2+}-Mg order–disorder reaction in opx can be described by means of the distribution coefficient (K_D) between

Table 2 Mössbauer hyperfine parameters of the orthopyroxenes from São João Nepomuceno (IVA) meteorite obtained at room temperature (300 K), 80 K and 4.2 K

Temperature (±1 K)	Sites	IS (±0.03 mm/s)	QS (±0.03 mm/s)	W (±0.02 mm/s)	A (±2 %)	ω (MHz)	B_{hf} (±1 T)
300 K	Fe^{2+} (M1)	1.24	2.26	0.36	18	–	–
	Fe^{2+} (M2)	1.14	2.10	0.33	82	–	–
80 K	Fe^{2+} (M1)	1.44	2.62	0.43	14	–	–
	Fe^{2+} (M2)	1.26	2.13	0.31	86	–	–
4.2 K	Fe^{2+} (M1)	1.49	2.61	0.42	15	–	–
	Fe^{2+} (M2)	1.26	2.15	0.42	45	–	–
	Fe^{2+} (M2)—magnetic	1.29	1.88	1.10	40	2.6	12.3

IS isomer shift relative to alpha iron, *QS* quadrupole splitting, *A* relative area, ω relaxation rate, B_{hf} hyperfine field

M1 and M2 sites defined as: $K_D = [X_{Fe^{2+}(M1)}X_{Mg(M2)}]/[X_{Fe^{2+}(M2)}X_{Mg(M1)}]$. In this work we obtain a $K_D \sim 0.126 \sim 0.13$. To estimate the equilibrium temperature (or closure temperature) of the intracrystalline distribution of Fe^{2+} and Mg between M1 and M2 sites, we can use the geothermometric equation of Wang et al. (2005) [11]. In that equation, the distribution coefficients K_D taken from Mössbauer data are independent of opx composition and they can be related to the equilibrium temperature by: $T = -2205/[\ln K_D - 0.391]$, where T is temperature in K. According to our data, we found $T \sim 634\ °C$.

Measurement at 4.2 K clearly exhibits relaxation effects (Fig. 1c). This feature was previously observed in natural opx and it has been attributed to spin-spin relaxation of Fe^{2+} ions [10]. The 4.2 K spectrum was fitted with the doublets identified at RT and to take into account relaxation effects it was included a component assigned as Fe^{2+}(M2)-magnetic. The decreasing in the relative area of Fe^{2+}(M2) doublet at 4.2 K indicates that Fe^{2+} ions at M2 site mainly contributes for magnetic ordering (probably Fe^{2+} at M1 sites order at lower temperatures). These results are in agreement with previous works where low temperature Mössbauer measurements were carried out in natural and synthetic low iron opx samples with Fe^{2+} ions primarily populating M2 sites [10, 12]. It is important to highlight that even at 4.2 K it was not detected any magnetic component due to terrestrial weathering processes.

4 Discussion

These preliminary results show that, in the studied samples, we have not identified secondary phases related to oxidation processes (e.g., presence of Fe^{3+}). Consequently, the minerals in SJN represent a non-altered assemblage, at least from the latest thermal event that could have affected this meteorite. The resolution of the doublets at RT suggest that theses crystals have low iron content and that Fe^{2+} mainly populate M2 site (high order degree). Both 80 K and 4.2 K measurements argue in favor of high order degree for these samples. Mössbauer spectrum at 4.2 K indicates that magnetic interaction is mainly provided by Fe^{2+} at M2 site, in accordance with Mössbauer measurements carried out in natural and synthetic low iron opx samples with Fe^{2+} ions primarily populating M2 sites [10, 12].

Variable temperature Mössbauer and single-crystal diffractions experiments are in progress to evaluate the accuracy of the site populations and to obtain additional information regarding the thermal/pressure history of these samples.

Acknowledgements The authors are grateful to Centro Brasileiro de Pesquisas Físicas (CBPF/MCTI) for work support and M. E. Zucolotto (Museu Nacional/UFRJ) for providing São João Nepomuceno samples. E. dos Santos is indebted to CAPES for granting his Ph.D. fellowship at CBPF. R. B. Scorzelli acknowledges financial support from FAPERJ and CNPq. M. E. Varela acknowledges financial support from PIP 1645 (CONICET) and CNPQ during scientific visits to CBPF. P. Munayco is indebted to FAPERJ for granting his Postdoc fellowship at CBPF.

References

1. Scott, E.R.D., Haack, H., McCoy, T.J.: Core crystallization and silicate-metal mixing in the parent body of the IVA iron and stony-iron meteorites. Geochim. Cosmochim. Acta **60**, 1615–1631 (1996)
2. Dundon, R.W., e Hafner, S.S.: Cation disorder in shocked orthopyroxene. Science **174**, 581–582 (1971)
3. Greenwood, N.N., Gibb, T.C.: Mössbauer Spectroscopy. Champman and Hall Ltd, Londres (1971)
4. Skogby, H., Annersten, H., Domeneghetti, M.C., Molin, G.M., Tazzoli, V.: Iron distribution in orthopyroxene: a comparison of Mössbauer spectroscopy and X-ray refinement results. Eur. J. Mineral. **4**, 441–452 (1992)
5. Abdu, Y.A., Scorzelli, R.B., Varela, M.E., Kurat, G., Azevedo, I.S., Stewart, S.J., Hawthorne, F.C.: Druse clinopyroxene in D'Orbigny angritic meteorite studied by single—crystal X-ray diffraction, electron microprobe analysis, and Mössbauer spectroscopy. Meteorit. Planet. Sci. **44**, 581–587 (2009)
6. Brand, R.A.: Normos-90: Mössbauer Fitting Package User's Guide. Universität Duisburg, Duisburg (1994)
7. Valenzuela, M., Munayco, P., Abdu, Y.A., Scorzelli, R.B., Dos Santos, E., Morata, D.: Fe-bearing minerals in weathered ordinary chondrites from the Atacama desert. Meteorit. Planet. Sci. **42**, A152 (2007)
8. Bancroft, G.M., Maddock, A.G., Burns, R.G.: Applications of the Mössbauer effect to silicate mineralogy—I. Iron silicates of known crystal structure. Geochim. Cosmochim. Acta **31**, 2219–2246 (1967)
9. Bancroft, G.M., Burns, R.G, Howie, R.A.: Determination of the cation distribution in the orthopyroxene series by the Mössbauer effect. Nature **25**, 1221–1223 (1967)
10. Srivastava, K.K.P.: Electronic spin relaxation in Mössbauer spectra of Fe^{2+} in mineral hypersthene. J. Phys C: Solid State Phys. **19**(6407), 6416 (1986)
11. Wang, L., Moon, N., Zhang, Y., Dunham, W.R., Essene, E.J.: Fe-Mg order-disorder in orthopyroxenes. Geochim. Cosmochim. Acta **69**, 5777–5788 (2005)
12. Pasternark, M.P., Taylor, R.D., Jeanloz, R., Bohlen, S.R.: Magnetic ordering transition in $Mg_{0.9}Fe_{0.1}SiO_3$ orthopyroxene. Am. Mineral. **77**, 901–903 (1992)

DOI 10.1007/978-94-007-6482-8_29
Reprinted from *Hyperfine Interactions*, DOI 10.1007/s10751-012-0755-8

^{57}Fe Mössbauer spectroscopy studies of chondritic meteorites from the Atacama Desert, Chile: Implications for weathering processes

P. Munayco · J. Munayco · M. Valenzuela · P. Rochette · J. Gattacceca · R. B. Scorzelli

Published online: 5 February 2013

Abstract Some terrestrial areas have climatic and geomorphologic features that favor the preservation, and therefore, accumulation of meteorites. The Atacama Desert in Chile is among the most important of such areas, known as dense collection areas. This desert is the driest on Earth, one of the most arid, uninhabitable locals with semi-arid, arid and hyper-arid conditions. The meteorites studied here were collected from within the dense collection area of San Juan at the Central Depression and Coastal Range of Atacama Desert. ^{57}Fe Mössbauer spectroscopy was used for quantitative analysis of the degree of weathering of the meteorites, through the determination of the proportions of the various Fe-bearing phases and in particular the amount of oxidized iron in the terrestrial alteration products. The abundance of ferric ions in weathered chondrites can be related to specific precursor compositions and to the level of terrestrial weathering. The aim of the study was the identification, quantification and differentiation of the weathering products in the ordinary chondrites found in the San Juan area of Atacama Desert.

Keywords Weathering · Meteorite · Atacama desert

Thirteenth Latin American Conference on the Applications of the Mössbauer Effect, LACAME 2012, Medellín, Colombia, 11–16 November 2012.

P. Munayco (✉) · J. Munayco · R. B. Scorzelli
CBPF, Rua Xavier Sigaud 150, 22290-180 Rio de Janeiro, Brazil
e-mail: mpablo@cbpf.br

M. Valenzuela
Dpto. Geología, Universidad de Chile, Chile Plaza Ercilla 803, Santiago, Chile

P. Rochette · J. Gattacceca
CEREGE (CNRS/ Aix-Marseille University), BP 80, 13545 Aix-en-Provence Cedex 4, France

J. Gattacceca
Department of Earth, Atmospheric, and Planetary Sciences, Massachusetts Institute of Technology, 77 Massachusetts Avenue, Cambridge, MA 02139, USA

1 Introduction

The study of meteorites depends largely on finding samples that may have been under the influence of the environment for over hundred years. An interdisciplinary study is under way with meteorite samples from the Atacama Desert, one of the oldest and driest desert of the world [1], in an attempt to understand the weathering processes acting on these primitive materials and the conditions of the accumulation surfaces that have preserved them. Those samples may have undergone weathering processes that may result in large changes of their original phases.

Oxidation is certainly the most important reaction to consider and their products are usually crystalline oxides associated to the iron matrix. The degree of weathering in ordinary chondrites (OC) can be quantitatively measured using Mössbauer spectroscopy to determine the proportions of the Fe-bearing phases occurring mainly as Fe^0 in the Fe-Ni metal, Fe^{2+} in the ferromagnesian silicates and troilite, and Fe^{3+} in the terrestrial alteration products. In newly fallen equilibrated OC the amounts of Fe^0 (kamacite and taenite) and Fe^{2+} (olivine, pyroxene, and troilite) are known within narrow limits. Thus, the abundance of oxidized iron in weathered chondrites can be related to specific starting compositions and to the level of terrestrial weathering

The aim of this study is the identification and quantification of the weathering products of OC found in San Juan areas of the Atacama Desert (AD) [2].

2 Materials and methods

^{57}Fe Mössbauer spectroscopy in transmission geometry was performed at room temperature (RT) and 4.2 K in a 512 channels Halder spectrometer. The drive velocity was calibrated using a ^{57}Co in Rh matrix source and an iron foil both at RT. All measurements were performed at high velocity (12 mm/s), average recording time of 24 h per sample. Mössbauer absorbers were prepared with 90 mg/cm^2 of the bulk meteorite sample. NORMOS code [3] was used for the spectrum analysis based on least-square fitting routine, assuming each spectrum to be a sum of Lorentzian lines grouped into quadrupole doublets and magnetic sextets. The isomer shifts was reported in relation to α-Fe.

The Mössbauer measurements have been performed on 9 meteorites collected in the San Juan area. The meteorites are weathered ordinary chondrites (OC) and include the chemical groups H, L and LL.

3 Results and discussion

The Mössbauer spectra at 300 K show an overlapping of paramagnetic and magnetic phases as observed in Fig. 1 for two selected samples from San Juan area, showing a low (San Juan 013) and high (San Juan 025) weathering degree. In San Juan 012 magnetic components associated with iron oxides are not observed indicating a low weathering degree. On the other hand, magnetic components associated to iron oxides indicate a high weathering degree in San Juan 025 sample. The spectra were fitted with three quadrupole doublets, two quadrupole doublets attributed to

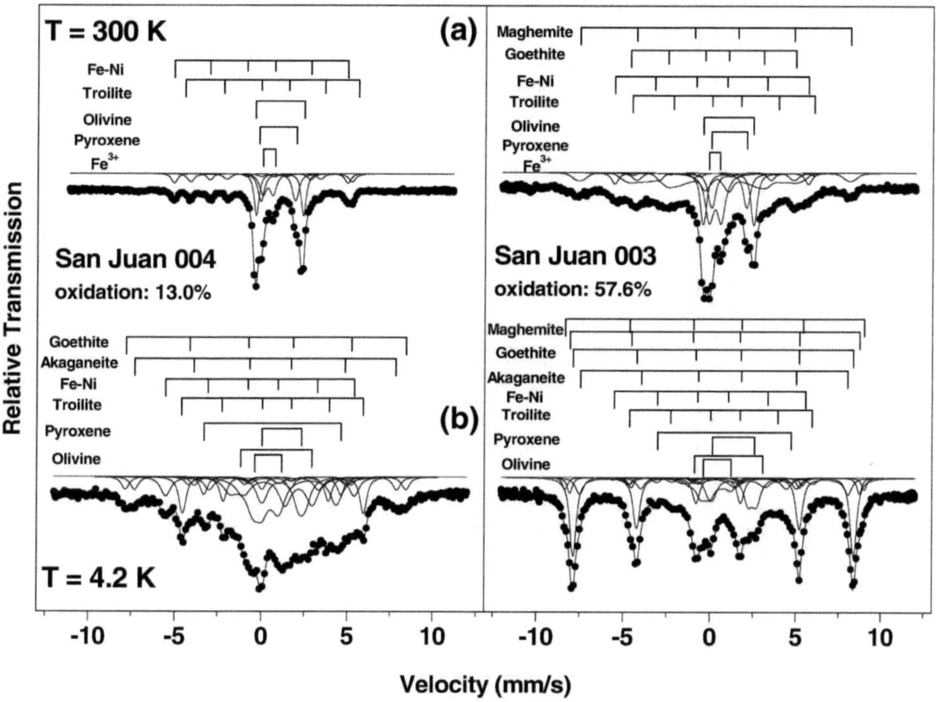

Fig. 1 Mössbauer spectra of some selected meteorite samples showing low (SJ 013) and high (SJ 025) degree of weathering from the San Juan area: **a** room temperature **b** low temperature (4.2 K)

Fe^{2+}: one associated with olivine (IS = 1.19 mm/s; QS = B3.0 mm/s) and the other to pyroxene (IS = 1.21 mm/s; QS = 2.14 mm/s). The third doublet is due to Fe^{3+}, whose Mössbauer parameters (IS = 0.36, QS = 0.77) are typical of a number of oxides and oxyhydroxides in a superparamagnetic state (e.g., ferrihydrite, small-particle goethite, lepidocrocite, akaganéite), and additionally, four magnetic sextets: two of them associated with the Fe-Ni primary phases (kamacite and taenite) and Troilite (FeS), the two remaining associated to goethite (B_{hf} = 38 T) and maghemite (B_{hf} = 50 T) [4].

Mössbauer measurements at low temperature (4.2 K) exhibit a too complex spectra exhibiting relaxation effects but apparently no magnetic ordering due to olivine and pyroxene and magnetic components associated to iron oxides/oxyhydroxides (Fig. 1b).

Mössbauer spectra at low temperature were fitted with the same components as in RT fitting: olivine, pyroxene and four magnetic components (troilite, goethite, maghemite, kamacite/taenite). In addition, it was observed one magnetic component (not observed at RT) with an hyperfine field of 48–49 T with nearly the same relative area of the Fe^{3+} doublet observed at RT. This magnetic component identified by their hyperfine parameters as akaganéite [5] can be associated to the magnetic splitting of the Fe^{3+} doublet at 4.2 K. In order to take into account the relaxation effects due olivine and pyroxene, it was necessary to use the spherical relaxation model of the fitting program Normos [3].

Table 1 Relative area for primary phases and weathering products obtained from the fitting of the Mössbauer spectra of meteorites recovered in the San Juan area at room and low temperature

Meteorite	Type	Primary mineralogy				Weathering products				Total
		Olivine (%)	Pyroxene (%)	FeNi (%)	Troilite (%)	Akaganéite (%)	Goethite (%)	Maghemite (%)	Fe^{3+}(%)	oxide (%)
$T = 300$ K										
San Juan 004	L	40.4	27.5	4.2	13.4	0.0	0.0	0.0	14.6	14.6
San Juan 008	LL	56.7	21.9	1.0	1.0	0.0	0.0	0.0	19.4	19.4
San Juan 006	H	26.0	16.4	5.1	8.7	0.0	16.5	4.8	22.6	43.8
San Juan 007	H	25.6	17.0	2.9	7.6	0.0	17.7	13.4	15.7	46.9
San Juan 012	H	24.2	15.7	4.3	7.7	0.0	17.9	13.0	17.2	48.1
San Juan 011	H	23.6	17.7	3.2	4.8	0.0	22.1	10.5	18.1	50.7
San Juan 010	H	21.6	16.6	2.3	3.6	0.0	27.7	4.9	23.4	55.9
San Juan 003	H	17.2	10.6	2.7	12.7	0.0	31.3	8.2	17.5	56.9
San Juan 005	H	18.3	11.0	2.6	10.3	0.0	20.0	13.5	24.3	57.8
$T = 4.2$ K										
San Juan 004	L	41.6	25.5	6.5	13.4	7.4	5.5	0.0	0	13.0
San Juan 008	LL	55.0	22.8	1.0	1.0	10.6	9.6	0.0	0	20.2
San Juan 006	H	27.7	16.8	5.3	9.4	14.5	20.9	5.5	0	40.8
San Juan 007	H	26.7	16.2	2.9	9.7	9.0	21.7	13.7	0	44.4
San Juan 012	H	25.7	17.3	4.3	7.8	7.7	25.0	12.2	0	44.9
San Juan 011	H	23.8	14.6	3.6	5.7	20.7	21.7	9.9	0	52.3
San Juan 010	H	21.7	15.8	2.4	3.7	14.4	35.1	6.9	0	56.4
San Juan 003	H	18.9	11.0	2.2	10.2	8.9	40.2	8.5	0	57.6
San Juan 005	H	20.2	11.2	1.2	11.6	17.1	26.7	12.0	0	55.8

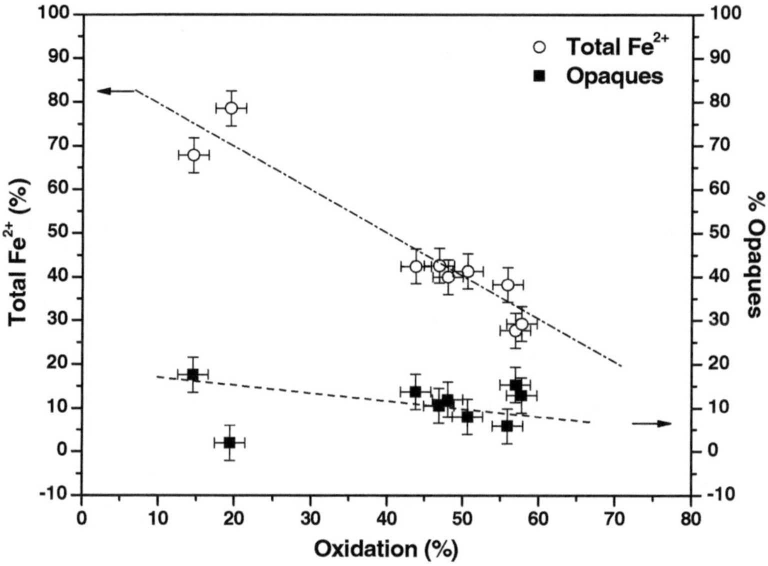

Fig. 2 Spectral area of total Fe^{2+} (%) and opaques against total ferric oxidation derived from Mössbauer spectra recorded from weathered samples of San Juan area, showing that both are weathered almost at a constant rate

The phase quantification was done using the relative areas taken from fitted Mössbauer spectrum (see Table 1) assuming that all of the phases have the same recoil-free fraction.

The samples show a wide range in total oxidation between 10 % and 58 %, given by the presence of Fe-oxides appearing as magnetically ordered Fe^{3+} (at low temperature) or as paramagnetic Fe^{3+} (at room temperature).

In order to study the weathering effect on the primary mineralogy of the samples we compare spectral areas of Fe^{2+} total (olivine and pyroxenes) and opaque (Fe-Ni and troilite) with the total amount of oxidation. Thus it was possible to determine which phases are the most susceptible to weathering (Fig. 2). The decrease in primary phases with increasing oxidation suggest that all iron-containing minerals within the meteorite are affected by weathering to some degree, as previously reported for OCs from other hot deserts [6].

4 Conclusions

From ^{57}Fe Mössbauer spectroscopy results we are able to conclude that the weathering degrees of San Juan OC samples vary in the range of ~10 to 57 %. The observed iron oxides are present as magnetically ordered Fe^{3+} (magnetite, maghemite, goethite) and paramagnetic Fe^{3+}. The last one, identified as akaganéite, the first product of Fe-Ni oxidation found in all samples in the range of 7 to 22 %. The decrease in primary phases with increasing oxidation suggests that all iron-containing minerals within the meteorite are affected by weathering to some degree

Acknowledgements R. B. Scorzelli would like to thank FAPERJ and CNPq for financial support; P. Munayco is grateful to FAPERJ and J. Munayco is grateful to CNPq for his fellowship.

References

1. Dunai, T.J., González López, G.A., Juez-Larré, J.: Oligocene–Miocene age of aridity in the Atacama Desert revealed by exposure dating of erosion-sensitive. Geology **33**, 321–324 (2005)
2. Gattacceca, J., Valenzuela, M., Uehara, M., Jull, A.J.T., Giscard, M., Rochette, P., Braucher, R., Suavet, C., Gounelle, M., Morata, D., Munayco, P., Bourot-Denise, M., Bourles, D., Demory, F.: The densest meteorite collection area in hot deserts: the San Juan meteorite field (Atacama Desert, Chile). Meteorit. Planet. Sci. **46**, 1276–1287 (2011)
3. Brand, R.A.: Normos Mössbauer Fitting Program, User's Guide. Wissenschaftlich Elektronik GmbH, Starnberg (1995)
4. Stevens, J.G., Khasanov, A.M., Miller, J.W., Pollak, H., Li, Z.: Mössbauer Mineral Handbook. Mössbauer Effect Data Center. The University of North Carolina, Asheville (1998)
5. Barrero, C.A., García, K.E., Morales, A.L., Kodjikian, S., Greneche, J.M.: New analysis of the Mössbauer spectra of Akaganéite. J. Phys., Condens. Matter **18**, 6827–6840 (2006)
6. Bland, P.A., et al.: Climate and rock weathering: a study of terrestrial age dated ordinary chondritic meteorites from hot desert regions. Geochim. Cosmochim. Acta **62**, 3169–3184 (1998)

DOI 10.1007/978-94-007-6482-8_30
Reprinted from *Hyperfine Interactions*, DOI 10.1007/s10751-013-0874-x

Mössbauer study of the inorganic sulfur removal from coals

F. Reyes Caballero · S. A. Martínez Ovalle

Published online: 6 June 2013

Abstract Mössbauer Spectroscopy (MS) was applied to study the occurrence and behavior of the iron-sulfur-containing minerals in coal and coal fractions obtained by different separation methods: hydrocyclonic, flotation and chemical removal process. Samples of one high sulfur coal from Guachinte mine (Valle, Colombia) and three low sulfur coals from the El Salitre zone (Paipa-Boyacá, Colombia) were analyzed. MS evidenced only the presence of pyrite in Esmeralda and Las Casitas coals, while it identified pyrite and siderite on Cerezo coal. MS and SEM- EDX confirm the inorganic sulfur removal on Guachinte coal submitted to hydrocyclonic removal process. MS of the precipitated coal fraction from Las Casitas mine obtained by flotation in water showed the presence of ferrous sulfate because of coal-weathering process. Treatment with hot diluted HNO3 equal to 27 acid on raw coal sample from Las Casitas mine showed that almost all of the pyrite in raw coal was removed.

Keywords Coal · Inorganic sulfur removal · Mössbauer spectroscopy

1 Introduction

Coal is mainly used for power worldwide. It is estimated that about 41 % of global elcctricity is generated from coal. In Colombia are the major reserves of coal from Latin American, 6.7 Gton dispersed in different zones in the country.

Proceedings of the thirteenth Latin American Conference on the Applications of the Mössbauer Effect, (LACAME 2012), Medellín, Colombia, 11–16 November 2012.

F. Reyes Caballero · S. A. Martínez Ovalle (✉)
Grupo de Física Nuclear Aplicada y Simulación,
Universidad Pedagógica y Tecnológica de Colombia, UPTC, Tunja, Colombia
e-mail: s.agustin.martinez@uptc.edu.co

As it is known, sulfur in coal occurs in two principal forms [1, 2]: organic sulfur and inorganic sulfur. The inorganic sulfur includes pyritic-sulfur (monosulfides and disulfides) and sulfate-sulfur (in combination with Ca or Fe). Free sulfur is present only sporadically. The presence and amounts of these sulfur forms vary considerably in coals from diverse origin. ASTM standard test method for forms of sulfur in coal, D 2492, is very laborious. Sulfate sulfur is extracted from the analysis sample with dilute hydrochloric acid (HCl). The sulfate sulfur in the extract is determined gravimetrically. Sulfates are soluble in HCl acid, but pyritic and organic sulfur are not. Pyritic sulfur is calculated from an analysis of the iron content of a nitric acid (HNO_3) solution. The iron is determined by atomic absorption techniques. The organic sulfur is then determined by the difference between the inorganic sulfur (sulfate + pyritic sulfur) and the total sulfur in the original material. The major objection to this standard method is the fact that an error in any one step of the procedure can affect the values determined in other portions of the method.

In our work, ^{57}Fe transmission Mössbauer Spectroscopy (MS) was applied in order to study the occurrence and behavior of the iron-sulfur-containing minerals in coal and coal fractions obtained by different separation methods: hydrocyclonic, flotation and chemical removal processes. MS is applied specifically to obtain a better characterization of the iron-sulfur-containing minerals of the analyzed samples.

The iron-sulfur-containing minerals play an important role in all coal-conversion processes (beneficiation, carbonization, combustion, gasification, liquefaction and pyrolysis): severe pollution caused by sulfur in coal during its combustion is considered to be responsible for acid rain, ozone layer depletion and several other harmful effects on the environment [3, 4]; previous work has indicated that pyritic-sulfur is the most important source of trace-elements such as As and Se in coal [5–7]; problems in technological performance (problems of fouling and slagging) [8, 9]; health impacts (respiratory problems of humans and animals [10, 11]; organic sulfur is chemically bound to the coal and very difficult to remove by physical methods, while inorganic sulfur is not bound chemically and could be removed to certain degrees from the raw coal. It is essential to reduce inorganic sulfur contents in particular and mineral matter contents in general from coal prior to its utilization. The study of different physical and chemical properties of the coal has contributed to various sulfur removal methods which involve a large variety of physical, chemical and biological processes [8, 12–16].

2 Experimental

Samples of beneficiated and precipitated coal obtained in a hydrocyclone separation unit with grain size between 600 to 1000 μm from Guachinte mine (Valle, Colombia) were selected for SEM-EDX (Scanning Electron Microscopy—Energy Dispersive X-ray analysis) studies. SEM-EDX analysis was carried out using a JEOL 8900R system. Mössbauer and XRD characterization of the mineral matter of coal from Guachinte mine were reported before [17]. The mineral phases identified by XRD were kaolinite, quartz, pyrite, gypsum, dolomite and calcite. MS confirm the presence of pyrite and kaolinite, besides evidence the additional presence of jarosite which was not detected by XRD due their low amount. The Guachinte mine coals are used

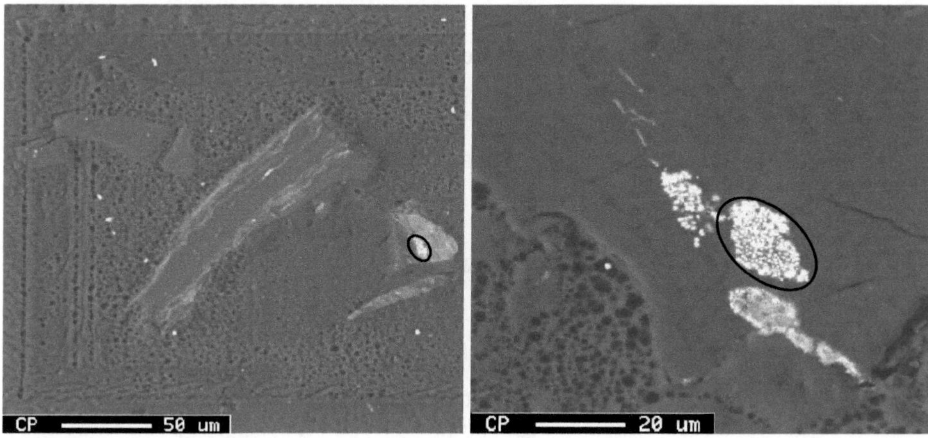

Fig. 1 SEM images of a general view of beneficiated coal product (*left side*) and precipitated coal product (*right side*) by hydrocyclonic separation process. In the *circles* were identified clusters of grains of pyrite

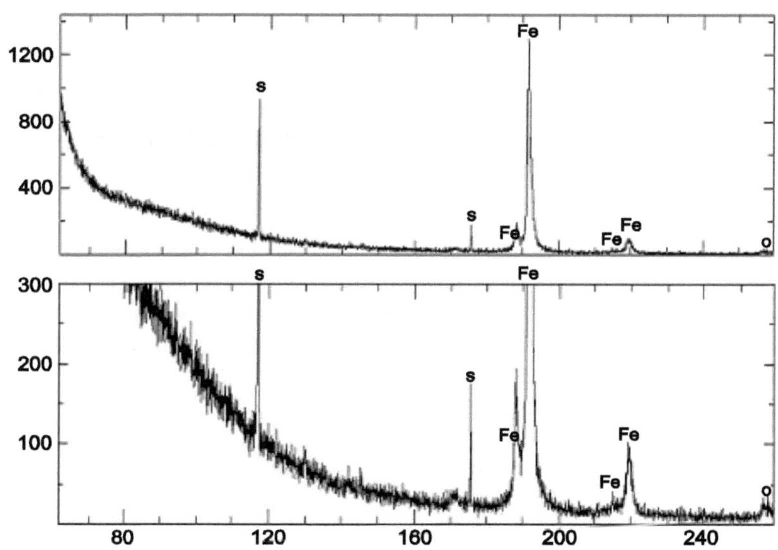

Fig. 2 EDX spectra associated with the grains of pyrite (FeS_2) identified in the Fig. 1

for combustion in the power generation plants for sugar, paper and cement local industry.

Run of mine coal samples from Cerezo, Esmeralda and Las Casitas mines located in El Salitre zone (Paipa-Boyacá, Colombia) were prepared according to ASTM standard procedure, D 2013 (pulverized to pass No. 60 (250 μm) sieve). These coals are used in the thermal power station TERMOPAIPA located to 40 Km of our laboratory.

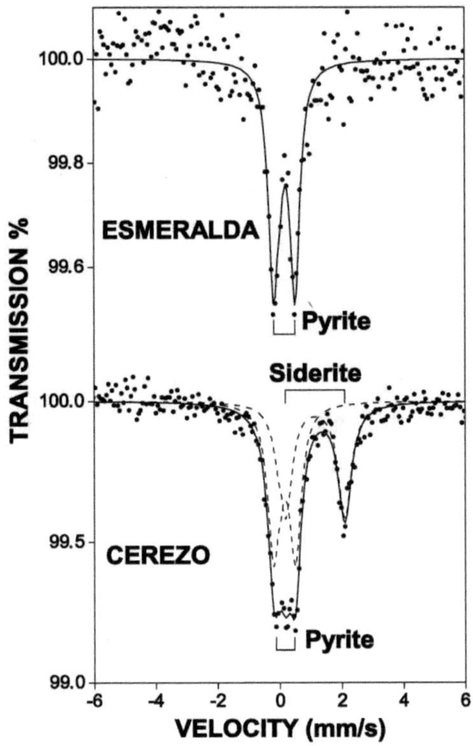

Fig. 3 Mössbauer spectra of the Cerezo and Esmeralda coals, El Salitre zone, Paipa, Boyacá, Colombia

Proximate analysis, total sulfur and ash contents were determined according to ASTM standard procedures. Sulfur removal (*SR*) or ash removal (ASHR) are defined as:

$$SR\,(or\,ASHR) = \frac{X_o - X}{X_o} \times 100\ \%,$$

where X_o is equal to initial value of sulfur or ash, and X is equal to corresponding residual value after removal process.

The flotation method in water was used for to study the removal of iron-sulfur-containing minerals on coal from Las Casitas mine with grain size 250 μm. Raw coal from Las Casitas mine was desulfurized and demineralized by treatment with hot dilute nitric acid. The experimental procedure was carried out with 100 g coal and 400 ml dilute nitric acid. The mixture in a glass beaker was heated in a stove for 8 min. The coal residues were filtered and dried in air.

For Mössbauer analysis the grain size of samples used was 250 μm. Mössbauer spectra of raw coal or coal fractions were recorded at room temperature using a conventional spectrometer with a ^{57}Co/Rh source and calibration with respect to iron metal. The spectra were fitted using the VARFIT and MOSSFIT programs.

Table 1 Mössbauer parameters of phases identified of the coals from El Salitre zone

Sample	IS [$mm \cdot s^{-1}$]	QS [$mm \cdot s^{-1}$]	H (KOe)	Phase
Cerezo	0.28	0.64	0	Pyrite
	1.23	1.90	0	Siderite
Esmeralda	0.30	0.62	0	Pyrite
Las Casitas	0.27	0.50	0	Pyrite

Table 2 Ash and total sulfur analysis of coals from El Salitre zone

Sample	Ash [wt. %]	Total sulfur [wt. %]
Cerezo	19.83	0.76
Esmeralda	14.25	0.69
Las Casitas	12.63	1.10

Table 3 Proximate analysis of coal Las Casitas mine

Analysis	ASTM standard	Air-dried	Dry
Moisture [wt. %]	D 7582	10.61	
Volatile matter [wt. %]	D 7582	40.26	45.04
Ash [wt. %]	D 7582	11.29	12.63
Fixed Carbon [wt. %]	D 3172	37.84	42.33
Calorific Power [BTU/Lb]	D 5865	10952	12252
Total sulfur [wt. %]	D 4239	0.98	1.10

3 Results and discussion

SEM-EDX analysis confirms the presence of the pyrite phase in the samples of the hydrocyclonic removal process applied to the Guachinte coal. Figure 1 show SEM images of a general view of beneficiated coal product (left side) and precipitated coal product (right side). In the circles were identified clusters of grains of pyrite. An increase in the pyrite content is evident in the precipitated coal fraction. These images have revealed the presence of pyrite in the form of clusters of framboidal particles. These images are representative of a common way which pyrite crystals occur in coal. It is well known that the pyrite in the coal is present in different forms [18–21]. Figure 2 shows EDX spectra associated with the grains of pyrite identified in the Fig. 1.

Figure 3 shows the Mössbauer spectra of raw coal samples from Cerezo and Esmeralda mines. The Esmeralda coal produced the simplest spectrum where the only identified phase was pyrite. In the spectrum of the Cerezo coal one can note the presence of pyrite and siderite. Table 1 shows Mössbauer parameters of the identified phases for coals from El Salitre zone. The identified phases in Cerezo and Esmeralda coals by Non-Rietveld XRD single analysis were kaolinite, quartz and pyrite. Table 2 shows ash and total sulfur analysis of coals from El Salitre zone.

Proximate analysis of the Las Casitas coal is presented in Table 3. This is a high-volatile B Bituminous coal in according to ASTM standard procedure, D388. Figure 4 (Top) shows Mossbauer spectrum of Las Casitas coal, this spectrum was fitted with one only doublet corresponding to pyrite.

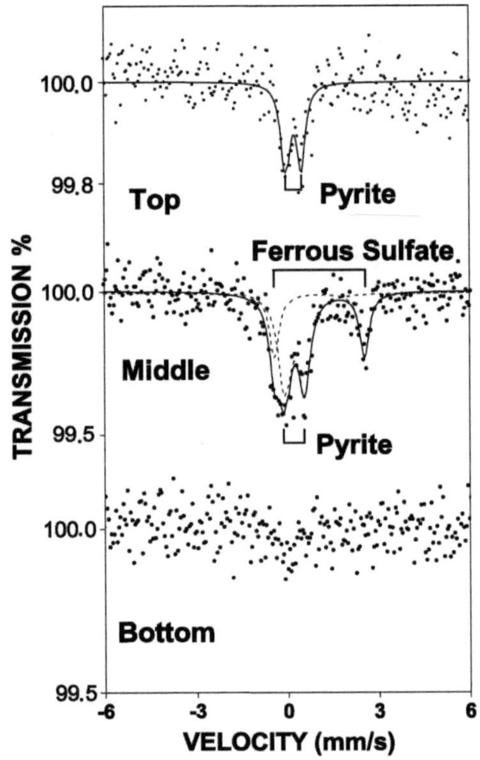

Fig. 4 (*Top*) Mössbauer spectrum of the Las Casitas coal. (*Middle*) Mössbauer spectrum of the precipitated coal fraction in water. (*Bottom*) Collapsed spectrum of the coal treated with hot dilute nitric acid

Table 4 Mössbauer parameters of the precipitated coal fraction in water

Sample	IS [$mm \cdot s^{-1}$]	QS [$mm \cdot s^{-1}$]	H (KOe)	Phase
Precipitated	0.27	0.63	0	Pyrite
	1.10	2.92	0	Ferrous sulfate

Coal from Las Casitas mine

Habitually, the coal is pretreated before use. When a coal is pulverized, its mineral components may be liberated from the coal particles, producing as pure mineral particles. The mineral grains in pulverized coal can be named included or excluded. Included mineral grains are associated with or contained within a carbonaceous coal particle. Excluded mineral grains do not have any carbonaceous material associated with them. The flotation process is a method for desulfurizing fine coal and is based on the assumption that coal is easily floatable whereas pyrite is not. In practice, included pyrite particles often float with the coal.

Figure 4 (Middle) shows Mössbauer spectrum of the precipitated coal fraction sample from flotation process in water. This spectra was fitted with a doublet corresponding to pyrite and a doublet corresponding to ferrous sulfate ($FeSO_4$, $FeSO_4 \cdot H_2O$, $FeSO_4 \cdot 4H_2O$, $FeSO_4 \cdot 7H_2O$). For revision of the values earlier reported in the literature was chosen $FeSO_4$ (with QS = 2.92 mm/s) [22]. The ferrous

Table 5 Total sulfur and ash analysis of the coal fractions from Las Casitas mine

ASTM standard	Sample	Analysis dry	
		Total sulfur [wt. %]	Ash [wt. %]
D 4239 – D 7582	Floated	1.15	11.78
	Precipitated	1.26	13.22
	Extract	0.50	7.15

Table 6 Comparative of the reductions in values of total sulfur (SR) and ash (ASHR)

Sample	Removal method	SR [%]	ASHR [%]	Resultant effect
Guachinte	Hydrocyclonic	23.8	73.8	High removal
Cerezo	Hydrocyclonic	–	65.9	High removal
Esmeralda	Hydrocyclonic	–	21.5	High removal
Las Casitas	Flotation	No change	6.73	Coal-weathering
Las Casitas	HNO_3 Acid	54.5	43.4	Desulphurization and demineralization

sulfate is found as a result of weathering and oxidation of pyrite. The iron-bearing minerals in general are particularly sensitive to weathering [23]. Table 4 shows the Mössbauer parameters of the precipitated coal fraction from Las Casitas mine. Table 5 shows total sulfur and ash analysis of floated and precipitated coal fractions and of the extract product treatment with hot dilute nitric acid on the original coal sample.

The MS evidenced the presence of only pyrite in Las Casitas coal. Then our aim was remove this pyrite. Results of experimental procedure carried out over a sample of Las Casitas coal treated with hot dilute nitric acid indicate that the coal was desulfurized and demineralized by removal of the original pyrite. In Fig. 4 (Bottom) shows the collapsed spectrum of coal sample treated with hot dilute nitric acid. The reductions in values of total sulfur and ash of all the removal processes are compared in Table 6.

4 Conclusions

There exist differences in occurrence and contents of minerals for coals from diverse origin. SEM-EDX allows one to confirm the presence of mineral phases and elemental compositions in coals and by-products. There is evidence of an increment of the pyrite content in precipitated coal fraction obtained by hydrocyclonic removal process.

The only iron-sulfur phase identified by Mössbauer spectroscopy on samples of coal from Salitre zone (Paipa, Boyacá-Colombia) was pyrite. The pyrite was not removed in the flotation process. Besides, it promotes the presence of sulfate sulfur by coal-weathering process. It can be concluded that the coal from Las Casitas mine treated with hot dilute nitric acid was desulfurized and demineralized. MS plays an important role in the characterization of the iron-sulfur-containing minerals present in coals.

References

1. Calkins, W.H.: Fuel **73**, 475–484 (1994)
2. Gryglewicz, G.: Fuel **74**, 356–361 (1995)
3. Bemirbas, A.: Energ. Convers. Manage. **43**, 885–895 (2002)
4. Baruah, B.P., Khare P.: Energ. Fuels **21**, 2156–2164 (2007)
5. Clemens, A.H., Deely, J.M., Gong, D., Moore, T.A., Shearer, J.C.: Fuel **79**, 1781–1784 (2000)
6. Kolker, A., Huggins, F.E., Palmer, C.A., Shah, N., Crowley, S.S., Huffman, G.P., Finkelman, R.B.: Fuel Proces. Tech. **63**, 167–178 (2000)
7. Spears, D.A., Manzanares-Papayanopoulus, L.I., Booth, C.A.: Fuel **78**, 1671–1677 (1999)
8. Bryers, R.W.: Progr. Energ. Combust. Scien. **22**, 29–120 (1996)
9. Shirazi, A.R., Börtin, O., Eklund, L., Lindqvist, O.: Fuel **74**, 247–251 (1995)
10. Finkelman, R.B., Gross, P.M.K.: Internat. J. Coal Geol. **40**, 91–101 (1999)
11. Ambedkar, B., Chintala, T.N., Nagarajan, R., Jayanti, S.: Chem. Engin. Proces: Proc. Intens. **50**, 236–246 (2011)
12. Borah, D., Baruah, M.K.: Fuel **78**, 1083–1088 (1999)
13. Rubiera, F., Steve, T., Hall, T., Shah, C.L.: Fuel **76**, 1187–1194 (1997)
14. Acharya, C., Kar, R.N., Sukla, L.B.: Fuel **80**, 2207–2213 (2001)
15. Karavaiko, G.I., Lobyreva, L.B.: Fuel Procces. Tech. **40**, 167–182 (1994)
16. Thoms, T.: Fuel Proces. Tech. **43**, 123–128 (1995)
17. Reyes, F., Pérez Alcázar, G.A., Barraza, J.M., Bohórquez, A., Tabares, J.A., Speziali, N.L.: Hyperfine Interact. **148/149**, 39–46 (2003)
18. Creelman R.A., Ward, C.R.: Int. J. Coal Geol. **30**, 249–269 (1996)
19. Shirazi, A.R., Eklund, L., Lindqvist O.: Fuel **73**, 193–198 (1994)
20. Waanders, F.B., Vinken, E., Mans, A., Mulaba-Bafubiandi, A.F.: Hyperfine Interact. **148/149**, 21–29 (2003)
21. Wigley, F., Williamson, J., Gibb, W.H.: Fuel **76**, 1283–1288 (1997)
22. Stevens, J.G., Khasanov, A.M., Miller, J.W., Pollak, H., Li, Z.: Mössbauer Minerals Handbook. Baltimore Press, North Carolina, USA (1998)
23. Gracia, M., Gancedo, J.R., Barrero, M.L., Garcia, A.B., Martínez-Alonso, A., Tascon, J.M.D.: Nucl. Instrum. Phys. Res. B. **76**, 191–194 (1993)

DOI 10.1007/978-94-007-6482-8_31
Reprinted from *Hyperfine Interactions*, DOI 10.1007/s10751-013-0873-y

Mössbauer analysis of coal coke samples from Samacá, Boyacá, Colombia

W. A. Pacheco Serrano · D. Quintão Lima · J. D. Fabris

Published online: 6 June 2013

Abstract Three samples of coke produced from coal from a mine in the municipality of Samacá, department of Boyacá, Colombia, were studied essentially with Mössbauer spectroscopy. The samples were treated with NaOH 5 mol L^{-1} in order to increase the proportion of iron oxides, by selectively dissolving silicate minerals and any remaining gibbsite, before the physical analysis. Room temperature Mössbauer data revealed that all samples do contain major proportions (>50 % of the relative subspectral area) of hematite along with (super) paramagnetic species as iron-bearing chemical compounds. The superparamagnetic contribution may be due to very fine grains of iron oxides, including nanometric hematite.

Keywords Iron in coal · Coke · Hematite · Chemical dissolution

Proceedings of the thirteenth Latin American Conference on the Applications of the Mössbauer Effect, (LACAME 2012), Medellín, Columbia, 11–16 November 2012.

W. A. Pacheco Serrano (✉)
Escuela de Física, Universidad Pedagógica y Tecnológica de Colombia, UPTC, Tunja, Boyacá, Colombia
e-mail: william.pacheco@uptc.edu.co

D. Quintão Lima · J. D. Fabris
Departamento de Química, ICEx, Universidade Federal de Minas Gerais, Belo Horizonte, MG, Brasil

D. Quintão Lima
e-mail: dianaquintao@yahoo.com.br

J. D. Fabris
Universidade Federal dos Vales do Jequitinhonha e Mucuri (UFVJM), 39100-000 Diamantina, MG, Brasil
e-mail: jdfabris@ufmg.br

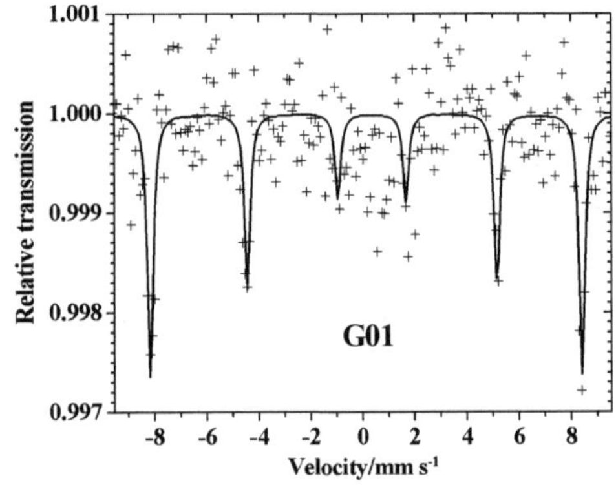

Fig. 1 Mössbauer spectrum of coal coke sample G 01 without chemical treatment at RT

1 Introduction

Coal production is experiencing a growing economical importance in Colombia. A significant part of the coal production is direct to supply the industrial demand in the country; part is exported after being processed to form coke. In Boyacá, especially in Samacá (5° 29′ 60.00″ N 73° 30′ 0″ E), the coal activity has been taking a particular economical importance. The extraction of coal mineral and its industrial processing to produce coke have been playing a very significant role in the whole economy of Boyacá. However, many concerns on any potential threats to the natural environment are permanent issues regarding national policies to mining and other industrial activities. Knowing more about the chemical and mineralogical nature of such an ore requires efforts of ample scientific researches in order to allow planning for its rational exploitation, while preventing uncontrolled harm to the natural environment.

The relationship between coal and iron oxides is a fundamental matter in the various application processes in industrial scale, especially if the coke is to be used for energy generation.

Mössbauer spectroscopy is a nuclear physical technique that allows assessing the intrinsic properties of the iron-bearing mineral phases in coal as well as following the subsequent stages of industrial processing, such as in the coking process. An example of this state of affairs is the correlation between the coal aromaticity and chemical state of iron in the ore [1].

In order to improve the Mössbauer analysis, these three coal coke samples were treated with NaOH 5 mol L^{-1}, following the method originally proposed by Norrish and Taylor [2, 3], to selectively remove silicates and, eventually, any remaining gibbsite.

2 Experimental

Prior to coking the coal production, as a first industrial procedure, in the Samacá industrial plant, the material is usually subjected to a cleaning process with water; in some cases magnetite is added intending to remove iron phase inclusions of the mineral coal. Magnetite is mixed in water at different densities and during gravimetric

Fig. 2 RT - Mössbauer spectra for samples after chemical treatment

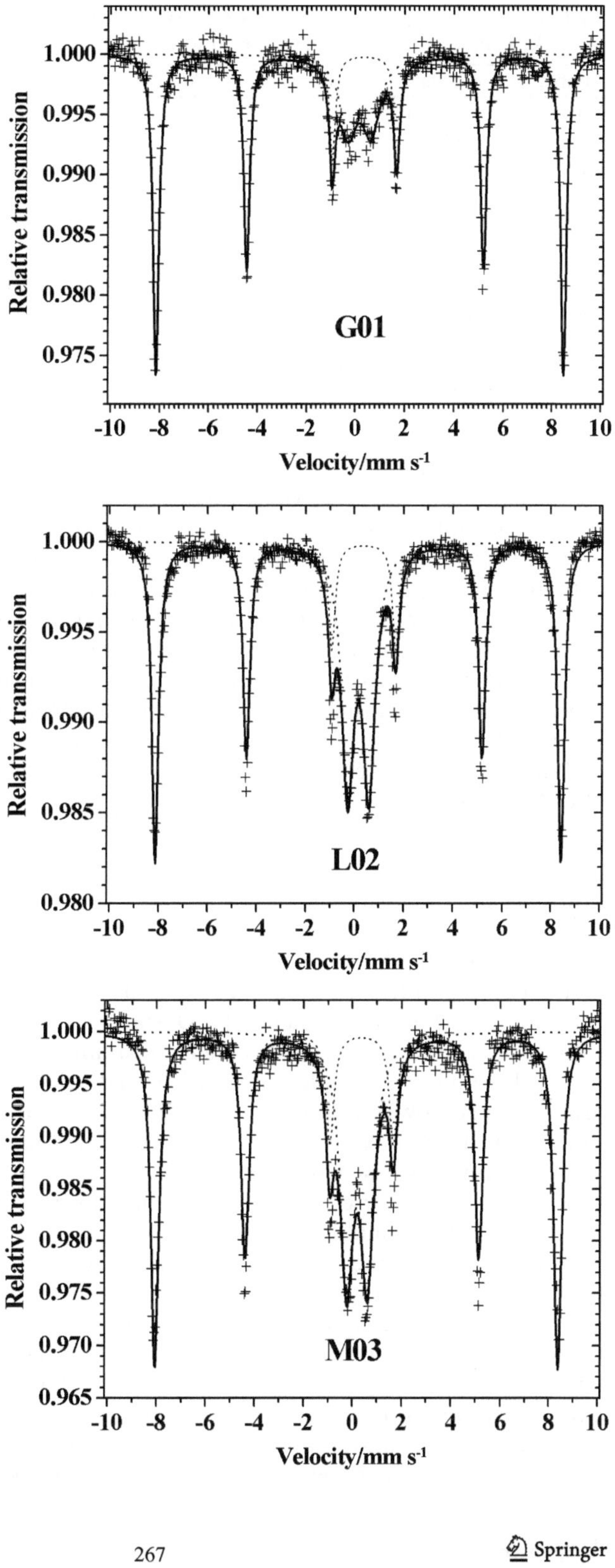

Table 1 Hyperfine parameters measured by ^{57}Fe Mössbauer spectroscopy at room temperature for coal coke samples after chemical treatment with NaOH 5 mol L^{-1}

Sample	δ/mms^{-1}	ε/mms^{-1}	Δ/mms^{-1}	B_{hf}/T	RA/%	Assignment
G01	0.371(2)	–		51.57(1)	76(1)	Hm
	0.28(2)	0.208(3)	0.92(3)		24(1)	
L02	0.368(1)	–		51.349(7)	60(1)	Hm
	0.283(3)	0.237(2)	0.867(4)		40(1)	
M03	0.369(2)	–		50.98(1)	63(1)	Hm
	0,306(4)	0.216(3)	0.847(7)		37(1)	

δ = isomer shift relative to αFe; Δ = quadrupole splitting, ε = quadrupolar shift; B_{hf} = hyperfine magnetic field; RA/% = relative subspectral area; Hm = hematite. Numbers in parentheses are uncertainties over the last significant digit corresponding to the value, as outputted by the least squares-fitting computer program

steps some minerals like pyrite, marcasite, gibbsite and others are removed from raw coal mineral. But the iron oxides are difficult to be removed from the raw coal.

Three samples from different parts of the region were subjected to oven heating to 900 °C in separate stages. Mössbauer measurement were performed on samples without any chemical treatment, and, depending on the observed results, with chemical treatment with NaOH 5 mol L^{-1} at 90 °C for 2 hours; such a treatment made it possible to concentrate the magnetic phases in other mineralogical components of the coke [3–5]. Mössbauer measurements were performed at RT on a conventional spectrometer in a transmission configuration, using a ^{57}Co/Rh source with a nominal activity of ~20 mCi. Chemical formulae were allocated through balance of charges and masses based on chemical composition data, following the algorithm described elsewhere [6].

3 Results and discussions

Figure 1 shows the Mössbauer spectrum for G01 ore sample without chemical treatment. The absorption is relatively low; ~ 0.3 %. The magnetic hyperfine structure is characteristic of that for hematite. Despite of the scattered points on the pattern, fitting spectrum with just a sextet leads to parameters in Table 1.

For the other two samples, namely L02 and M03, Mössbauer spectra (not shown) also show very low absorption (~ 0.2 %), again due to the low concentration of iron in the coal. In both cases, a larger data scattering made it more difficult to fit and assign parameters to spectra.

Figure 2 shows corresponding Mössbauer spectra obtained after chemical treatment with NaOH 5 mol L^{-1}.

Table 1 presents the Mössbauer corresponding hyperfine parameters for samples G01, L02 and M03 after chemical treatment with NaOH 5 mol $L^{-}1$.

All samples were found to contain hematite (ideal formula, αFe_2O_3) and Fe^{3+} (super) paramagnetic component. The iron oxides obtained through the coking process at 900 °C may be actually regarded as being a relatively complex mixture of decomposition product of hydrated iron(III) oxide – oxides in varying proportions . Parameters for the sextets are characteristic for hematite. The doublets are generally assignable to (super) paramagnetic ferric chemical species. Any superparamagnetic contribution would be due to iron oxides in very fine particles, including nanosized hematite. To confirm this hypothesis further Mössbauer measurements at low temperatures down to 4.2 K are still needed.

The chemical treatment allowed us to clearly observe the six lines of the hyperfine structure due to hematite at room temperature, along with few details of the whole doublet on spectra for these three samples.

4 Conclusion

Hematite and (super) paramagnetic Fe^{3+} species form the iron-containing fractions in this coal, the starting material to obtain coke in Samacá, Colombia.

The current steps to clean the raw coal in the industry, to remove iron oxides, is relatively optimized, but still far from ideal. The coke product from this coal may be more safely used in industrial metallurgical furnaces but this is less evident for applications in energy generation as the remaining iron forming hematite and other Fe^{3+} chemical species can produce outcomes that may be harmful to the natural environment.

The selective chemical treatment of the raw coal samples with NaOH 5 mol L^{-1} at 90 °C for 2 hours has proved to be effective enough to allow a clear Mössbauer identification of hematite, the dominant iron oxide in these coal samples.

Acknowledgements Work supported by CNPq (grant # 302479/2010-4), and FAPEMIG (grant # PPM 00419/10) (Brazil), and DIN – UPTC (Colombia). JDF is Visiting Professor at UFVJM, under the PVNS/CAPES program (Brazil).

References

1. Ahmed, M.A., Vandenberghe, R.E., De Grave, E., Eissa, N.A., Ibarra J.V.: Fuel. **78**(4), 453–457 (1999)
2. Norrish, K., Taylor, R.M.: J. Soil Sci. **12**, 294–306 (1961)
3. da Costa Couceiro, P.R.: Características quimico – mineralógicas de óxidos de ferro e de caulinita de perfis de solos da regiǎo de Manaus, Amazônia, Brasil. Tese de Doutorado. ICEx, UFMG, (2004), p. 68. In portuguese
4. de F.F. Lelis, M., Gonçalves, C.M., Fabris, J.D., da N. Mussel, W., Pacheco Serrano, W.A.: J. Braz. Chem. Soc. **15**(6), 884–889 (2004)
5. Gonçalves, C.M., Fabris J.D., Pacheco Serrano, W.A.: Hyperfine Interact. **122**, 171 (1999)
6. Fabris, J.D., Coey, J.M.D., da N. Mussel, W.: Hyperfine Interact. **113**, 249 (1998)

DOI 10.1007/978-94-007-6482-8_32
Reprinted from *Hyperfine Interactions*, DOI 10.1007/s10751-013-0876-8

ERRATUM

Erratum to: Mössbauer analysis of coal coke samples from Samacá, Boyacá, Colombia

W. A. Pacheco Serrano · D. Quintão Lima · J. D. Fabris

Published online: 24 July 2013

Erratum to: Hyperfine Interact DOI 10.1007/s10751-013-0873-y

This article was published with an erroneous version of Table 1. Please find the correct Table 1 on this page that should be regarded by the reader as the final version.

Table 1 Hyperfine parameters measured by ^{57}Fe Mössbauer spectroscopy at room temperature for coal coke samples after chemical treatment with NaOH 5 mol L^{-1}

Sample	δ/mm s^{-1}	ε/mm s^{-1}	Δ/mm s^{-1}	B_{hf}/T	RA/%	Phase
G01	0.371(2)	−0.208(3)		51.57(1)	76(1)	Hm
	0.28(2)		0.92(3)		24(1)	Fe^{3+}
L02	0.368(1)	−0.237(2)		51.349(7)	60(1)	Hm
	0.283(3)		0.867(4)		40(1)	Fe^{3+}
M03	0.369(2)	−0.216(3)		50.98(1)	63(1)	Hm
	0.306(4)		0.847(7)		37(1)	Fe^{3+}

δ = isomer shift relative to αFe; Δ = quadrupole splitting, ε = quadrupole shift; B_{hf} = hyperfine magnetic field; RA/% = relative subspectral area; Hm = hematite; Fe^{3+} = (super) paramagnetic high spin ferric iron oxide(s). Numbers in parentheses are uncertainties over the last significant digit corresponding to the value, as outputted by the least squares-fitting computer program

The online version of the original article can be found under http://dx.doi.org/10.1007/s10751-013-0873-y.

W. A. Pacheco Serrano (✉)
Escuela de Física, Universidad Pedagógica y Tecnológica de Colombia, UPTC, Tunja, Boyacá, Colombia
e-mail: william.pacheco@uptc.edu.co

D. Quintão Lima · J. D. Fabris
Departamento de Química, ICEx, Universidade Federal de Minas Gerais, Belo Horizonte, MG, Brasil

D. Quintão Lima
e-mail: dianaquintao@yahoo.com.br

J. D. Fabris
Universidade Federal dos Vales do Jequitinhonha e Mucuri (UFVJM), 39100-000 Diamantina, MG, Brasil
e-mail: jdfabris@ufmg.br

DOI 10.1007/978-94-007-6482-8_33
Reprinted from *Hyperfine Interactions*, DOI 10.1007/s10751-013-0826-5

Indoor atmospheric corrosion of conventional weathering steels in the tropical atmosphere of Panama

Juan A. Jaén · Josefina Iglesias · Olga Adames

Published online: 28 February 2013

Abstract One year indoor atmospheric corrosion examinations have been carried out on two conventional weathering steels for a year, at two test sites, Tocumen and Sherman Breakwater in Panama. They are environmentally classified by ISO 9223 as $S_1P_0\tau_4$ and $S_3P_0\tau_5$, respectively. In this humid-tropical marine climate corrosion rates are rather high, especially at Sherman Breakwater test site, mainly due to the high deposition of chloride, among other environmental conditions. Our results indicate that indoor corrosion is highly determined by the time of wetness and chloride ions. A-588 weathering steel corroded at a generally lower rate than COR-420 weathering steel. Rust characterization was performed by XRD, FTIR, and Mössbauer spectroscopy. Lepidocrocite, goethite, maghemite and akaganeite were found as corrosion products. Akaganeite is only detected when high chlorides deposition rates are obtained, and no washing effect occurs. This phase, together with maghemite, is obtained when there is greater aggressiveness in the environment.

Keywords Weathering steel · Mössbauer spectroscopy · X-ray diffraction · IR spectroscopy · Indoor corrosion

Proceedings of the Thirteenth Latin American Conference on the Applications of the Mössbauer Effect, (LACAME 2012), Medellín, Colombia, 11–16 November 2012.

J. A. Jaén (✉)
Depto. de Química Física, CITEN, Edificio de Laboratorios Científicos-VIP,
Universidad de Panamá, Panamá, Panamá
e-mail: jjaen@ancon.up.ac.pa

J. Iglesias
Laboratorio de Química y Física Aplicada, Universidad Tecnológica de Panamá, Panamá, Panamá

O. Adames
Escuela de Química, Universidad de Panamá, Panamá, Panamá

1 Introduction

Weathering steels belong to the group of steels of high strength low alloy (HSLA), in which certain alloying elements contribute to the formation of a stable, adherent and protective rust layer called patina. They are materials with very good resistance to certain corrosive atmosphere without being painted or protected by another type of coating.

The atmospheric corrosion of weathering steels initially occurs in a manner similar to the corrosion of carbon steel, but decreases significantly with increasing exposure time because of the patina. The protective properties of the patina, depends on the chemical composition of metal, the humid–dry cycles, pH and composition of rainwater and atmospheric pollutants [1]. The salinity of aerosols may play an adverse role in the formation of protective patina accelerating the process of atmospheric corrosion, even in indoor conditions [2, 3]. Such was the case of the Luling Bridge, which crosses the Mississippi River in New Orleans, Louisiana [4, 5]. The protective characteristics of patinas also depend on the porosity and adherence of the different phases of rust formed on the steel surface. The knowledge of the morphology and growth characteristics provides information on the composition and protective potential of the patina [6].

Centennial Bridge, the new cable-stayed bridge over the Panama Canal, was built with steel delta frames made out of high performance steel (HPS) ASTM A709, Grade HPS-70W. This sparked a renewed interest in studying the performance of weathering steels in the humid-tropical marine climate of Panama. This paper presents the results of exposing conventional weathering steels in indoor /interior/ conditions for a period of one year, compared with outdoor exposure. The steels were exposed in a urban test site near the Pacific Ocean, and at the entrance of the Panama Canal in the Caribbean coast of Panama.

2 Materials and methods

Weathering steels (A-588 and CSN COR 420) used in this study have been described elsewhere [7, 8]. Coupons of 150 mm $\times$ 100 mm were used after blasting, cleaning and degreasing with acetone. In outdoor conditions, the samples were exposed for a year at 45° to the horizontal in the Tocumen site (T) (9° 4′ 2″ N 79° 24′ 16″ W) and at 30° in Fort Sherman Breakwater site (SB) (9° 22′ 25″ N 79° 56′ 50″ W). On the other hand, for indoor conditions metallic samples in vertical position and pollutant collectors were located inside booths made of "Fiber-Cement", avoiding heat trap effect. Cabinets had windows that allowed natural ventilation. Because of practical reasons, exposure of the A-588 samples started during dry season (January), whereas exposure of steel COR 420 started during the rainy season (July). During the dry season (from the middle of December through the end of April) there is little or no precipitation; however in the wet season (from May to the beginning December) there is frequent and heavy precipitations.

Environment is characterized at each site for meteorological and atmospheric pollution data [9], both outside and inside cabins, but in this paper we set out indoor results. The chloride deposition rate and SO_2deposition rate were determined monthly. Data for temperature and relative humidity (RH) were processed in order to determine time of wetness (TOW). After exposure, corrosion rates (corrosion penetration p) were calculated by weight loss according to standard methods [10].

The corrosion products were characterized by XRD using Cu (Kα) radiation with a Bruker D8 Advance diffractometer, equipped with a detector angle Vantek 2000 at the Materials Institute of Universidad Autónoma de México (UNAM). IR spectra were recorded using a Nicolet Avatar 360 FTIR spectrometer. The resolution of the spectra was 4 cm^{-1}. Mössbauer spectra (MS) were recorded using a conventional spectrometer of constant acceleration with a 57Co(Rh) source of nominal activity of 10 mCi (370 MBq). A closed-cycle cryostat (CCS 850 Janis) was employed for low temperature measurements. Calibrations were done with a standard α-iron foil absorber at room temperature. The Mössbauer data was evaluated with the Recoil software (University of Ottawa, Canada) using Voigt base fitting.

3 Results and discussion

Panama is a tropical country, very humid and warm, with temperatures averaging 27 °C all year long. The two test sites, Tocumen and Sherman Breakwater, are environmentally classified by ISO 9223 [9] as $S_1P_0\tau_4$ and $S_3P_0\tau_5$, respectively. In Sherman Breakwater test site, being a splash zone, conditions are particularly harsh due to the high deposition of chloride, among other environmental conditions. As shown in Fig. 1, deposition of chlorides inside the booths is very high in Sherman Breakwater station, though significantly diminishes compared to outside [8]. During indoor exposures the temperatures and relative humidity were practically the same as outside the booths, and SO_2 deposition rate is negligible. TOW is also higher in indoor conditions (given in Fig. 2) than outdoors in both sites [8], though always higher in Sherman Breakwater. This is an important factor influencing indoor corrosion aggressivity, since adsorbed water is present longer times, affecting the wet-dry cycles.

We have found [8] that the outdoor corrosivity in Tocumen and Sherman Breakwater test sites, according to ISO 9223, is C3 and >C5, respectively. Nonetheless, metals exposed to indoor conditions presented lower corrosion rates compared to outdoors, as shown in Fig. 3. Vera et al. [11] demonstrated that the exposure angle has some influences in outdoor corrosion rate. It is estimated that the difference in outdoor corrosion rates between 45° and 90° is ca. 1.3. The fact that in outdoor conditions, chloride and other pollutants deposit on the surface, but rain and other precipitations periodically clean the surface and that in indoor conditions there is no washing of the contaminants, so they accumulate on the metal surface compel us to look for different explanation for the significance of the observed corrosion rates. Even though the temperatures and relative humidity were practically the same as outside the booths, surface drying is slower than in outdoors due to the influence of air and sun radiation and rainfall (thicker water layers). This explains why indoor vertical corrosion rates are significantly lower than outdoors. In indoor conditions, the chloride deposition rate is significantly lower than outdoors contributing to lower corrosion rates. Both steels exposed to indoor conditions had not tendency of corrosion rates to decrease with the time, in contrast to the usual parabolic diminishing behavior observed in outdoor conditions. This can be interpreted as a

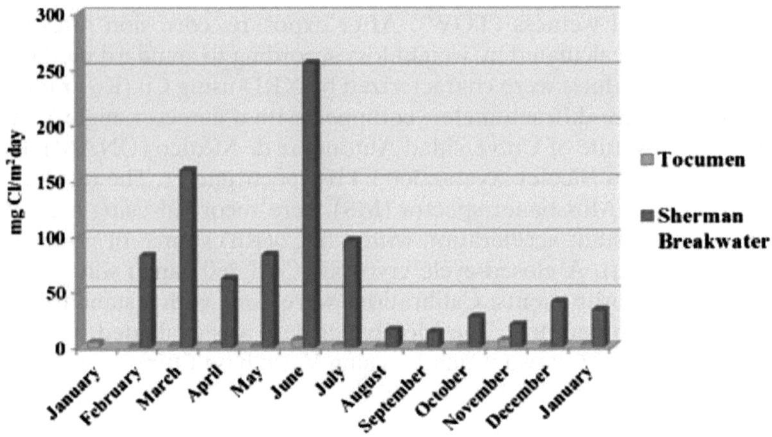

Fig. 1 Monthly average of deposition rate of chlorides inside booths for Tocumen and Sherman Breakwater sites

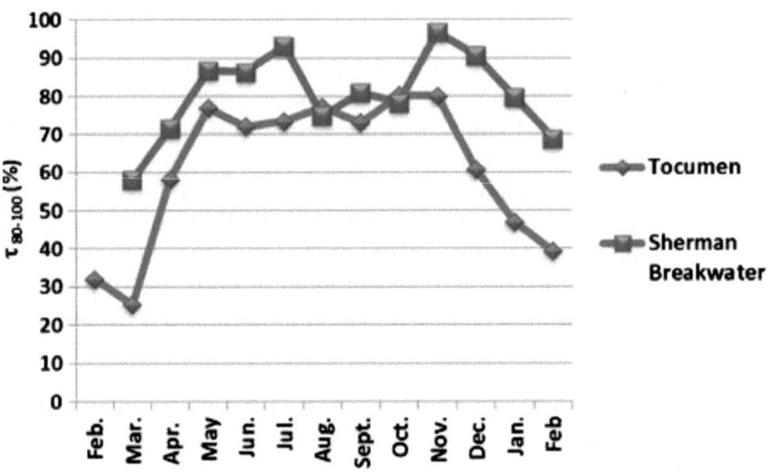

Fig. 2 Time of wetness (RH 80–100 %) under indoor conditions obtained at the Tocumen and Sherman Breakwater areas

lack of protective effect of the corrosion products. In Table 1 the indoor corrosion aggressivity results is summarized.

X-ray powder diffraction pattern for all samples were recorded to obtain a basic identification of the oxides present due to corrosion. As in the case of outdoor corrosion for these steels [8], the main phases constituting the rust layers formed after a year of indoor exposure are lepidocrocite γ-FeOOH, goethite α-FeOOH, and the iron oxide maghemite γ-Fe_2O_3. Interestingly, no akaganeite β-FeOOH is observed in Tocumen site (due to low chloride), whereas in Sherman-Breakwater site the formation of akaganeite is observed for both steels (see Fig. 4). In indoor conditions there is no washing effect on the metallic surface by precipitations, and

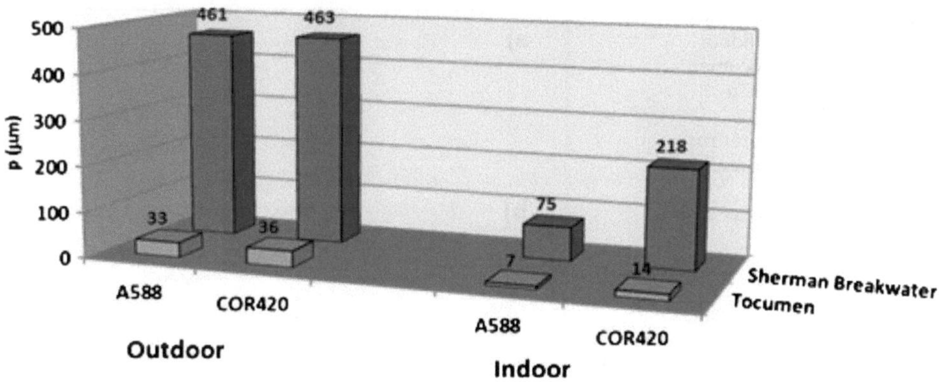

Fig. 3 Average corrosion rate (µm) depending on exposure condition and location of test site

Table 1 Classification of corrosivity for indoor samples

Classification norm	Test site	Type of steel	
		A-588	COR 420
ISO 9223 [9]	Tocumen	C2 low	C2 low
	Sherman-Breakwater	C4 high	>C5 very high
ISO/CD 11844 [12]	Tocumen	IC2 low indoor	IC2 low indoor
	Sherman-Breakwater	IC2 low indoor	IC3 medium indoor

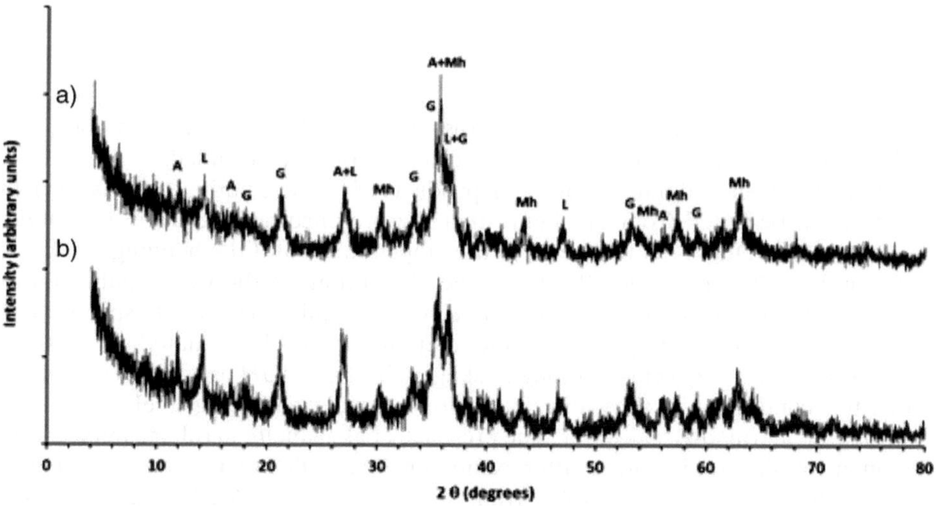

Fig. 4 XRD patterns of some characteristic samples of indoor corrosion products from Sherman-Breakwater site **a** A-588 and **b** COR 420 for 1 year. A, G, L, and Mh stand for akaganeite, goethite, lepidocrocite, and maghemite, respectively

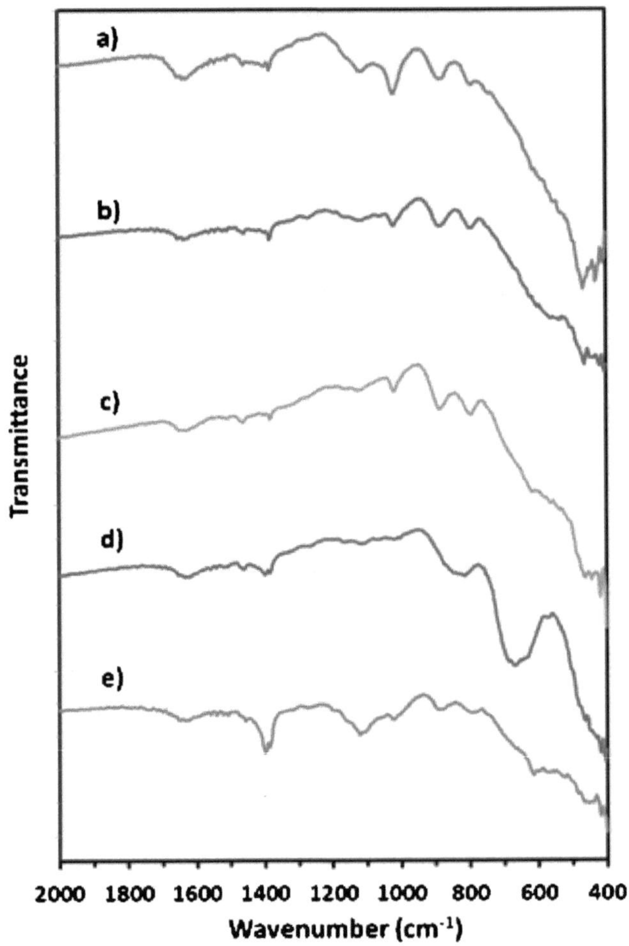

Fig. 5 Fourier transformed infrared spectra of indoor rusts formed at **a** Tocumen site A-588, **b** Tocumen site COR 420, **c** Sherman-Breakwater rust A-588, **d** Sherman-Breakwater internal rust A-588, **e** Sherman-Breakwater rust COR 420

chloride ions are retained in the metal-rust interface. In outdoor conditions, this phase is only observed in regions in which chloride ions are occluded.

The FTIR spectra of indoor corrosion products of the weathering steels is given in Fig. 5 It can be noted from this figure that the presence lepidocrocite γ-FeOOH, goethite α-FeOOH, is confirmed by the appearance of OH bending peaks at 1021 cm^{-1}, at 887 cm^{-1} and 795 cm^{-1}. The absorptions in the wavenumber range 550–670 cm^{-1} (Fe-O range) are attributed to maghemite. The FTIR spectrum of the corrosion product scrapped from the internal parts of delaminated sheets show peaks at 887 cm^{-1} y 795 cm^{-1} hinting at akaganeite β-FeOOH as a main constituent under this conditions, as shown in Fig. 5d for internal weathering steel A-588 after exposure in Sherman-Breakwater. The presence of akaganeite in other samples from Sherman-Breakwater is evident after examining the profile of FTIR spectra of Fig. 5.

Mössbauer analysis of samples from weathering steel A-588 and COR 420 confirmed the compositions of the indoor corrosion products identified with XRD and FTIR. Transmission Mössbauer measurement performed at room temperature and liquid nitrogen are presented in Figs. 6 and 7.

Fig. 6 Transmission Mössbauer spectra of rust of weathering steels formed after one year indoor exposure at Tocumen site at 295 K: **a** A-588, **b** COR 420 **c**, and at 80 K: **c** A-588, **d** COR 420

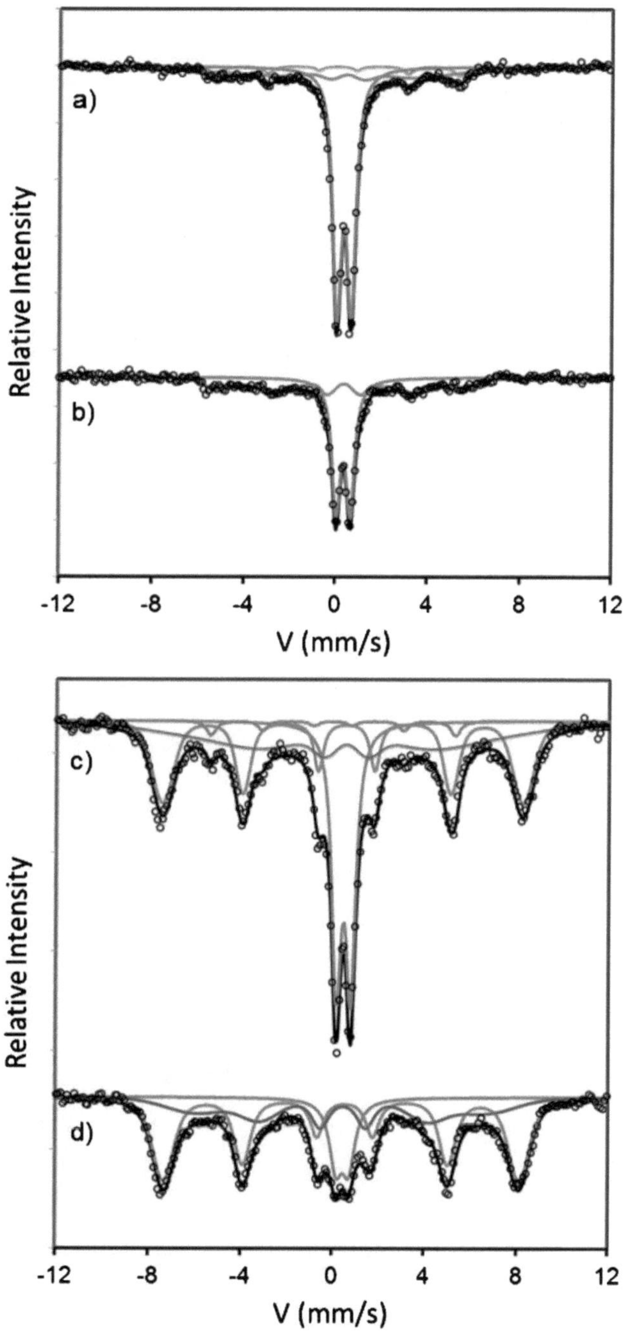

The RT (295 K) spectra of corrosion products of weathering steels formed after one year indoor exposure at Tocumen (Fig. 6) consist of broadened magnetic component and a dominant paramagnetic doublet. A small contribution of an extra

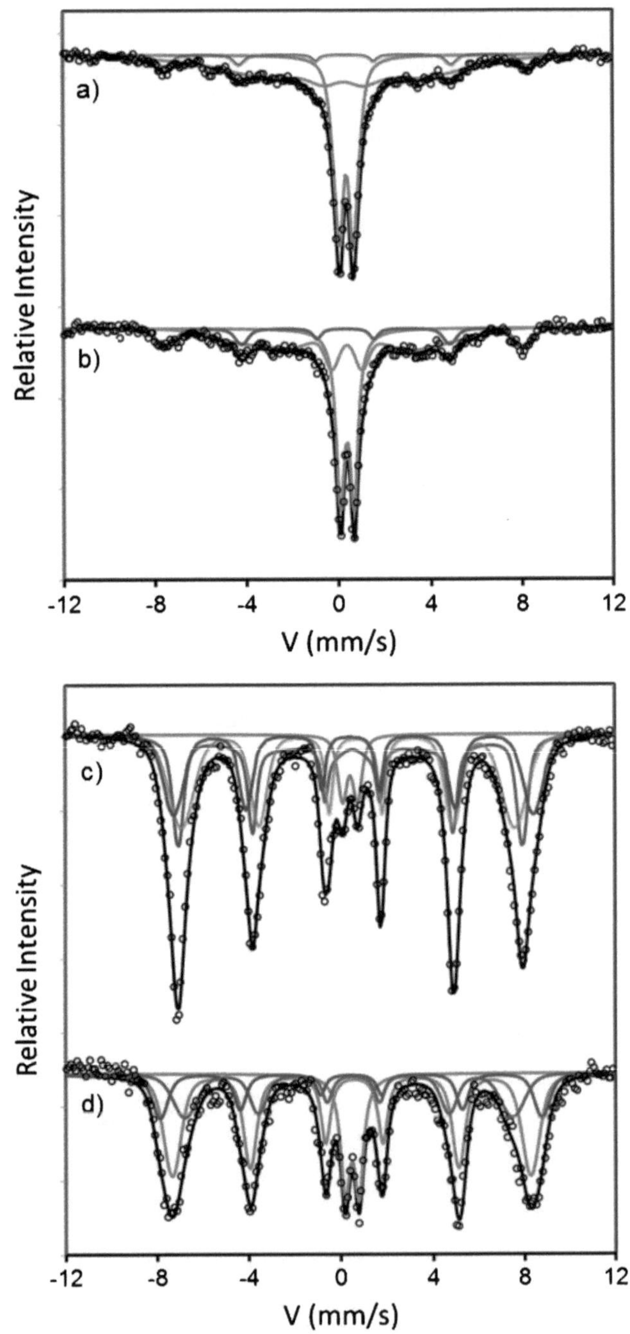

Fig. 7 Transmission Mössbauer spectra of rust of weathering steels formed after one year indoor exposure at Sherman Breakwater site at 295 K: **a** A-588, **b** COR 420 **c**, and at 80 K: **c** A-588, **d** COR 420

sextet of α-Fe from the scraping contamination of steel plates is observed. The Mössbauer parameters of the doublet ($\delta = 0.37$ mm/s and $\Delta = 0.62$ mm/s) are in good agreement with those published for lepidocrocite and superparamagnetic component

Table 2 Area percentages of corrosion products calculated from 78 K Mössbauer spectra

Type of steel	Test site	α-FeOOH	γ-FeOOH	γ-Fe_2O_3	β-FeOOH
A-588	Tocumen	70	30	–	–
	Sherman Breakwater	37	5	25	33
COR 420	Tocumen	85	15	–	–
	Sherman Breakwater	46	14	17	23

of goethite and maghemite at RT [13]. The magnetic component of each spectrum were fitted to a distribution with magnetic field in the range of 22–28 Tesla and are assigned to goethite with particle size >15 nm. At 80 K there is remarkable change compared to the RT spectra with an important increase in the relative intensity of the magnetic component (see Fig. 6), fitted using two distributions with magnetic fields around 48 Tesla and 37 Tesla. These subspectra are assigned to goethite of particle size ca. 17 nm and 9 nm, respectively. The remaining doublet is attributed to lepidocrocite, phase detected using XRD and FTIR.

Rather similar spectra of corrosion products of weathering steels formed after one year indoor exposure at Sherman Breakwater, but with two clear differences. In first place, the magnetic field of the goethite component is now in the range of 25–32 Tesla. Second and more important, another magnetic component is produced, having a magnetic field of 48–49 Tesla. This phase is ascribed to maghemite γ-Fe_2O_3. The magnetic hyperfine field of the well crystallized maghemite is in the range of 48–51 Tesla [13]. For both weathering steels, the goethite component has magnetic fields aprox. 48 Tesla, thus the particle size should be ca. 17 nm.

Area percentages of all coupons exposed in indoor conditions calculated from 78 K Mössbauer spectra are collected in Table 2. The ratio (β-FeOOH $+\gamma$-Fe_2O_3) to (β-FeOOH $+\gamma$-FeOOH $+\gamma$-Fe_2O_3) suggest that the rust obtained after one year exposure under indoor conditions is non-protective, in agreement with the observations in the corrosion rate measurements.

4 Conclusion

It is concluded that corrosion rates of weathering steels significantly decrease in the tropical atmosphere of Panama atmosphere when exposed in indoor environments. A-588 weathering steel corrodes at a generally lower rate than COR-420 weathering steel.

The interaction between the chloride depositions rates with the time of wetness are the most significant variables influencing the indoor corrosion of weathering steels. Because of that, corrosivity is higher Sherman Breakwater site as compared to Tocumen.

The nature of corrosion products is correlated with the observed corrosion rates. The major compounds are lepidocrocite, goethite, followed by maghemite. Akaganeite is only detected under high chlorides deposition rates, and no washing effect occurs. Maghemite and akaganeite were identified as prominent components obtained in the most aggressive conditions of Sherman Breakwater.

References

1. Pourbaix, M.: Tournée d'etde de la CEFA, Corrosion atmosph'erique des aciers patinables. Rapport Technique 217, Ed. Cebelcor, Bruxelles, pp. 18–32 (1973)
2. Mendoza, A.R., Corvo, F.: Outdoor and indoor atmospheric corrosion of carbon steel. Corros. Sci. **41**(1), 75–86 (1999)
3. Corvo, F., Torrens, A.D., Betancourt, N., Pérez, J., Gonzalez, E.: Indoor atmospheric corrosion in Cuba. A report about indoor localized corrosion. Corros. Sci. **49**, 418–435 (2007)
4. Cook, D.C., Van Orden, A.C.: The Luling bridge: an inside story. Corrosion 2000, paper no. 449, NACE International, Houston, TX, pp. 1–11 (2000)
5. Cook, D.C.: Spectroscopic identification of protective and non-protective corrosion coatings on steel structures in marine environments. Corros. Sci. **47**, 2550–2570 (2005)
6. Raman, A., Nasrazadani, S., Sharma, L.: Morphology of rust phases formed on weathering steels in various laboratory corrosion tests. Metallography **22**, 79–96 (1989)
7. Jaén, J.A., Muñóz, A., Justavino, J., Hernández, C.: Characterization of initial atmospheric corrosion of conventional weathering steels and a mild steel in a tropical atmosphere. Hyperfine Interact. **192**(1), 51–59 (2009)
8. Jaén, J.A., Iglesias, J., Hernández, C.: Analysis of short-term steel corrosion products formed in tropical marine environments of Panama. Int. J. Corr. vol. 2012, Article ID 162729, p. 11 (2012). doi:10.1155/2012/162729
9. ISO 9223: Corrosion of Metal and Alloys - Classification of Corrosivity of Atmospheres. Geneve: International Standards Organization (1992)
10. ISO 9226: Corrosion of Metals and Alloys. Method for Determination of Corrosion Rate of Standard Specimens for the Evaluation of Corrosivity. Geneve: International Standards Organization (1991)
11. Vera, R., Rosales, B.M., Tapia, C.: Effect of the exposure angle in the corrosion rate of plain carbon steel in a marine atmosphere. Corros. Sci. **45**(5), 321–37 (2003)
12. ISO/CD 11844: Corrosion of metals and alloys—classification of corrosivity of indoor atmospheres. Determination and estimation of indoor corrosivity (2006)
13. Cook, D.C., Oh, S.J., Balasubramanian, R., Yamashita, M.: The role of goethite in the formation of the protective corrosion layer on steels. Hyperfine Interact. **122**, 59–70 (1999)

DOI 10.1007/978-94-007-6482-8_34
Reprinted from *Hyperfine Interactions*, DOI 10.1007/s10751-013-0831-8

Structural and magnetic characterization of the "GASPAR" meteorite from Betéitiva, Boyacá, Colombia

L. M. Flor Torres · G. A. Pérez Alcazar

Published online: 13 March 2013

Abstract A structural and magnetic characterization has been performed of a plate obtained from the "Gaspar" meteorite from the Otengá region of the Betéitiva municipality, Boyacá, Colombia. The sample was provided by Ingeominas (Colombian Geological Agency). After the studies the sample was classified as an octahedral iron meteorite, due the Fe and Ni concentrations and the Widmanstätten pattern which was observed on the surface of the sample. The plate shows a crack which divides the sample in two regions (side A and B, respectively). Both sides were studied using techniques like X-rays diffraction (XRD), Mössbauer spectrometry, optical microscopy, and scanning electronic microscopy (with EDAX). On both sides an iron Fe-Ni matrix (kamacite) was found; a large quantity of carbon in the form of graphite and in two types: nodular and laminar; and different preferential orientation in both sides of the sample. The studies permit to prove that Gaspar is a fragment of the registered Santa Rosa de Viterbo meteorite.

Keywords Fe-Ni iron meteorite · X-ray diffraction · Mössbauer spectrometry · Optical microscopy · Scanning electron microscopy

1 Introduction

There are extraterrestrial bodies of different sizes traveling in space named meteors. These are produced due the degradation of the surface of bodies like asteroids, comets or inclusive planets different to the earth.

Proceedings of the thirteenth Latin American Conference on the Applications of the Mössbauer Effect, (LACAME 2012), Medellín, Columbia, 11–16 November 2012.

L. M. Flor Torres (✉) · G. A. Pérez Alcazar
Departamento de Física, Universidad del Valle, A.A. 25360, Cali, Colombia
e-mail: lauren.flor.univalle@gmail.com

Table 1 Textural classification of metallic meteorites based in the Ni percentage [2]

Class	Symbol	Strip thickness (mm)	%Ni
Hexahedrite	H	>50	4.5–6.5
Octhahedrite	O		
Very thick	Ogg	3.3–50	6.5–7.2
Thick	Og	1.3–3.3	6.5–7.2
Medium	Om	0.5–1.3	7.4–10.3
Fine	Of	0.2–0.5	7.8–12.7
Very fine	Off	<0.2	7.8–12.7
Plassitic	Opl	<0.2	12.7–16.0
Ataxite	D	Ausente	>16.0

When these bodies enter in contact with the earth atmosphere, before they strike the earth surface, they lose mass due the strong friction with the atmosphere, and in this way they lose kinetic energy which is transformed in light and heat, forming in this way the as called fugal start or meteor. Those meteors which have a mass above some tens of kilograms can survive the travel across the atmosphere and impact the earth surface and are called meteorites [1].

Meteorites can be classified in three groups, depending on their composition [1]: stony, iron, and stony-iron meteorites. The stony ones principally consist of olivine, pyroxene, feldspar, and in some cases they can present iron and nickel in small proportions. This group is divided in two subgroups: the stony meteorites which contain chondrules, called chondrites and those without chondrules called achondrites.

The second group is also named metallic or siderites. They consist for more than 90 % of Fe-Ni alloys like Kamacite (4–7 at. % of Ni) and Taenite (20–45 at. % of Ni), which present the so called Widmanstätten pattern. This type of meteorites can also contain other minerals like troilite, graphite, coenite and some silicates, in non-considerable amounts.

The textural classification of the iron meteorites is based in the width of the Kamacite plates in the Widmanstätten texture [2]. This width depends on the Ni content of the meteorite (Table 1).

The third group or stony-iron meteorites are also known as sideralites. They are bodies composed by iron-nickel phases and silicates, in approximately similar fractions.

The meteorite object of present study (see Fig. 1), was found in the Otengá region, Betéitiva municipality, Boyacá Department, Colombia, by peasants of the region during their land preparation labors for corn sowing. They collected this sample because it was a different material than those normally found in the zone, and it was transported to INGEOMINAS (Colombian Geologic Agency) of Boyacá. This meteorite was named "Gaspar".

The aim of this paper is to report the analysis performed on the "Gaspar" meteorite, by using X-ray diffraction, Mössbauer spectrometry, optical microscopy, and scanning electron microscopy (including EDAX). These results will be compared with previous ones reported for other meteorites found in neighboring zones in order to know if this is a part of the same sample or is a body of different origin.

Previous studies conducted in some bodies found in neighboring zones of the site in which the Gaspar meteorite was found, like Santa Rosa de Viterbo, Rasgatá, and others [2–4], reported that the samples are part of the registered Santa Rosa de

Fig. 1 "Gaspar" meteorite found in Betéitiva. Boyacá

Fig. 2 Widmanstätten pattern of the plate cutted from the GASPAR meteorite. A crack divides the sample in two regions

Viterbo meteorite [5], which presents an elemental composition near 91 at. % Fe, 7 at. % Ni and nearly 2 at. % of silicates.

2 Experimental

Initially a plate of nearly 0,6 cm width was cut, using thc *IsoMet® 5000 Family of Linear Precision Saws* cutter,of the Universidad del Valle. This plate was polished with sandpapers of different grades (220, 320, 360, 400, 600, 1200, and 2000), until a specular surface was obtained. Then it was attacked with nital and the Widmanstätten pattern was obtained. Also the crack which divides the plate in two zones was present. These two zones were studied and analyzed (see Fig. 2).

2.1 Optical microscopy results

The optical microscopy analysis was realized by using the equipment of the Materials School of Universidad del Valle, and permitted to observe the different grain sizes on both sides of the sample, and also the different orientations of the Widmanstätten pattern (Fig. 3).

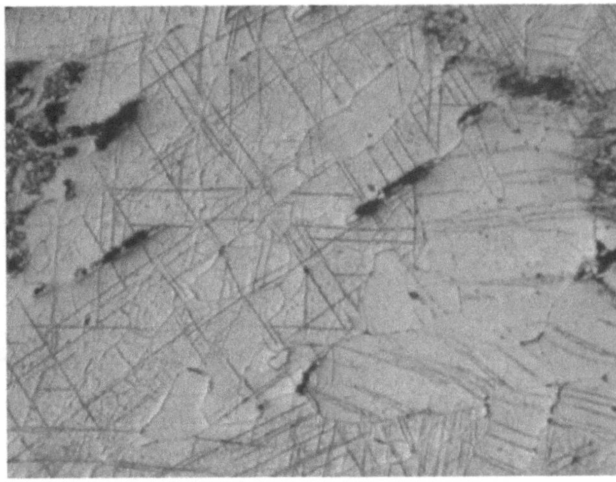

Fig. 3 Optical microscopy picture of the A side of the GASPAR meteorite showing the different orientations of the Widmanstätten pattern

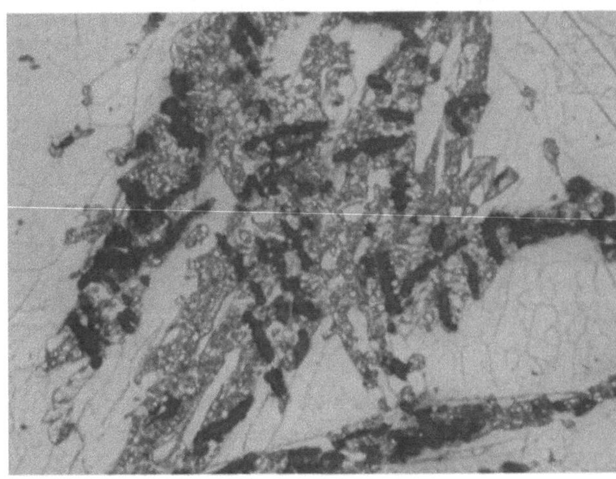

Fig. 4 Optical microscopy picture of bright, dark, and clear zones of side B

Different microstructures (see Fig. 4) were found of bright, dark, and clear zones. These microstructures were found in different zones of both sides of the sample.

2.2 Scanning Electron Microscopy results (SEM and EDAX)

The JEOL JSM-6490LV scanning electron microscopy of the materials school of Universidad del Valle, with an INCA PENTAFE×3 OXFORD prove, was used.

With this equipment the Widmanstätten pattern was observed with bigger resolution and some particles were observed inside the lines of the pattern (Fig. 5). Besides these particles, some dark zones were additionally found, as shown in Fig. 6.

The analysis by EDAX shows that both regions of the dark zone (1 and 2 of Fig. 6), present a similar composition. Fe is present with ~45 wt. %, C presents a weight of

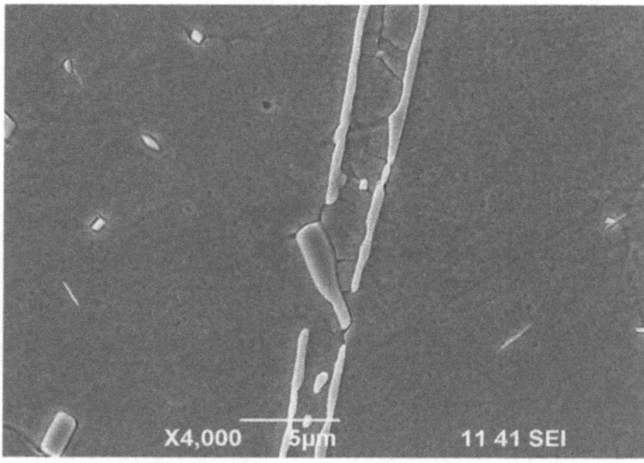

Fig. 5 SEM micrograpy of side A of the sample. It shows a magnification of the Widmanstätten pattern, with some particles inside

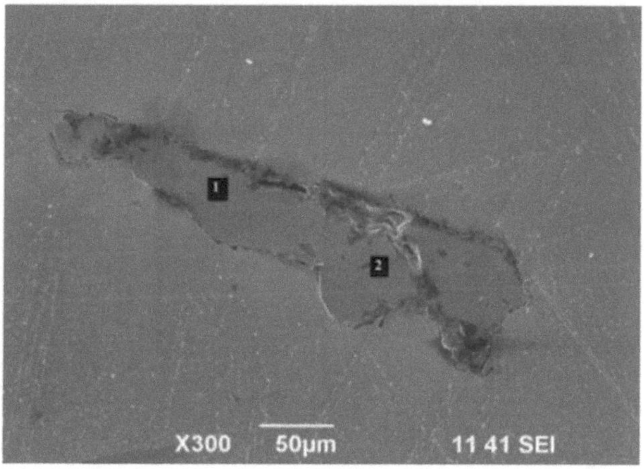

Fig. 6 Dark zone found in the B side of the sample

~23 %, Ni presents a weight of ~20 %, and P is present with a weight of ~10 %. These dark regions were found in both sides of the sample.

Using EDAX it was possible to find the chemical composition of the bright and dark zones detected by optical microscopy of Fig. 4.

The dark zones present a mean composition, which is similar in both sides, with ~94 wt. % of C and ~5 wt. % of O. This composition indicates that these zones may correspond to graphite. The bright zones present a composition similar in both sides of the sample with ~55 wt. % Fe, ~25 wt. % of C, and small quantities of Ni, P, and O. The morphology and composition of these zones permit to attribute them to perlite. However, as can be seen in Fig. 7a and b, the shape of the graphite is not the same at both sides of the sample. It is present as laminar (side B) and nodular (side A) types. According with the metallurgical process followed for heat treatment of iron-carbon steels, the lamellar graphite can be formed if the furnace temperature is maintained between 900 and 955 °C, followed by cooling down to 300 °C during two hours (higher cooling rate). The conditions for the formation

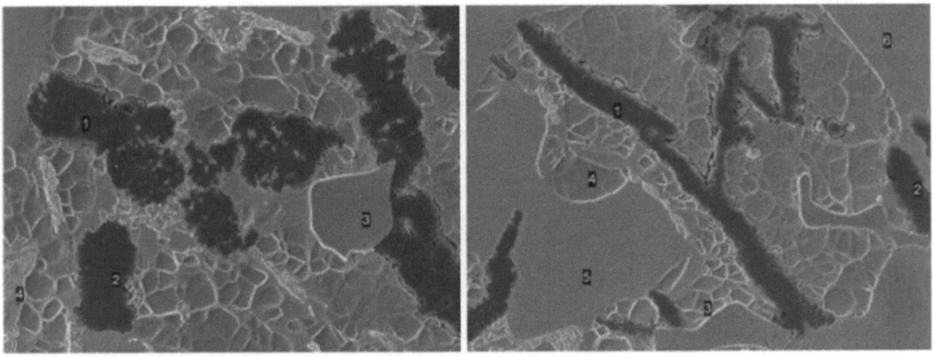

Fig. 7 SEM pictures of **a** bright and dark zones of side A (×1000), and **b** bright and dark zones of side B of the sample (×1000)

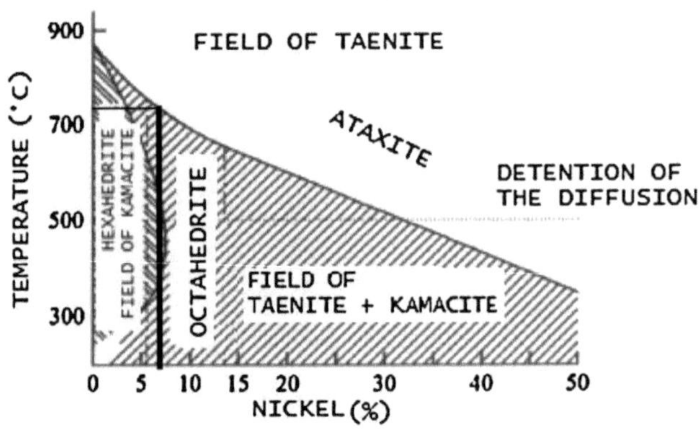

Fig. 8 Phase stability diagram of the different Fe-Ni phases [2]

of nodular graphite are: temperature of the furnace maintained between 915 and 930 °C, and followed by cooling down to 700 °C during two hours (smaller cooling rate). Taking into account that meteorites maintain a very low temperature until they reach the upper layers of the atmosphere, when they come into contact with these upper layers, an increase in temperature is produced due to the friction between the meteorite and the atmosphere, volatizing partially or completely if the meteorite is very small. The higher temperature corresponds to that side which is in friction with the atmosphere, producing a different heating and cooling regime to that of the other side of the meteorite, explaining in this way the formation of the nodular graphite in side A and the laminar one in side B. When the meteorite falls on the earth it cools further, reaching its coldest temperature at the earth.

Compositional analysis was also performed to the matrix of both sides of the meteorite. The same three elements were found in both sides, Fe with near 70.65 wt. %, C with near 24.69 wt. %, and Ni with near 4.67 wt. % in the side A and 70.07 wt. % of Fe, 24.54 wt. % of C and 5.39 wt % of Ni in the side B. According to this composition the meteorite can be classified as iron or metallic, and the composition is nearly the same at both sides.

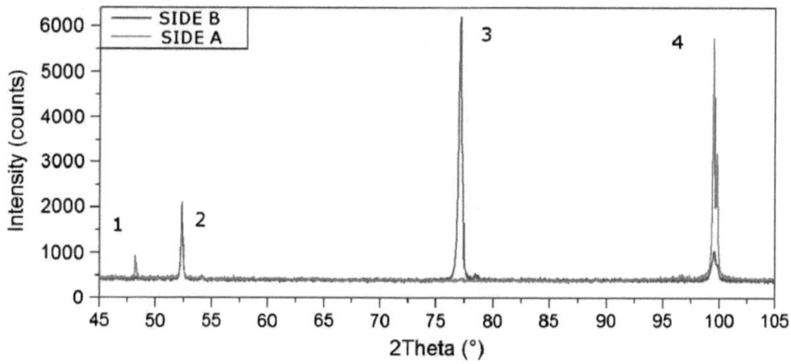

Fig. 9 XRD patterns of both sides of the GASPAR meteorite. *Red line* corresponds to the Side A and *blue line* to the side B

Table 2 Results of the crystallographic analysis of side A of the "GASPAR" meteorite

Parameters	Phase 1 - C	Phase 2 - αFe10.8Ni
Crystallite size $\perp$ [nm]	280.11	93.12
Crystallite size $\|\|$ [nm]	280.11	93.12
a – b – c [Å]	2.470 – 2.470 – 6.800	2.872
$\alpha - \beta - \gamma$ [°]	90 – 90 – 120	90
Lattice	Hexagonal	Cubic
Symmetry	6/mmm(6/m2/m2/m)	m3m (4/m32/m)

Table 3 Results of the crystallographic analysis of side B of the "GASPAR" meteorite

Parameters	Phase 1 - C	Phase 2 - αFe10.8Ni
Crystallite size $\perp$ [nm]	1142.45	67.30
Crystallite size $\|\|$ [nm]	5.80	199.63
a – b – c [Å]	2.524 – 2.524 – 4.471	2.871
$\alpha - \beta - \gamma$ [°]	90 – 90 – 120	90
Lattice	Hexagonal	Cubic
Symmetry	6/mmm(6/m2/m2/m)	m3m (4/m32/m)

The approximate weight concentration values, in the matrix, for nickel and iron are 5.04 and 70.36 %, respectively. The total nickel concentration, for comparison and classification of the samples in the different subgroups of metallic meteorites, is obtained by the equation

$$Ni_T = \frac{Ni}{Fe + Ni}\% \tag{1}$$

The obtained total weight value for nickel was ~6.68 %. Comparing this value to those required in each subgroup of meteorites, the "GASPAR" meteorite is classified as an octahedrite (Og) IIIAB type.

In the stability phase diagram of the different Fe-Ni phases (Fig. 8), it can be observed that a material with Ni between ~7 and 14 wt. %, can be classified in the

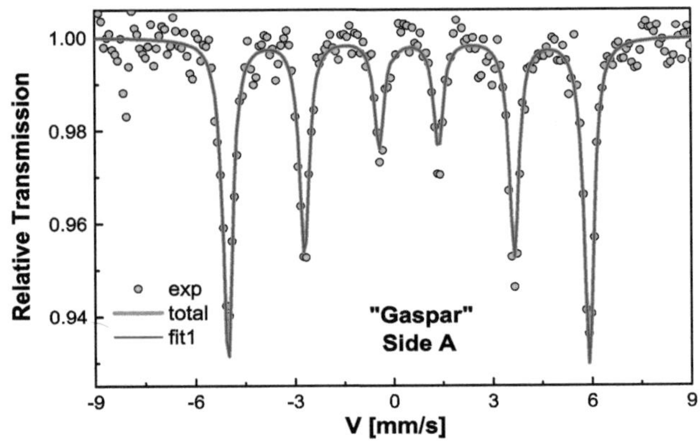

(a)

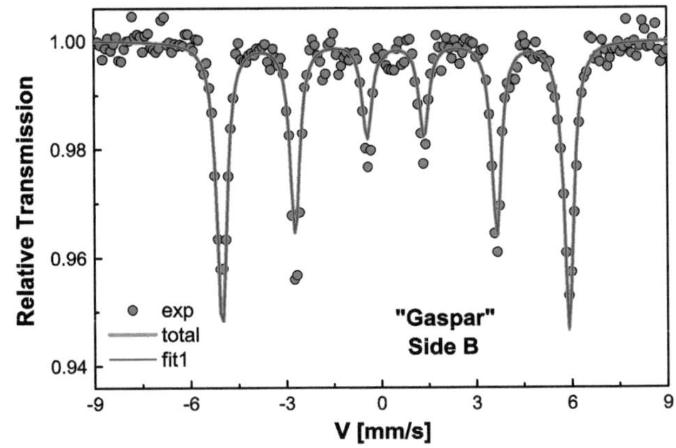

(b)

Fig. 10 Mössbauer spectra of both sides of the "GASPAR" meteorite, **a** side A of the sample, and **b** side B of the sample

octahedrite meteorite group with Kamacite presence. According to its composition GASPAR meteorite is part of this group [6]. Besides, the associated temperature is less than 900 °C.

2.3 X-ray diffraction (XRD) results

The XRD patterns of both sides of the sample were refined by using the Rietveld method with the GSAS program [7], and the line sequence of the α-$Fe_{10.8}Ni$ (BCC) and carbon phases like Lonsdaleite (side A) and Graphite (side B). The patterns are shown in Fig. 9 and the refined parameters are shown in Tables 2 and 3.

Two differences in the peak intensities can be observed. The first one is in the intensities of the peaks 2 and 4 of the pattern, which correspond to the (110) and (211) of the α-$Fe_{10.8}Ni$ phase. In phase B the contribution of peak 2 is larger and for peak 4 the larger contribution is found on side A. The second one is that the intensity of the peak 1, which corresponds to the carbon phases, is dominating on side B. These results show that the different thermal regimens of both sides induce

Table 4 Mössbauer parameters obtained for both sides of the GASPAR meteorite

	DI (mm/s)	SQ (mm/s)	H (kOe)
Side A	$0.009 \pm 0.667E\text{-}02$	$-0.022 \pm 0.916E\text{-}02$	337.1 ± 0.3
Side B	$0.004 \pm 0.567E\text{-}02$	$-0.022 \pm 0.719E\text{-}02$	337.1 ± 0.3

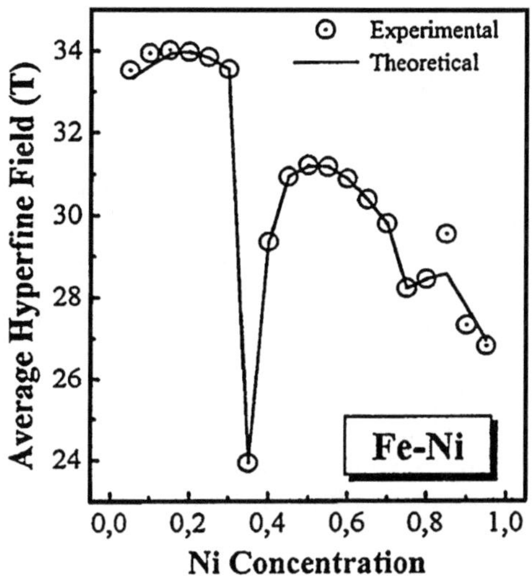

Fig. 11 Dependence of the mean hyperfine field with the Ni concentration (figure taken from Ref. [8])

different preferential orientations in them. After the refinement of the two patterns the following crystallographic parameters were obtained:

It can be noted that graphite (Phase 1-C) of side A presents crystallite sizes parallel and perpendicular with similar values (≈280 nm), in according with the nodular morphology detected by SEM. Besides, the obtained graphite in size B presents a perpendicular crystallite size (≈1142 nm) much bigger than that of the parallel size (≈6 nm) showing its lamellar shape in according with the SEM results.

2.4 Mössbauer spectrometry studics

The Mössbauer spectra of both sides of the sample are shown in Fig. 10. In Table 4 the hyperfine parameters are listed obtained after the fitting of the spectra. It can be observed that the isomer shift and the quadrupole splitting of the obtained sextet correspond to those of iron, however the hyperfine field obtained is near 337 kOe, then it is bigger than that of pure iron.

This larger hyperfine field can be attributed to the substitution of Fe atoms by Ni atoms, because this substitution increase the mean hyperfine field as was reported by Restrepo et al. [8]. They proved that the mean hyperfine field of Fe-Ni alloys increases when the Ni content increases between 0 up to 20 at. % (see Fig. 11). Our material presents a Ni concentration around 5 wt. %, and is the region of larger mean hyperfine fields (Fe and Ni present nearly the same weight).

3 Conclusions

The GASPAR meteorite was classified as an iron meteorite or metallic octahedrite (Og) IIIAB type, due to the presence of a Widmanstätten pattern typical of these meteorites, a concentration of nickel of 6.6 wt. %, and the Kamacite and Taenite phases. Besides, its matrix contains a majority content of Fe, C, and Ni in both sides of the sample.

With the help of the stability phase diagram of the Fe-Ni alloys, it was possible to observe that a material with 6.6 wt. % is inside the field of the octahedrite with a temperature below 900 °C, and these conditions are those of the GASPAR meteorite.

The different types of graphite found in both sides of the meteorite and the different texture of then, permit to conclude that both sides were exposed to different heating and cooling regimes when one side was exposed more directly to the friction with the earth's atmosphere before the impact.

The increase in the hyperfine field (370 kOe) compared to that of pure iron is due to the percentage of Ni that is in the BCC material.

Comparing our results with previous studies, it can be said that this meteorite has very similar characteristics to those found and studied in other fragments collected in nearby areas, so it is possible that Gaspar meteorite is part of the Santa Rosa of Viterbo meteorite, found in Boyáca, Colombia, near the site in which Gaspar was found.

Acknowledgements The authors would like to thank Mr. Jaime H. Toro, owner of GASPAR meteorite, from La Pradera community, and Dr. Jose Arenas, Director of the Ingeominas Geological Museum in Bogotá who permitted the GASPAR study. We also thank Universidad del Valle and the CENM-Univalle by the financial support.

References

1. Lamus, C.M., Medina, B.: Meteoritos: petrografía, geoquímica y origen astrofísico. Medellín, Thesis, Colombia (2002)
2. Gil, J., Concha, A.: Caracterización petrográfica y clasificación textural del meteorito de Santa Rosa de Viterbo (Boyacá), Colombia. Geol. Colomb. No. 31, ISSN 0072-0992, (2006)
3. Ramirez, J.E.: Meteors and meteorites. Contributions of the Meteoritical Society. The Meteorites of Santa Rosa de Viterbo, Boyacá, Colombia (730,059), vol. LVII, pp. 29–36. Instituto Geofísico de los Andes Colombianos, Bogotá, Colombia (1949)
4. Buchwald, V.F., Wasson, J.T.: The two colombian iron meteorites, Santa Rosa and Tocavita. Analecta Geol. **3**, 5–29 (1968)
5. Meteoritical Society: "Santa Rosa". Year found: 1810. Page on the Internet: http://www.lpi.usra.edu/meteor/metbull.php?code=23167
6. Norton, O.R.: The Cambridge Encyclopedia of Meteorites, pp. 354. Cambridge University Press, UK (2002)
7. Larson, C.A., Von Dreele, R.B.: GSAS, General Structure Analysis System. Los Alamos National Laboratory, Los Alamos (2004)
8. Restrepo, J., Pérez Alcázar y, G.A., Bohórquez, A.: Description in a local model of the magnetic field distributions of $Fe_{1-x}Ni_x$ disordered alloys. J. Appl. Phys. **81**, 4101 (1997)

DOI 10.1007/978-94-007-6482-8_35
Reprinted from *Hyperfine Interactions*, DOI 10.1007/s10751-013-0821-x

Iron nano-clusters in ytterbium films: a ^{57}Fe Mössbauer spectroscopic study

C. Rojas-Ayala · W. T. Herrera · I. S. Dinóla · M. Kraken · E. C. Passamani · E. Baggio-Saitovitch · F. J. Litterst

Published online: 13 February 2013

Abstract We have performed a Mössbauer study on iron clusters that are formed in ytterbium films prepared by vapor co-deposition onto kapton substrates kept at room temperature. The film thicknesses were chosen in a range between 2.5 and 2.8 μm. XRD of the films reveals a mixture of fcc- and hcp-like ytterbium. Iron concentrations were between 0.3 and 5 at %. All samples reveal hyperfine spectra attributed to only two types of iron clusters with well defined hyperfine parameters. The clusters are supposed to be formed at boundaries of hcp- and fcc-like grains. In addition there is found a small contribution from monomeric iron. Spectra taken at 4.2 K reveal a complex distribution of magnetic hyperfine fields. The dynamic origin of the spectral shape is supported by susceptibility data revealing spin-freezing at temperatures below ca. 10 K. This proves that the iron clusters have sizes on the order of nm.

Keywords Mössbauer spectroscopy · Iron nano-clusters · Ytterbium films · Vapor deposition · Spin-freezing

---

Thirteenth Latin American Conference on the Applications of the Mössbauer Effect, LACAME 2012, Medellín, Colombia, 11–16 November 2012.

C. Rojas-Ayala (✉) · W. T. Herrera · I. S. Dinóla · E. Baggio-Saitovitch · F. J. Litterst
Centro Brasileiro de Pesquisas Físicas,
Rua Dr. Xavier Sigaud 150, 22290-180, Rio de Janeiro, RJ, Brazil
e-mail: chachi@cbpf.br

C. Rojas-Ayala
Facultad de Ciencias Físicas, Universidad Nacional Mayor de San Marcos,
P.O. Box 14-149, Lima 14, Lima, Peru

M. Kraken · F. J. Litterst
Institut für Physik der Kondensierten Materie, Technische Universität Braunschweig,
38106 Braunschweig, Germany

E. C. Passamani
Departamento de Física, Universidade Federal do Espírito Santo,
29075-910, Vitória, ES, Brazil

1 Introduction

Dilute magnetic impurities with concentrations ranging from parts-per-million up to parts-per-hundred in non-magnetic metallic matrices have been intensively studied in the past for investigating, firstly, the site occupation of the impurities and its influence on magnetism, and secondly, Kondo effect and competing spin-glassy behavior [1–3]. Based on this previous work, we have started some years ago investigations on the formation of iron nanoclusters in non-magnetic metallic hosts with low solubility among impurity and host elements [4, 5].

Earlier Mössbauer studies on iron in ytterbium [6, 7] had initially been devoted to investigate the preferential site occupation of Fe and its relation to magnetism. Further going investigations deal with the formation of magnetic Fe-clusters by diffusion of Fe atoms [8, 9]. The presence of iron magnetic moments in isolated sites and clusters could be shown. However, Mössbauer spectra are complex [8, 9], showing a superposition of several components, and a satisfying interpretation could not yet be presented. The main aim of the present work is to contribute to a better systematics and to propose some new simple assignments to spectral components for dilute Fe in Yb films prepared by deposition onto substrates kept at room temperature.

2 Experimental details

Iron-doped Yb films were prepared by vapor co-deposition of high purity Fe (95 % enriched in ^{57}Fe) and Yb with nominal compositions between 0.3 and 5.0 at. % of Fe. The sample preparation procedure and the evaporation details have been described previously [5, 9].

Depositions were performed onto Kapton substrates mounted tilted 45° relative to the metal vapor flux direction. The substrate temperature was kept at 290 K. The base pressure in the deposition chamber was 2×10^{-8} mbar, increasing to 2×10^{-7} mbar during the evaporation. The deposition rate was monitored using piezo-crystals and typical values were 2–7 Å/s for Yb and 0.02 Å/s for Fe. The total film thicknesses were typically 2.5 to 2.8 μm.

Samples were characterized by X-ray diffraction (XRD) and ^{57}Fe transmission Mössbauer spectroscopy. We used a standard type Mössbauer spectrometer with sinusoidal velocity sweep. Low temperature Mössbauer experiments have been performed in a variable-temperature ^{4}He cryostat. The ^{57}Co/Rh source and the Mössbauer absorbers were kept at the same temperatures. X-ray diffraction measurements were performed on a Rigaku Miniflex using Cu -Kα radiation.

3 Results and discussion

From XRD we find that all of our films possess a mixture of fcc- and hcp-like Yb phases with a ratio of fcc:hcp of about 3:2. This result agrees with a recently reported one obtained for an undoped Yb film [10], which also suggests that the fcc-like Yb phase is dominant. We propose that the structure of our films is consisting of fcc-domains surrounded by an interconnecting hcp phase. A detailed consideration of the influence of the mixture of structural phases of the films on the formation of specific clusters taking into account the very sensitive Martensitic transition between hcp and fcc Yb [11, 12] will be given elsewhere. The XRD Bragg peaks of both crystalline ytterbium phases of our iron-doped films reveal significant broadening, indicating the formation of small grains. Using the Scherrer relation, we estimate mean grain sizes of the fcc-like Yb of about 25 nm; this is very close to previously obtained values for Fe/Yb films [9]. From the relative intensities of Bragg peaks there are indications for varying degrees of texture of the ytterbium films.

RT Mössbauer absorption spectra of Yb films with different iron concentrations are shown in Fig. 1. All samples reveal paramagnetic behavior. These spectra were fitted using a superposition of three components: one single absorption line labeled S and two doublets (D1, D2). For all films studied, these components have very similar hyperfine parameters. From our fits there are no indications for an asymmetry in areas between the two lines of each doublet. This means, there is no preferred orientation of the electric field gradient, despite the XRD data of the films reveal some texture. From this we conclude that the doublet spectra D1 and D2 should be attributed to aggregates or clusters that are not affected by the texture of the ytterbium matrix.

The variation of the relative spectral areas for the three components is presented in Fig. 2. Whereas the relative spectral area for the D1-component slightly increases with iron concentration, the area for the D2-component slightly decreases. The area for the S-component remains constant within experimental error.

The introduction of the singlet (S) in the fitting procedure is based on results obtained earlier from films deposited at low substrate temperatures [8]. For these films this component dominates the Mössbauer spectra. Its spectral fraction is reduced when increasing the temperature of the films and instead there is found an increase of the area of doublet D1 [8]. But even after annealing the films up to 300 K some spectral contribution of the single line S remains present. We attribute S to iron monomers in interstitial sites of fcc-Yb [8, 9]. The negative isomer shift (see Fig. 3b) indicates that the electron density is high in comparison to metallic iron, which can be understood to result from a charge transfer from ytterbium host atoms. Since Yb, being a rare-earth element, is highly electropositive and iron is highly electronegative, the latter tends to acquire electrons from the ytterbium atoms. This behavior is in good agreement with earlier work [6, 7].

The isomer shift of the D1-component is higher than the isomer shift of the monomers (S), indicating different surroundings with higher number of iron neighbors. However, its low value of quadrupole splitting (see Fig. 3a) may suggest an almost cubic arrangement. We attribute D1 to iron clusters formed upon Fe atomic diffusion during the deposition process (respectively during annealing after low temperature deposition due to diffusion of monomers) at grain boundaries of the fcc-like Yb phase, where the growth of clusters is supposed to be favorable.

The doublet D2 has a bigger value of quadrupole splitting (see Fig. 3a), which indicates a surrounding with lower symmetry. We attribute D2 to very small iron aggregates (dimers, trimers, etc) formed mainly at hcp-like grain boundaries.

4.2 K Mössbauer spectra of the iron doped ytterbium films prepared at RT are plotted in Fig. 4. While for low Fe-concentrations (< 3.5 at. %), superparamagnetic contributions dominate the spectra even at 4.2 K, the spectra for high Fe-contents display a broad magnetic component. The complex spectral shapes found for all concentrations are supposed to be influenced by fluctuations of the magnetic cluster moments on a time scale comparable to nuclear Larmor precession.

Fig. 1 Room temperature ^{57}Fe Mössbauer absorption spectra of ytterbium films with different iron concentrations

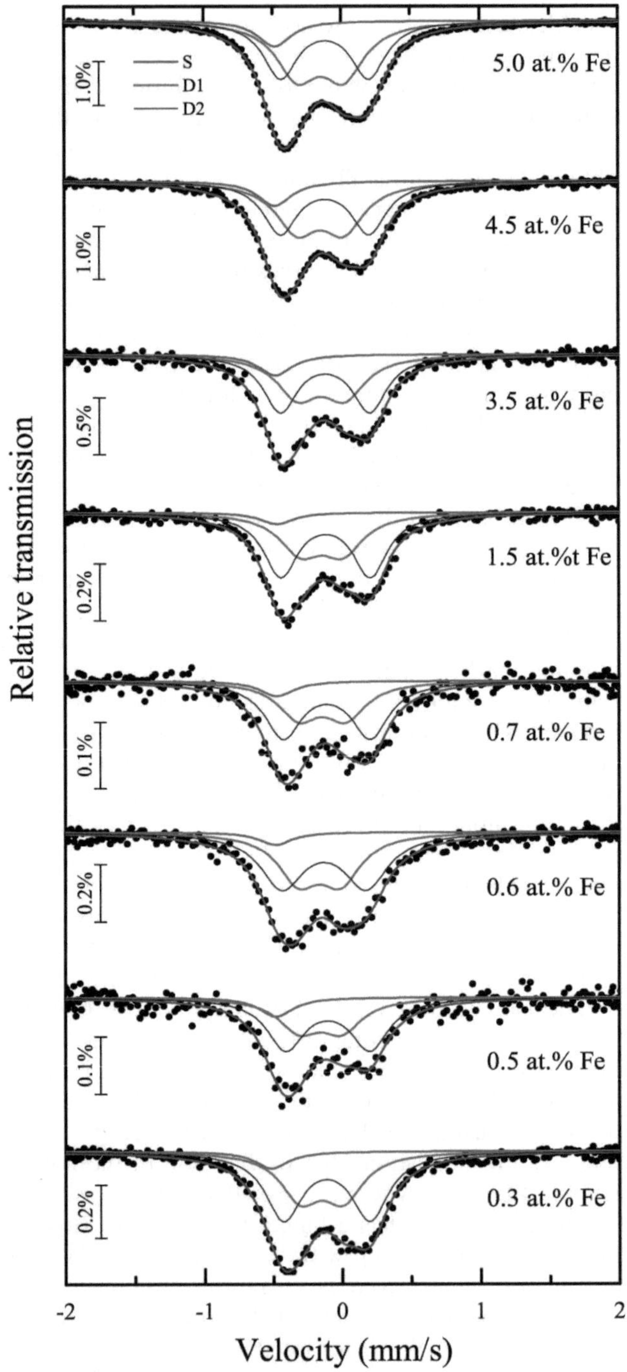

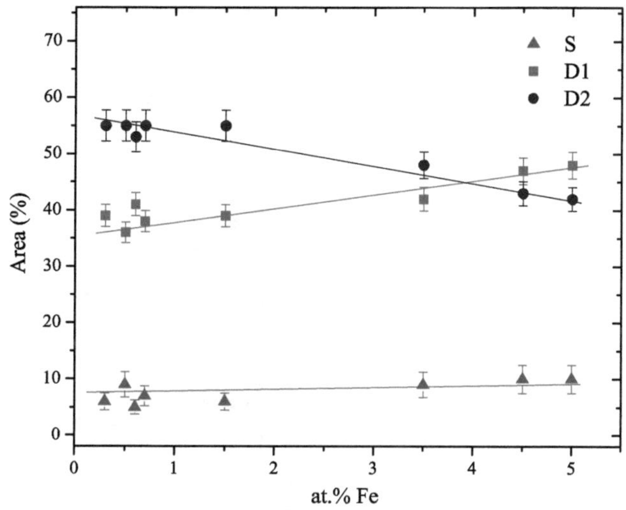

Fig. 2 Variation of relative spectral area of S, D1 and D2 as a function of iron concentration in iron doped ytterbium films deposited at RT

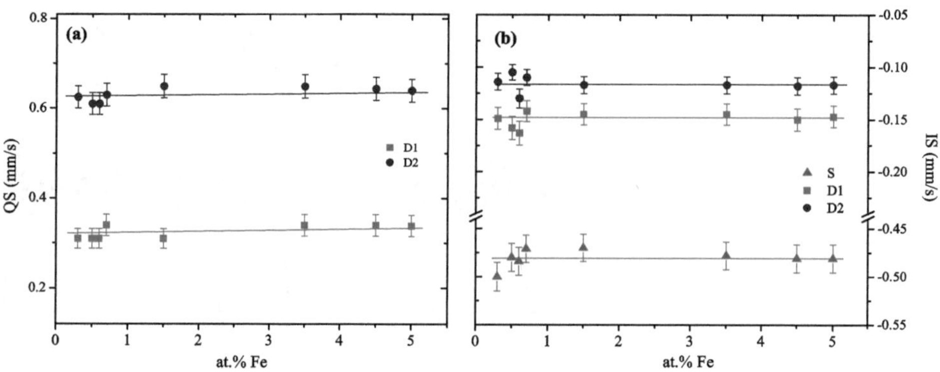

Fig. 3 Hyperfine parameters for iron doped ytterbium films prepared at RT. **a** Quadrupole splitting, QS, and **b** isomer shift, IS. Values of IS are relative to the ^{57}Co/Rh source at RT

We have omitted here a rigorous dynamic fit to the complex spectral shapes. The drawn fit lines in Fig. 4 come from a simplistic static magnetic hyperfine field distribution, which is certainly inadequate for the low iron concentrations and therefore should serve only as a guide to the eye. It should be noted that according to the hyperfine parameters there are no indications for the presence of clusters of metallic iron or iron oxides.

The increased concentration of clusters for samples with 4.5 at. % and 5.0 at. % Fe leads to considerable slowing-down of fluctuations and an increase of magnetic interactions between clusters. As consequence nearly static distributions of magnetic hyperfine fields are found. This freezing of cluster dynamics is in agreement with dc magnetic susceptibility revealing spin-glassy behavior below about 10 K for these Fe concentrations, whereas for lower concentrations there are no indications for spin-freezing down to 4.2 K. From this we conclude that the iron clusters in our samples must have sizes on the order of few nanometers.

Fig. 4 ^{57}Fe Mössbauer absorption spectra of ytterbium films with different iron concentrations taken at 4.2 K

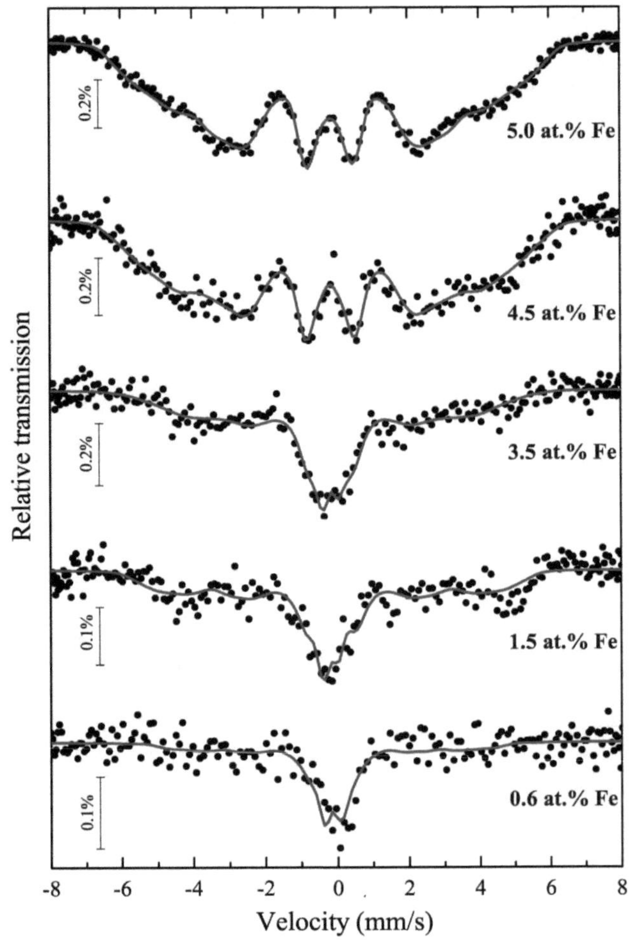

The superparamagnetic fraction visible in the spectra for lowest iron concentrations is dominated by doublet D2, i.e. the spectral area of doublet D1 is mainly comprised in the broad magnetically split background. The clusters associated with D1 should therefore be bigger than those associated with D2.

A detailed interpretation of these spectra, taking into account dynamic magnetic fluctuations including cluster-cluster interactions and a comparison of the dynamical behavior with that derived from susceptibility data, will be presented elsewhere.

4 Summary

We have shown by Mössbauer spectroscopy that two types of iron clusters are formed in ytterbium films prepared by vapor co-deposition onto kapton substrates kept at room temperature. The films possess a mixture of fcc- and hcp-like ytterbium. In contrast to earlier studies [8, 9], which have shown that the formation of iron clusters is strongly dependent on film thickness, preparation and annealing conditions, we took special care of nearly identical conditions of preparation. The film thicknesses were chosen in a narrow range between 2.5 and 2.8 μm. Iron concentrations were between 0.3 and 5 at %. All samples reveal hyperfine spectra attributed to only two types of iron clusters with well defined hyperfine parameters. The clusters are supposed to be formed at boundaries of hcp- and fcc-like grains. In addition there is found a small contribution from monomeric iron.

Spectra taken at 4.2 K reveal a complex distribution of magnetic hyperfine fields. The spectral shape is clearly influenced both by inhomogeneous broadening and relaxation effects. The dynamic origin of the spectral shape is supported by susceptibility data revealing spin-freezing at temperatures below ca. 10 K. This also proves that the iron clusters possess sizes on the order of nm.

Additional experiments, as systematic magnetization measurements and Mössbauer spectroscopy under external magnetic field, will be presented elsewhere.

Acknowledgements This work has been made possible by financial supported from CAPES, FAPERJ, FAPES, Brazilian-German research program financed by CNPq and DFG 444BRA-113/55/0-1, and CLAF/CNPq.

References

1. Longworth, G., Jain, R.: J. Phys. F **8**, 993 (1978)
2. Larica, C., Baggio-Saitovitch, E., Xia, S.K.: Hyperfine Interact. **66**, 339 (1991)
3. Passamani, E.C., Baggio-Saitovitch, E., Micklitz, H.: Phys. Rev. B **55**, 12382 (1997)
4. Herrera, W.T., Dinóla, I., Baggio-Saitovitch, E., Kraken, M., Litterst, F.J.: Hyperfine Interact. **203**, 149 (2011)
5. Herrera, W.T., Xing, Y.T., Ramos, S.M., Munayco, P., Fontes, M.B., Baggio-Saitovitch, E.M., Litterst. F.J.: Phys. Rev. B **84**, 014430 (2011)
6. Sawicka, B.D., Letocha, E., Sawlcki, J.A., Tyliszczak, T., Binczycka, H.: Phys. Lett. A **100**, 45 (1984)
7. Kapoor, J., Riegel, D., Li, Y., Polaczyk, C., Andres, J., Mezei, F., Sielemann, R., Yoshida, Y., Brewer, W.D., de Mello, L.A., Frota-Pessoa, S.: Phys. Rev. Lett. **78**, 1279 (1997)
8. Passamani, E.C., Morales, M.T., Baggio-Saitovitch, E.: Phys. Rev. B **60**, 1111 (1999)
9. Herrera, W.T., Rojas-Ayala, C., Dinóla, I.S., Passamani, E.C., Kraken, M., Engelke, J., Baggio-Saitovitch, E., Litterst, F.J.: Hyperfine Interact. **203**, 143 (2011)
10. Jerap, M., Sakurai, K.: J. Phys.: Condens. Matter **22**, 474010 (2010)
11. Lide, D.: CRC Handbook of Chemistry and Physics, 88th edn. Taylor & Francis Group, Boca Raton (2008)
12. Bucher, B., Schmidt, P., Jayaraman, A., Andres, K., Maita, J., Nassau, K., Dernier, P.: Phys. Rev. B **2**, 3911 (1970)

DOI 10.1007/978-94-007-6482-8_36
Reprinted from *Hyperfine Interactions*, DOI 10.1007/s10751-013-0813-x

Processing of gadolinium–iron garnet under non-equilibrium conditions

S. C. Zanatta · F. F. Ivashita · K. L. da Silva · C. F. C. Machado · A. Paesano Jr.

Published online: 1 February 2013

Abstract We have investigated the mechanosynthesis of gadolinium iron garnet (GdIG) by high-energy ball-milling of 3.(Gd_2O_3) + 10.(α-Fe) followed by thermal annealing conducted at moderate temperatures (1100 °C). The samples were characterized by X-ray diffraction and Mössbauer spectroscopy in order to determine the influence of the milling time on the final products. For as-milled samples the results revealed the enlargement of the magnetic component belonging to iron and a discrete paramagnetic component. The formation of a garnet phase was observed in all as-annealed samples treated at 1100 °C for 6 h in quantities proportional to the time of grinding the precursors. Evidently, high-energy ball milling of Gd_2O_3 + α-Fe powders is an important step in GdIG synthesis by a ceramic method. Single-phase garnet is observed for the samples milled for 12 and 24 h treated at 1100 °C for 6 h.

Keywords Garnet · Mechanosynthesis · Mössbauer spectroscopy

1 Introduction

$RE_3M_5O_{12}$ garnets (where RE = rare-earth, M = transition metal) present unique magnetic and magneto-optical properties such as ferromagnetic ordering at room temperature, with constant magnetization over a large temperature range, ferromagnetic resonance and Faraday rotation, among others. These features make this

Thirteenth Latin American Conference on the Applications of the Mössbauer Effect, LACAME 2012, Medellín, Columbia, 11–16 November 2012.

S. C. Zanatta (✉)
Universidade Estadual do Paraná – UNESPAR, Av. Gabriel Esperidião, s/n, Paranavaí, Brazil CEP 87703-000
e-mail: shalicaza@yahoo.com.br

F. F. Ivashita · K. L. da Silva · C. F. C. Machado · A. Paesano Jr.
Universidade Estadual de Maringá – UEM, Av. Colombo, 5790, Maringá, Brazil CEP 87029-900

system one of the most promising materials for application in the technology of passive not-reciprocal and magneto-optical recording devices [1, 2].

Garnets crystallize in the cubic structure (space group *Ia3d*) and can be synthesized with partial substitutions in the different cationic sites, e.g., with the ideal formula $(RE', RE'')_3(M', M'')_5O_{12}$. It is known that their properties can change significantly depending on the nature and amount of ion substitution and the synthesis conditions route.

The most common process for synthesis of polycrystalline garnets is by solid-state reaction (SSR) of precursors powders (i.e., RE_2O_3 and α-Fe_2O_3), mixed in stoichiometric amounts [3, 4]. In general, this method requires heat treatments at temperatures above 1400 °C and this high temperature inevitably leads to coarsened microstructures, which affect the sinterability of the synthesized powders [5, 6]. When obtained at lower temperatures, a similar garnet (i.e., $Gd_3Fe_3Al_2O_3$) demanded a prolonged heating of 2 weeks [7]. Certainly, this has implications for the production costs of garnets.

Aiming to overcome this constraint, some alternative routes for preparation of garnets such as sol-gel, co-precipitation, citrate gel process and high-energy ball-milling have been successfully used [8, 9]. We have investigated the mechanosynthesis of a gadolinium iron garnet and an yttrium iron garnet by high-energy ball milling the oxide $3.Gd_2O_3 + 5.Fe_2O_3$ and $3.Y_2O_3 + 5.Fe_2O_3$, respectively. However, the garnet phase was clearly observed only after heat treatments carried out at 1000 °C and 1100 °C, and besides, the relative amounts of the phases (perovskites and garnet) depend on the milling time and temperature. Perovskites have been described as transient phases in garnet formation [3]. The analysis of our results points to an inverse correlation between the amount of perovskite existing in the as-milled samples and the quantity of garnet observed in the heat treated samples. In other words, the perovskite phase appears in the as-milled samples. Once it is formed, it works as a reaction barrier for garnet crystallization, demanding higher annealing temperatures. It is plausible to speculate that in order to avoid the formation of perovskite by mechanosynthesis one could use metallic iron instead of iron oxide. Indeed, perovskite is not supposed to be formed by ball-milling α-Fe + RE_2O_3 in a poorly oxidizing atmosphere.

In spite of that, up to date no results were reported using metallic iron and a RE oxide as precursors in a mechanical milling aiming at the synthesis of the garnet phase.

In order to explore this possibility, in the present study we investigate the conditions of synthesis of the garnet from the milling system $3.(Gd_2O_3) + 10(\alpha\text{-}.Fe)$. The results suggest that high energy ball-milling of this system followed by thermal annealing leads to the synthesis of a single-phase garnet.

2 Experimental

The samples were prepared, firstly, by ball milling of Gd_2O_3 (99.9 %) with α-Fe (99.9 %), manually pre-mixed in nominal compositions of 3:10. The milling was conducted for 1, 8, 12 and 24 h in free atmosphere, in a planetary ball-mill, using a hardened steel vial of 80 cm^3 charged with hardened steel balls of 10 mm in diameter. The ball-to-powder mass ratio (20:1) and speed rotation of the supporting disc and vial (300 rpm) were kept constant throughout the experiments. Subsequent

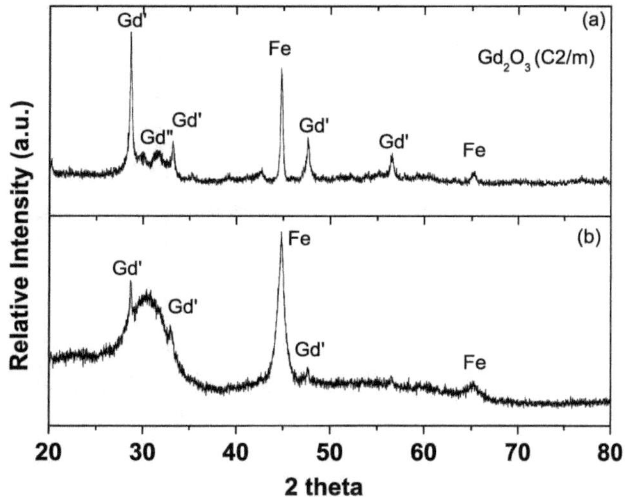

Fig. 1 X-ray diffraction patterns for 1 h (**a**) and 24 h (**b**) as-milled samples; Gd′ = Gd_2O_3 cubic; Gd″ = Gd_2O_3 monoclinic; Fe = metallic iron

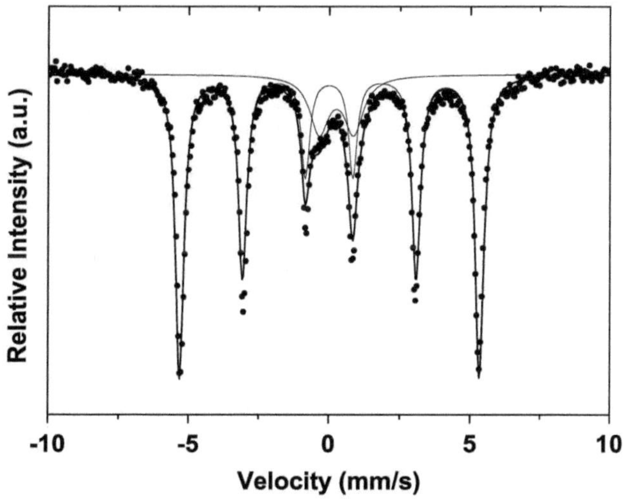

Fig. 2 Mössbauer spectrum for the 24 h as-milled sample

to the milling, the samples were heat treated in free atmosphere at 1100 °C for 3 h and 6 h. The XRD patterns of the as-milled and annealed products were obtained at room temperature (RT), using a Siemens D500 X-ray diffractometer in Bragg-Brentano geometry, with CuKα radiation, in the $20° \leq 2\theta \leq 80°$ range. The MS characterization was performed in transmission geometry, using a conventional Mössbauer spectrometer in constant acceleration mode. The spectra were analyzed with a non-linear least-square routine, with Lorentzian line shapes. The values of the isomer shift (IS) are referred to an α-Fe foil at RT.

3 Results and discussion

Figure 1 shows the diffraction patterns for some as-milled samples.

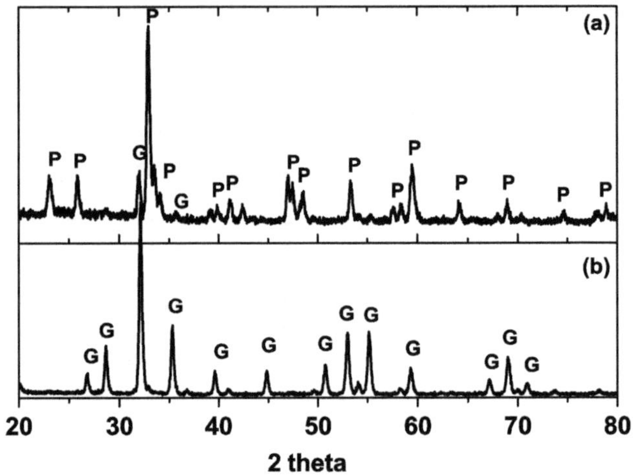

Fig. 3 X-ray diffractograms of treated samples, previously milled by 1 h (**a**) and 24 h (**b**)

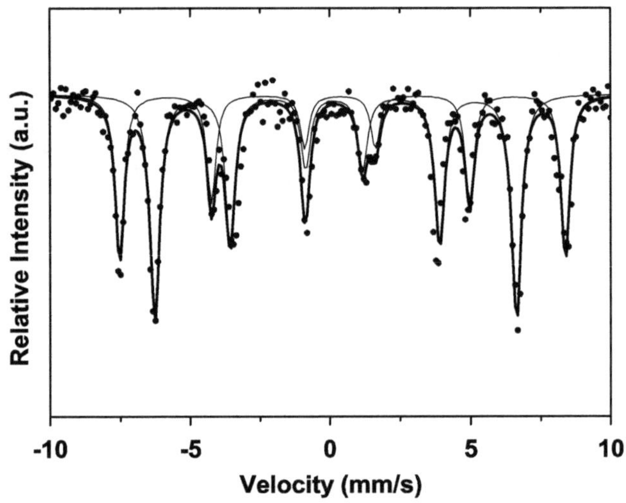

Fig. 4 Mössbauer spectrum of single-phase garnet

The X-ray diffraction pattern in Fig. 1a shows that the high-energy milling induces a partial transition for the gadolinia precursor, from cubic structure (space group *Ia3*) to monoclinic (space group C2/m), as seen, mainly, by peaks between 30° and 35°, whereas leaving unreacted most of the iron precursor. Increasing the milling time, the system tends to amorphization, with the iron grain strongly refined, as can be seen by the large broadening of the peak situated at 44.1° (Fig. 1b).

Figure 2 shows the Mössbauer spectrum for the 24 h as-milled sample. This and the other spectra for the as-milled samples were fitted with a discrete sextet, which hyperfine parameters belong to metallic iron, plus a paramagnetic component, with isomer shift compatible with Fe^{3+} and a subspectral area that slightly increases with grinding time. We believe that this doublet is the indication of some blend between the precursors produced by grinding, resulting in undefined iron-gadolinium oxide.

Diffractograms of some heat-treated samples are shown in Fig. 3. They reveal the formation of garnet and perovskite phases, in relative amounts that depend on the milling time. For 1 h of milling, e.g., perovskite is the major phase (Fig. 3a), whereas

Table 1 Hyperfine parameters for the sample milled for 24 h and heat-treated at 1100 °C for 6 h

Site	δ^a (mm/s) (±0.01)	$\Delta E_Q{}^b$ (mm/s) (±0.02)	$B_{hf}{}^c$ (T) (±0.3)	Γ^d (mm/s) (±0.01)	Area[e] (%) (±0.1)
16a	0.39	0.09	49.4	0.32	35.2
24d	0.16	0.01	40.2	0.37	52.8

[a] Isomer shift relative to α-Fe, at room temperature
[b] Quadrupole Splitting
[c] Magnetic Hyperfine Field
[d] Linewidth
[e] Subspectral area

for 8 h the garnet (diffractogram not shown) is the major phase. For milling times of 12 h or more, a monophasic (garnet) sample is obtained (Fig. 3b).

Figure 4 shows the Mössbauer spectra for the heat-treated sample, previously milled during 24 h. The spectrum was fitted with two discrete sextets, corresponding to the sites 16a and 24d of the garnet structure. Table 1 presents the hyperfine parameters for this sample. The values fitted are in good agreement with those reported in literature for GdIG [10].

4 Conclusions

The results suggest that the high-energy ball-milling system $Gd_2O_3 + \alpha$-Fe up to 24 h, does not promote the structure of garnet and/or perovskite. These phases are observed in the heat-treated samples. The quantity of garnet increases with milling time. Single-phase garnet can be observed in samples milled for 12 and 24 h (Table 1).

High-energy milling of $Gd_2O_3 + \alpha$-Fe is an important step in GdIG synthesis by ceramic method. Thus, the data favor the conclusion that the high energy milling system $Gd_2O_3 + \alpha$-Fe followed by heat treatment at 1100 °C during 6 h represents an alternative for synthesis of garnet.

References

1. Hudson, A.S.: Molecular engineering in the design of microwave ferrimagnetic garnets. J. Phys., D, Appl. Phys. **3**, 251–268 (1970)
2. Pardavi-Horvath, M.: Microwave applications of soft ferrites. J. Magn. Magn. Mater. **215–216**, 171–183 (2000)
3. Sánchez-De Jesús, F., Cortés C.A., Valenuela R., Ammar, S., Bolarín-Miró, A.M.: Synthesis of $Y_3Fe_5O_{12}$ (YIG) assisted by high-energy ball milling. Ceram. Int. **38**(6), 5257–5263 (2012)
4. Widatallah, H.M., Johnson, C., Al-Harthi, S.H., Gismelssed, A.M., Al-Rawas, A.D., Stewart, S.J., Elzain, M.E., Al-Omari, I.A., Yousif, A.A.: A Structural and Mössbauer study of $Y_3Fe_5O_{12}$ nanoparticles prepared with high energy ball milling and subsequent sintering. Hyperfine Interact. **183**, 87–92 (2008)
5. Kong, L.B., Ma, J., Huang, H.: Low temperature formation of yttrium aluminum garnet from oxides via a high-energy ball milling process. Mater. Lett. **56**, 344–348 (2002)
6. Vaqueiro, P., López-Quintela, M.A., Rivas, J., Greneche, J.M.: Annealing dependence of magnetic properties in nanostructured particles of yttrium iron garnet prepared by citrate gel process. J. Magn. Magn. Mater. **169**, 56–58 (1997)

7. Grasset, F., Mornet, S., Etourneau, J., Haneda, H., Bobet, J.-L.: Effects of ball milling on the grain morphology and the magnetic properties of $Gd_3Fe_3Al_2O_{12}$ garnet compound. J. Alloys Compd. **359**, 330–337 (2003)
8. Mergen, A., Qureshi, A.: Characterization of YIG nanopowders by mechanical synthesis. J. Alloys Compd. **478**, 741–744 (2009)
9. Paesano, A., Jr., Zanatta, S.C., de Medeiros, S.N., Cótica, L.F., da Cunha, J.B.M.: Mechanosynthesis of YIG and GdIG: a structural and Mössbauer Study. Hyperfine Interact. **161**, 211–220 (2005)
10. Lataifeh, M.S., Lehlooh, A.F.D.: Mössbauer spectroscopy study of substituted yttrium iron garnets. Solid State Commun. **97**(9), 805–807 (1996)

DOI 10.1007/978-94-007-6482-8_37
Reprinted from *Hyperfine Interactions*, DOI 10.1007/s10751-013-0865-y

Effect of boron in $Fe_{70}Al_{30}$ nanostructured alloys produced by mechanical alloying

M. M. Rico · G. A. Pérez Alcázar · J. M. Greneche

Published online: 18 April 2013

Abstract The substitution of aluminum by boron in the $Fe_{70}Al_{30}$ system prepared by high energy ball milling is studied when the B content ranged from 0 up to 20 at. %, and the milling times were 24, 48 and 72 h. X-ray diffraction (XRD) patterns of $Fe_{70}Al_{30}$ showed a predominant bcc structural phase with a lattice parameter larger than that of α-Fe. A second (tetragonal) phase arose with the addition of boron. It is associated to the existence of $(Fe, Al)_2B$, although the values of the lattice parameters are slightly different from those found in the literature. This phase shows high stability; its lattice parameters and the Mössbauer parameters do not show notable variations, either with milling time or composition. It was also evidenced that an increase of boron content and of milling time produced a decrease of the lattice parameter of the Fe-Al bcc structure. This is in agreement with the small atomic radius of boron in comparison with that of aluminum. This also allows boron to occupy interstitial sites in the lattice, increasing the grain size and giving rise to the ductile character of the alloy. On the other hand, 300 K transmission Mössbauer spectra (TMS) were fitted, for low boron concentrations (<8 at.%), with a hyperfine field distribution (HFD) associated with the bcc phase. For high boron content (≥8 at.%), a magnetic component related to the tetragonal phase is added and its broadened lines are attributed to the disordered character of Fe_2B, probably induced by the milling process.

Proceedings of the 13th Latin American Conference on the Applications of the Mössbauer Effect, (LACAME 2012), Medellín, Colombia, 11–16 November 2012.

M. M. Rico (✉)
Departamento de Física, Universidad Autónoma de Occidente, A.A. 2790, Cali, Colombia
e-mail: mrico@uao.edu.co

G. A. Pérez Alcázar
Departamento de Física, Universidad del Valle, A.A. 25360, Cali, Colombia

J. M. Greneche
Laboratoire de Physique de l'Etat Condensé, UMR CNRS 6087, Université du Maine, 72085 Le Mans Cedex 9, France

Keywords FeAlB · Mössbauer spectrometry · Mechanical alloying

1 Introduction

It has been well established that high-energy mechanical milling is one of the best techniques for producing nanostructured powders [1, 2]. This method consists in the mixing of powders (of different metals or alloys/compounds), which are milled together to get a homogeneous alloy through successive welding and fracture processes [3]. The nanostructured powders obtained consist thus of particles composed of nanometer size crystalline grains, linked to one another through grain boundaries [4]. The reduction of the grain size improves the physical, mechanical and magnetic properties of the material [3]. For this reason nanostructured materials are potentially attractive for many applications [5].

This is particularly evident in the case of Fe-Al intermetallic systems. The dependence of the magnetic properties on the state of disorder and the microstructure makes this system interesting to study when it is produced by the mechanical alloying (MA) process.

The Fe-Al intermetallic compound possesses advantageous properties, in particular a high specific strength (strength-to-density ratio), good strength at intermediate temperature and an excellent corrosion resistance at elevated temperatures under oxidizing, carburizing and sulfidizing atmospheres. These features make the Fe-Al intermetallic compounds very attractive materials for structural and coating applications at elevated temperatures in hostile environments [6]. The addition of small amounts of boron to Fe-Al can change the fracture mode from intergranular failure to transgranular fracture. This change in fracture mode is believed to result from the segregation of boron to grain boundaries. Studies on the microstructure of these alloys gave evidence that the addition of boron can also result in the formation of Fe_2B precipitates and a tetragonal phase which is likely coherent with the matrix [7].

A recent study [4] showed that the addition of boron (up to 20 at. %) to the Fe-Al system produces growth of the grain boundaries giving rise to a low hyperfine magnetic field contribution that can be attributed to iron atoms located in them. However, there are few reports to our knowledge on the magnetic properties of ternary FeAlB alloys [4, 8, 9]. The purpose of the present investigation is to obtain a better understanding of the evolution of mechanically alloyed $Fe_{70}Al_{30}$ system with milling time and the effect of the addition of boron in its structural and magnetic properties. Both variations in particle size and in lattice parameter have been determined through X-ray diffraction studies. The magnetic hyperfine field distributions obtained from ^{57}Fe Mössbauer spectra have been used to identify the different magnetic phases that appear.

2 Experimental section

Fe, Al and B powders with 99.9 % purity were used to prepare samples of $Fe_{70}Al_{30-x}B_x$ ($0 \leq \times \leq 20$) by mechanical alloying. The powders were milled for 24, 48 and 72 h in a planetary milling (Fristch Pulverisette 5) using hardened stainless steel vials together with balls of the same material, under vacuum. The ball-to-powder mass ratio was 15:1. A structural characterization was carried out with X-ray diffraction (XRD) using a Philips X'pert diffractometer (with monochromatized Cu $K\alpha$ radiation; $\lambda = 1.54056$ Å). The patterns were then refined by the Rietveld method using the MAUD program, which is combined with Fourier analysis to describe the broadening of the Bragg peaks [10]. The magnetic characterization was carried out by transmission Mössbauer spectrometry (TMS) by means of a conventional constant acceleration spectrometer while the spectra were fitted by using the MOSFIT program involving discrete distributions of magnetic and/or quadrupolar components with lorentzian lines [11].

3 Results and discussion

Figure 1 shows XRD patterns for samples of $Fe_{70}Al_{30}$ and $Fe_{70}Al_{10}B_{20}$ milled for 24 and 72 h. The sample without B evidences characteristic lines of a bcc structure, coming from the bcc character of pure iron, whereas the samples with B concentration higher than 4 at. % exhibit both bcc and tetragonal phases. This last phase appears as a consequence of Fe and B alloying in a Fe_2B–like phase. The XRD patterns were fitted by means of the Maud program [10] and the results are presented in Fig. 2a (bcc phase) and b (tetragonal phase). The lattice parameter for the bcc phase has a higher value than that of bcc Fe, because of the presence of Al, which has a larger atomic radius. However, this parameter decreases when the B concentration increases, since some Al atoms are substituted by B (which is much smaller than both Al and Fe atoms). This substitution, together with milling time, helps to bring the diffraction planes nearer, which in turn contributes to decrease the lattice parameter. In the case of the tetragonal phase, there is no apparent dependence of lattice parameters on either boron concentration or milling time. Nevertheless, a slight variation of these parameters was observed in comparison with those of pure Fe_2B. It can be attributed to the inclusion of Al atoms in the positions of Fe (0.1649; 0.1649; 0). The grain size for the bcc phase ranges from about 10 to 29 nm, and decreases when the B concentration increases, because B atoms occupy interstitial sites in the bcc phase, producing a growth of grain boundaries [12]. On the other hand, there is no clear behavior with milling time. Nonetheless, for B concentrations between 15 and 20 at. %, there is a substantial reduction of the grain size between 24 and 48 h, as it is observed in Fig. 2a (middle). This is a consequencc of the increase of grain boundaries with milling time; this effect tends to stabilize for 15 and 20 at. % of B, starting from 48 h, where the grain size remains rather constant. For the tetragonal phase, no tendency is found for low concentrations of boron. For higher B contents (15 and 20 at. %), closely similar values for grain size were obtained.

The bcc and tetragonal phase contents are shown at the bottom of Fig 2a and b, respectively. An increase of the tetragonal phase is observed with B content, being higher for 48 and 72 h. Despite that this value is strongly increased with milling time, it is not higher than that reported for $Fe_{60}Al_{20}B_{20}$ alloys at the same milling time [4].

Figure 3 shows 300 K Mössbauer spectra for samples milled for 24 and 72 h. For the sample without B a characteristic spectrum of a disordered magnetic system is observed. The appearance of this spectrum suggests a fit with a hyperfine field

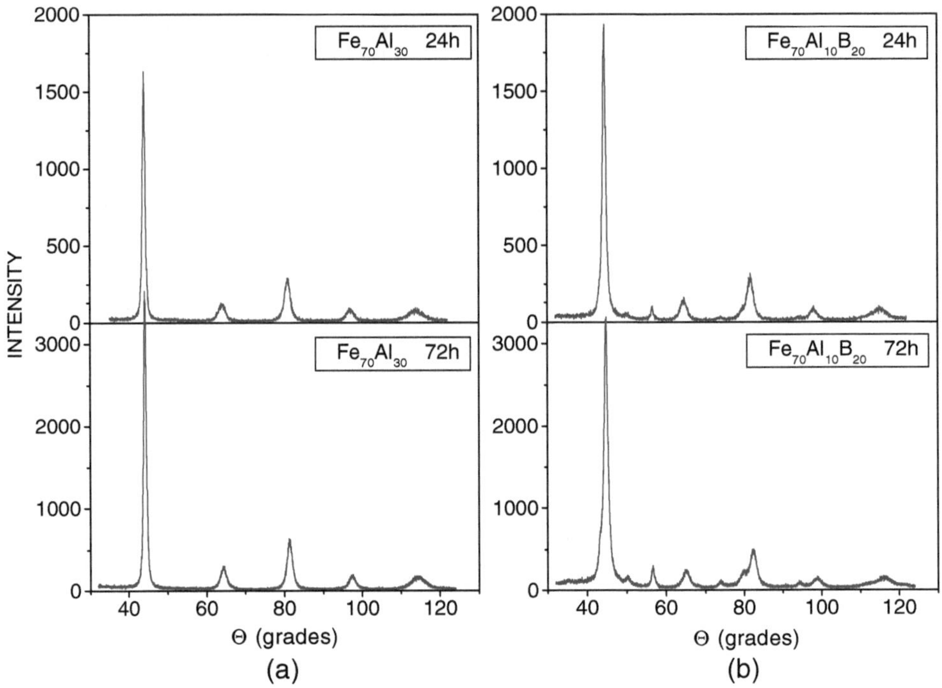

Fig. 1 X- Ray diffraction patterns for samples of $Fe_{70}Al_{30}$ (**a**) and $Fe_{70}Al_{10}B_{20}$ (**b**) milled for 24 and 72 h

distribution (HFD) associated to the bcc-disordered phase. It can be noted that all the spectra show an asymmetry which was reproduced in the spectra by fitting them with a distribution of IS. When milling time increases, the spectral lines are more separated and narrowed. This means that the HFD reaches more elevated field values with higher probabilities as illustrated in Fig. 4a, which shows the HFDs for samples without B (pure $Fe_{70}Al_{30}$ samples) for different milling times.

This enhancement of the HF values and the decrease of the lattice parameter (Fig. 2a) is in agreement with the reported results for milled $Fe_{60}Al_{40}$ samples [13, 14]. It can also be observed that the effect of milling time is to increase both the probability and the field of the most intense peak, and to decrease the probability of the peaks at small fields, giving rise thus to an increase of the MHF. This behavior was also reported by different authors for this type of alloys [15, 16], who attributed it to the increase of the disorder. Besides, the increase of the ferromagnetic behavior with the disorder was also reported in arc-melting [17] and MA [13] produced FeAl disordered crystalline alloys.

When B is added (Fig. 3b) the spectra were also fitted with HFD, but some new lines appear like a sextet that are attributed to the ferromagnetic Fe_2B tetragonal phase. These lines become more intense when milling time increases. However, this second phase was detected at 48 and 72 h for 10 at. % of B content and from 24 h for higher concentrations. In the case of $Fe_{70}Al_{10}B_{20}$ milled for 72 h, a first naive but rather good description is obtained by means of only two magnetic components with broadened lorentzian lines and a doublet: the hyperfine field values of the magnetic

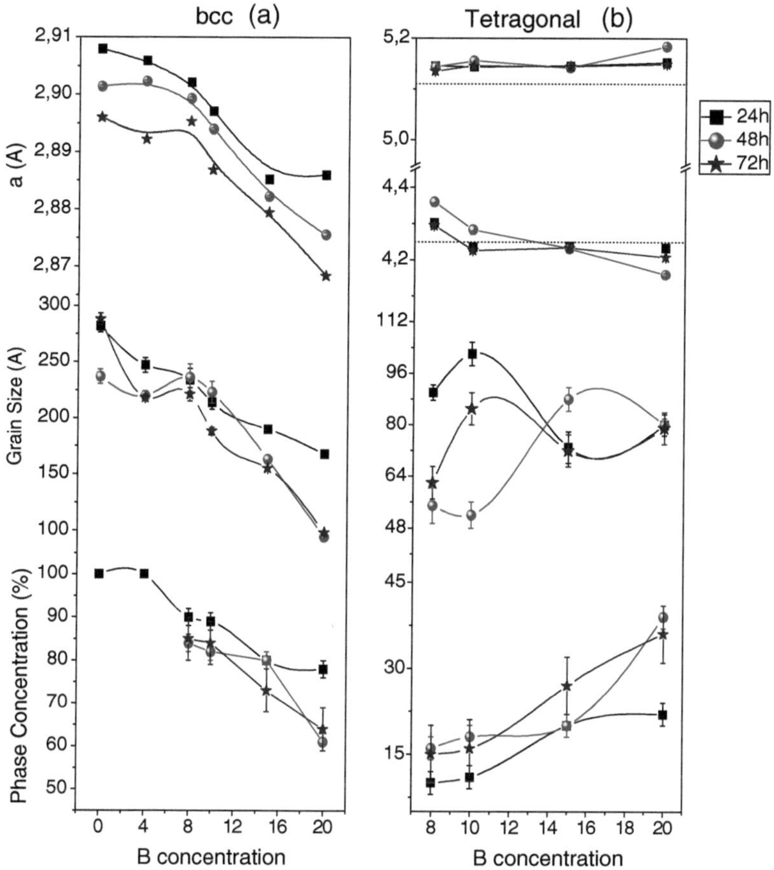

Fig. 2 Mean structural parameters as a function of B concentration obtained from XRD patterns for different milling times: lattice parameter, grain size and weight fraction of bcc (**a**) and tetragonal (**b**) phases

components are found close to 33 T and 23.4 T, and allow these 2 components to be assigned to bcc-Fe rich and tetragonal Fe_2B phases, respectively. Figure 3b-right corresponds to this fit. The modeling can be improved by considering a distribution of hyperfine fields for the first component. Then the second Mössbauer component was better described by means of two magnetic sextets associated with two hyperfine field values of 22,9 and 24,0 T, in rather good agreement with that of the pure Fe_2B phase [18]. Such a model was successfully achieved for all Mössbauer spectra with B concentration of 20 at. %. The doublet is attributed to those Fe sites of the bcc phase rich in Al atoms, which appear when some Fe atoms of the bcc phase migrate to form the Fe2B phase.

In order to see the effect of the increase of boron concentration and the effect of milling time in samples with and without B, the HFDs for 0 and 20 at. % B milled at different times are illustrated in Fig. 4. It can be noted that the addition of B tends to shift the HFDs to higher fields, and allows the observation of two regions with more clarity: the first one in the higher field range, which gets narrower and achieves higher

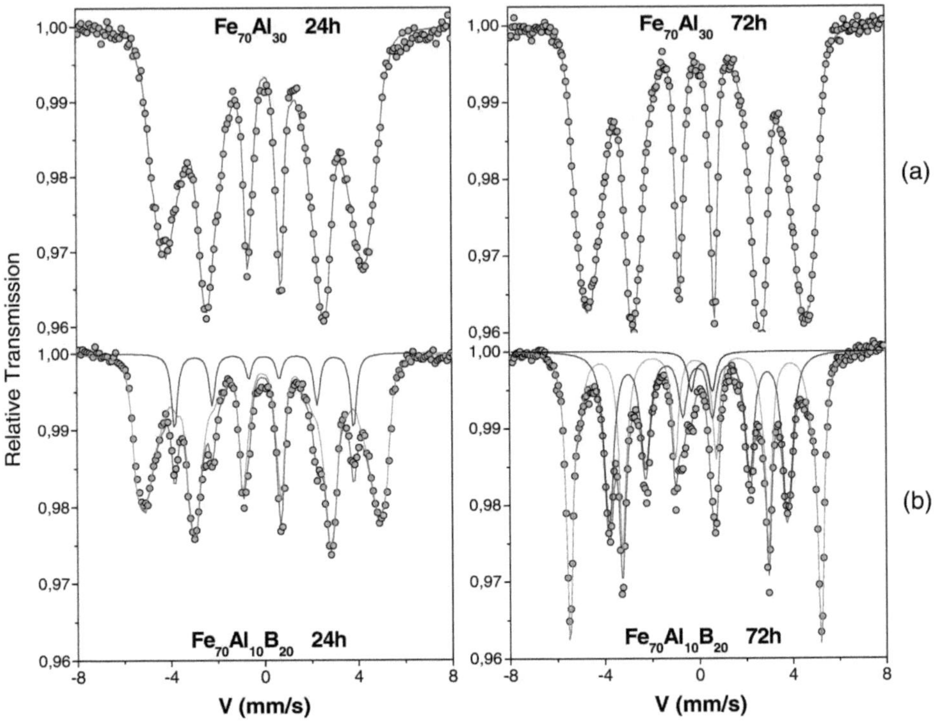

Fig. 3 Mössbauer spectra at room temperature for samples milled for 24 and 72 h

probabilities as B content increases, while the second one lies in the lower field range with lower probabilities. It should be noted that those low fields are slightly higher than those observed for the samples without B. Otherwise, the increase of milling time in both type of samples gives rise to the increase of the probability and the narrowing of the most intense peak, accompanied by the disappearing of one of the small peaks and the increase of the probability of the other. This narrowing is associated with the increase of the disorder of the samples due to the tendency to obtain chemical equilibrium. This chemical equilibrium homogenizes the samples decreasing in this way the concentration gradients and carrying the distribution around of only one more probable Fe site. It can be also seen that when the boron contents increases the HFDs become narrower, and this can be explained as a consequence of the formation of the Fe_2B phase. Then some iron atoms are lost from the bcc phase in order to form the new one and in this way there are less environments in the HFD of the bcc phase.

In Fig. 5 the behavior of Mössbauer parameters with boron content and milling time is observed. The isomer shift (IS) of the HFDs, at top of Fig. 5, diminishes with B concentration for 24 h of milling time. This diminution is associated with the increase of s electrons in the nuclei of Fe atoms due to penetration of boron atoms in the bcc phase. At 48 h and 72 h, the IS is approximately constant, with practically no influence of concentration. The meaning of this behavior is that for these alloys a saturation stage is reached which is not sensitive any more to longer synthesis times and there is consequently a saturation of the s electron density in the nuclei of the Fe atoms. The IS for the tetragonal phase of the sample milled during 24 h behaves

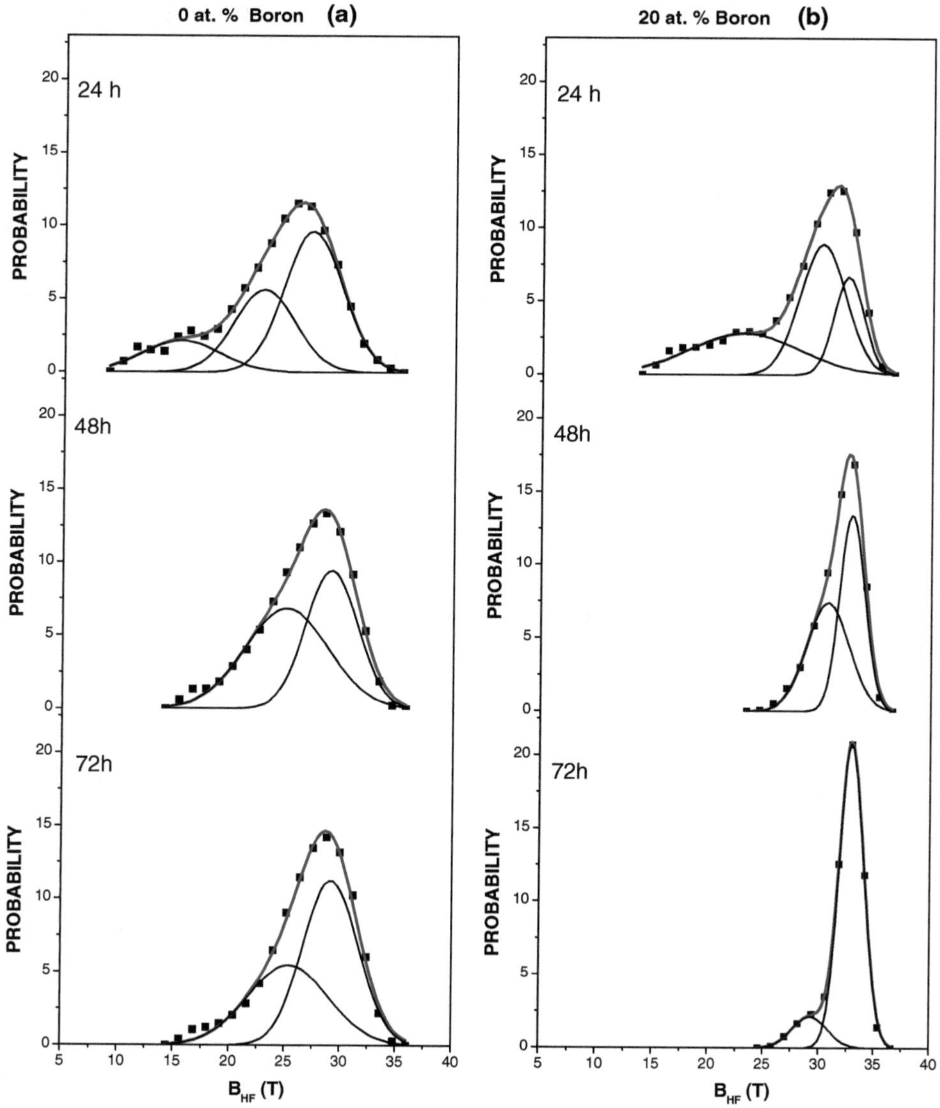

Fig. 4 Hyperfine field distribution and Gaussian fits for samples without B (**a**) and with 20 at. % of B (**b**), milled for 24, 48 and 72 h. All measures were taken at 300 K

in the range previously reported in literature for melted Fe_2B alloys [18]. For 48 h and 72 h this parameter shifts to values smaller than the lower limit, remaining nearly constant with B concentration.

Figure 5 (middle) shows the behavior of the MHF for the bcc phase and the HF of the Fe_2B phase. For the sample without boron the MHF increases with milling time tending to a constant value. At 24 h, the MHF is almost constant. For 48 h and 72 h when boron content increases these MHF values get lower as a consequence of the migration of the Fe atoms to the other phase and the decreasing of the local

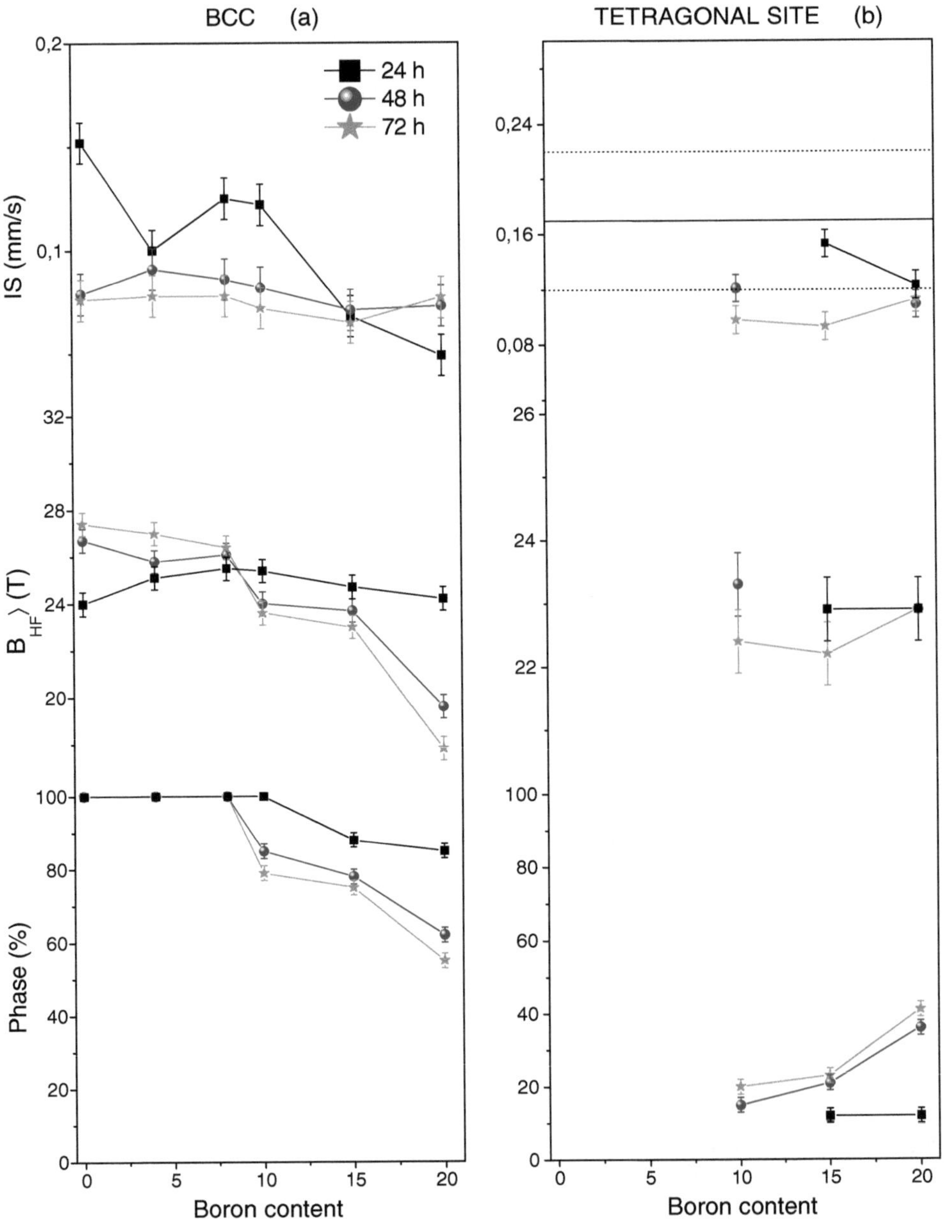

Fig. 5 Mössbauer parameters obtained. Isomer shift (**a**) MHF (**b**) and phase percent for the field distribution (**a**) and the tetragonal site (**b**)

magnetic moment of iron due to the increase of the nearest neighbors Al atoms [19, 20]. Besides, for these milling times, for each B concentration, the MHF values are very similar, which confirms that synthesis time has been reached at 48 h. The HF for the tetragonal phase for all milling times tends to coincide with the values found for 20 at. % of boron.

Figure 5 (bottom) depicts the proportion of Fe atoms that make part of the field distribution and the ferromagnetic site. It can be observed that the number of Fe atoms of the tetragonal phase increases with both milling time and boron concentration. This is in agreement with the XRD results shown in Fig. 2.

4 Conclusions

The effect of inclusion of boron and milling time in the FeAl mechanical alloying system was studied through XRD and MS. It was found that when the boron content increases, a bcc phase associated to Fe-Al tends to coexist with a tetragonal phase related to Fe_2B-like phase.

For B concentrations from 0 up to 10 at. % the Mössbauer spectra were fitted with a distribution that is characteristic of a disordered system which is associated with the bcc phase. For B contents higher than 10 at. %, the distribution becomes narrower and it was necessary to employ ferromagnetic sites with HF values close to those reported for pure Fe_2B.

Both, the increase of boron concentration and the milling time induce the narrowing of the HFD. The first one is a consequence of the loss of iron atoms from the bcc phase, which migrate to form the Fe2B phase; and the second one is due to the increase of the disorder. The paramagnetic site, which appears at 20 at. % B and 72 h of milling, corresponds to bcc Fe sites rich in aluminum.

References

1. Koch, C.C.: Nanostruct. Mater. **9**, 13–22 (1997)
2. Zhang, D.L.: Prog. Mater. Sci. **49**, 537–560 (2004)
3. Suryanarayana, C.: Prog. Mater. Sci. **46**, 1–184 (2001)
4. Rico, M.M., Greneche, J.M., Pérez Alcázar, G.A.: J. Alloy. Compd. **398**, 26–32 (2005)
5. Zhang, D.L.: Prog. Mater. Sci. **49**, 537–560 (2004)
6. Krasnowski, M. Grabias A. Kulik T.: J. Alloy. Compd. **424**, 119–127 (2006)
7. Pang, L., Kumar, K.S.: Mater. Sci. Eng. A **258**, 161–166 (1998)
8. De Jesus, D.R., Partiti, C.S.M.: J. Magn. Magn. Mater. **226–230**, 1527–1529 (2001)
9. McCally, R.L., Moorjani, K., Kistenmacher, T.J.: J. Appl. Phys. **61**, 3655 (1987)
10. Lutterotti, L., Scardi, P.J.: Appl. Crystallogr. **23**, 246 (1990)
11. Teillet, J., Varret, F.: unpublished Mosfit program, Université du Maine
12. Liu, C.T., George, E.P.: Scripta Metall. **24**, 1285 (1990)
13. Gialanella, X., Amils, M.D., Baró, P., Delacroix, G., Le Caer, Lutteroti, L., Suriñach, S.: Acta Mater. **46**, 3305 (1998)
14. Zamora Ligia, E., Pérez Alcázar, G.A., Vélez, G.Y., Betancur, J.D., Marco, J.F., Romero, J.J., Martínez, A., Palomares, F.J., González, J.M.: Phys. Rev. B **79**, 094418 (2009)
15. Hernando, A., Amils, X., Nogués, J., Suriñach, S., Baró, M.D., Ibarra, M.R.: Phys. Rev. B **58**, R11 864 (1998)
16. Nogués, J., Apiñaniz, E., Sort, J., Amboage, M., d'Astuto, M., Mathon, O., Puzniak, R., Fita, I., Garitaonandia, J.S., Suriñach, S., Muñóz, J.S., Baró, M.D., Plazaola, F., Baudelet, F.: Phys. Rev. B **74**, 024407 (2006)
17. Pérez Alcázar, G.A., Galvão da Silva, E.: J. Phys. F. Met. Phys. **17**, 2323 (1987)
18. Weisman, I.D., Swartzendruber, L.J., Bennett, L.H.: Phys. Rev. **177**, 465 (1969)
19. Arzhnikov, A.K., Dobysheva, L.V., Timirgazin, M.A.: J. Magn. Magn. Mater. **320**, 1904–1908 (2008)
20. Szymanski, K., Satula, D., Dobrzynski, L., Voronina, E., Yelsukov, E.P., Miyanaga, T.: Phys. Rev. B **72**, 104409 (2005)

DOI 10.1007/978-94-007-6482-8_38
Reprinted from *Hyperfine Interactions*, DOI 10.1007/s10751-013-0834-5

Mössbauer and X-ray study of the $Fe_{65}Ni_{35}$ invar alloy obtained by mechanical alloying

R. R. Rodriguez · J. L. Valenzuela · J. A. Tabares · G. A. Pérez Alcázar

Published online: 13 March 2013

Abstract $Fe_{65}Ni_{35}$ samples were prepared by mechanical alloying (MA) with milling times of 5, 6, 7, 10 and 11 h, using a ball mass to powder mass ratio of 20:1 and at 280 rpm. The samples were characterized by X-ray diffraction (XRD) and transmission ^{57}Fe Mössbauer spectrometry. The X-ray diffraction pattern showed the coexistence of one body centered cubic (BCC) and two face centered cubic (FCC1 and FCC2) structural phases. The lattice parameters of these phases did not change significantly with the milling time (2.866 Å, 3.597 Å and 3.538 Å, respectively). After 10 h of milling, the X-ray diffraction pattern showed clearly the coexistence of these three phases. Hence, Mössbauer spectrometry measurements at low temperatures from 20 to 300 K of this sample were also carried out. The Mössbauer spectra were fitted using a model with three components: the first one is a hyperfine magnetic field distributions at high fields, related to the BCC phase; the second one is a hyperfine magnetic field distribution involving low hyperfine fields related to a FCC phase rich in Ni, and the third one is a singlet related to a FCC phase rich in Fe, with paramagnetic behavior. As proposed by some authors, the last phase is related with the antitaenite phase.

Keywords FeNi · Mechanical alloying · Mössbauer spectrometry · Invar composition

Proceedings of the Thirteenth Latin American Conference on the Applications of the Mössbauer Effect, (LACAME 2012), Medellín, Colombia, 11–16 November 2012.

R. R. Rodriguez (✉)
Universidad Autónoma de Occidente, A.A. 2790, Cali, Colombia
e-mail: rubyrocio@gmail.com

J. L. Valenzuela · J. A. Tabares · G. A. Pérez Alcázar
Departamento de Física, Universidad del Valle, A.A. 25360, Cali, Colombia

1 Introduction

The investigation of the magnetic properties of Fe-Ni alloys has a long lasting tradition. New technological methods of preparation and treatment open possibilities to prepare materials with different structures and novel physical properties. This is also the case for the mechanical alloying process, which allows the preparation of nanocrystalline Fe-Ni alloys characterized by a grain size of a few nanometers [1, 2] with a high structural disorder [3–5]. The reduction of the grain size increases the saturation magnetization and decreases the coercivity. In this sense, nanocrystalline Fe-10 and Fe-20 wt.% Ni alloys exhibit a soft magnetic behavior [2].

The as-milled samples, near the invar composition, were characterized by Mössbauer spectrometry (MS) and showed to contain a mixture of α (bcc) and γ (fcc) phases after 48 h of milling [6, 7]. Scorzelli [8] and Valderruten et al. [9] observed a non-magnetic phase usually referred to as "paramagnetic phase". This phase was recently reported by Rancourt and Scorzelli [10] as a possible equilibrium phase in the Fe–Ni system and they propose that this is a low spin γ-Fe–Ni phase (γ_{LS}), with an estimated composition of 25–30 % Ni. Besides, it occurs in close microstructural association with tetrataenite. This phase is only observed by Mössbauer spectrometry because tetrataenite (ordered FeNi) and γ_{LS} (proposed to be called antitaenite) have practically indistinguishable lattice parameters. Therefore, the γ_{LS} phase is not readily observable as a distinct phase by TEM or X-ray diffraction. Additionally Valderruten et al. [9] proposed for this type of Fe-Ni samples a new fitting model, which includes two hyperfine magnetic field distributions (HMFD) corresponding to the two ferromagnetic phases, and a paramagnetic site ascribed to the antitaenite phase.

The aim of the present work is to produce the $Fe_{65}Ni_{35}$ alloy (invar melted composition) by the high-energy ball milling technique. Several alloying times ranging from 5 and 11 h were considered. The powders were investigated by means of X-ray diffraction and ^{57}Fe Mössbauer spectrometry at different temperatures from 20 K up to 300 K.

2 Experimental method

Samples were milled from pure elemental powders (Fe 99.9 % and Nickel 99 %) in a planetary high energy ball mill. The speed used was 280 rpm, and the milling was carried out using a sequence of 60 min milling and 30 min resting. The elemental powders were weighed in stoichiometric proportions, mixed in a small flask and then transferred into the vials under argon atmosphere. We used a 10:1 ball mass to powder mass ratio and the milling times were: 5, 6, 7, 10 and 11 h. XRD measurements were performed using $Cu_{K\alpha}$ radiation and the patterns were fitted using the GSAS software [11] which is based on the Rietveld method combined with Fourier analysis, well adapted for broadened diffraction peaks, and taking also into account instrumental parameters. The refinement of the patterns allows obtaining the lattice parameters, the average grain size and the relative fraction of the different phases. ^{57}Fe Mössbauer experiments were performed on the sample milled during 10 h at 20, 30, 60, 70, 80, 90, 150 and 300 K in transmission geometry using a ^{57}Co/Rh source. The analysis of the Mössbauer spectra was done by using the MOSFIT

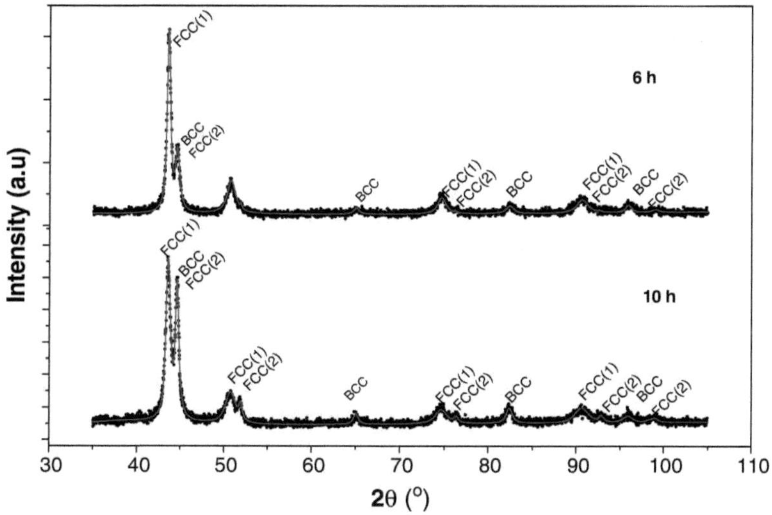

Fig. 1 XRD- patterns for $Fe_{65}Ni_{35}$ milled during 6 and 10 h

program [12] and the values of isomer shift are quoted relative to that of α-Fe at 300 K.

3 Results and discussion

3.1 X-ray diffraction

Figure 1 shows the $Fe_{65}Ni_{35}$ XRD patterns for 6 and 10 h; those corresponding to 5, 7 and 11 h are similar. The XRD patterns exhibit a coexistence of one Body Centered Cubic (BCC) and two Face Centered Cubic (FCC1 and FCC2) structural phases.

The estimation of the lattice parameter and the grain size was obtained from refinement of X-ray patterns using the GSAS procedure. Table 1 shows the observed phases and their respective proportions, and lattice parameters for the different milling times. The lattice parameter of the three phases are approximately constant for the different milling times, the average lattice parameter is 2.866 Å for BCC structure, 3.597 Å for FCC1 (the value reported for the bulk Fe-Ni is 3,5975 Å) and 3.538 Å for FCC2. The lattice parameter of BCC and FCC1 are well in accordance with those reported in Refs. [7, 13]. These two phases are well known in the literature as kamasite and taenite, respectively [1, 14]

Figure 2 shows the variation of Fe-Ni crystallite average size versus time. As can be seen in Fig. 2, during the earlier stages of milling, from 5 h to 7 h, the crystallite size increases, showing that the cold welding predominates over fracture. Afterwards fracture dominates, and the crystallites become smaller. After 11 h the crystallite size was about 15 nm for the three structural phases, according to Scorzelli results [8].

To the best of our knowledge, the FCC2 phase has not been observed by X-ray diffraction by other authors. However, Scorzelli [8], Rancourt and Scorzelli [10] and Valderruten et al. [9] have observed a non-magnetic phase usually referred to

Table 1 Phases, proportions and lattice parameters obtained for the $Fe_{65}Ni_{35}$ alloys milled for different times

Milling time (h)	Phase	Phase (%)	Lattice parameter (Å)
5	BCC	25	2,866
	FCC1	70	3,595
	FCC2	5	3,531
6	BCC	21	2,865
	FCC1	77	3,597
	FCC2	2	3,543
7	BCC	26	2,867
	FCC1	73	3,596
	FCC2	7	3,558
10	BCC	22	2,867
	FCC1	66	3,598
	FCC2	12	3,534
11	BCC	41	2,867
	FCC1	79	3,598
	FCC2	4	3,523

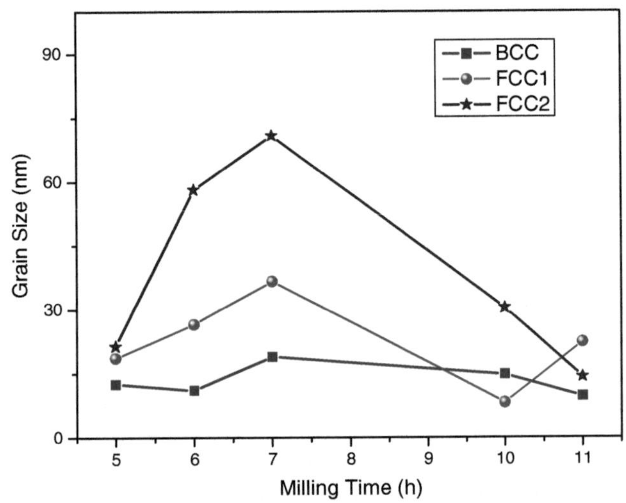

Fig. 2 Grain size of BCC, FCC1 and FCC2 structural phases vs. milling time

as a "paramagnetic phase", which was only observed by Mössbauer spectrometry because tetrataenite (ordered FeNi) and γ_{LS} (proposed to be called antitaenite) have practically indistinguishable lattice parameters.

By X ray-diffraction it was possible to observe that the highest weight percentage of the FCC2 phase was obtained for the sample milled during 10 h. For this reason we decided to study this sample by Mössbauer spectrometry at temperatures in the range from 20 to 300 K.

3.2 Mössbauer spectrometry

The Mössbauer spectra recorded at different temperatures, from 20 to 300 K, and their corresponding hyperfine magnetic field distributions (HMFD's) for the

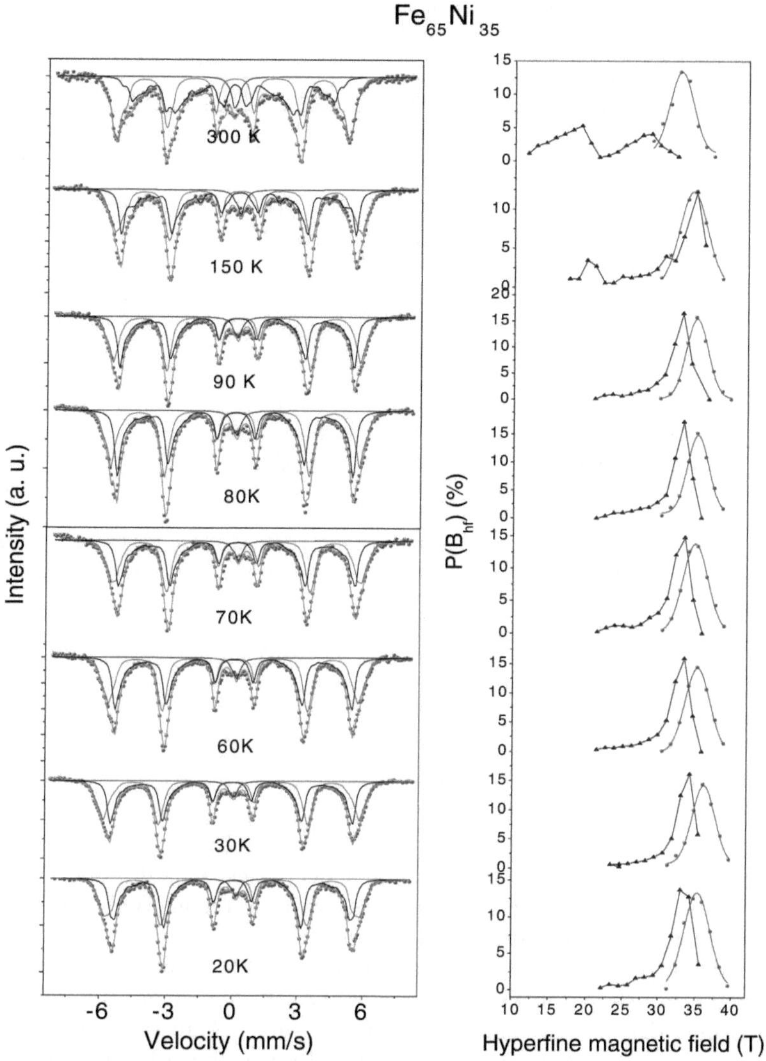

Fig. 3 Mössbauer spectra at different temperatures and their corresponding hyperfine magnetic field distributions, for $Fe_{65}Ni_{35}$ sample milled during 10 h

$Fe_{65}Ni_{35}$ alloys are presented in Fig. 3. Mössbauer spectra were fitted using a model with two hyperfine magnetic field distributions (HMFDs) and one singlet.

In Table 2 are shown the hyperfine parameters obtained from the fitting. The first HMFD, with higher fields, is related to the disordered BCC Fe-Ni phase rich in Fe, and its spectral area at room temperature (46 %) was maintained fixed for all temperatures. The second HMFD, with lower hyperfine fields, is related to the disordered FCC1 phase rich in Ni and the third one, the singlet, was related to the disordered FCC2 phase rich in Fe, with paramagnetic behavior.

Table 2 Hyperfine parameters obtained from the fitting of the Mössbauer spectra: isomer shift (IS), quadrupole splitting (QS), mean hyperfine field (MHF) and area percentage for $Fe_{65}Ni_{35}$ alloy at different temperatures

Temperature (K)	Component	IS (mm/s)	QS (mm/s)	MFH (T)	Area (%)
20	BCC1	0,14	−0,08	35,4	49,0
	FCC1	0,12	0,00	32,2	47,7
	FCC2	0,21			3,3
30	BCC1	0,14	−0,08	36,0	49,0
	FCC1	0,12	−0,02	32,6	47,6
	FCC2	0,19			3,4
60	BCC1	0,13	−0,07	35,2	49,0
	FCC1	0,11	−0,02	31,2	47,3
	FCC2	0,19			3,7
70	BCC1	0,12	−0,07	34,9	49,0
	FCC1	0,10	−0,03	31,5	47,4
	FCC2	0,17			3,6
80	BCC1	0,10	−0,07	35,0	49,0
	FCC1	0,10	−0,02	31,7	47,2
	FCC2	0,16			3,8
90	BCC1	0,09	−0,07	34,9	49,1
	FCC1	0,07	−0,02	31,4	46,9
	FCC2	0,15			4,0
150	BCC1	0,05	−0,07	34,2	49,0
	FCC1	0,05	−0,03	28,9	46,9
	FCC2	0,12			4,1
300	BCC1	−0,06	−0,07	33,3	49,0
	FCC1	−0,06	−0,06	21,4	46,5
	FCC2	−0,01			4,5

We note that the intensity of the singlet decreases with the decrease of temperature. Some Fe sites in the FCC2 phases, with paramagnetic behavior, become ferromagnetic when the temperature decreases.

The two hyperfine magnetic fields distributions, $P(B_{hf})$, obtained from the fitting of the spectra, are shown in Fig. 3. The HMFD with higher fields is related to the BCC structure, rich in Fe. When it was fitted with a Gaussian curve, its central value field was 32.6 T at room temperature, close to that of α-Fe. However, the HMFD contains fields larger than 32.6 T with appreciable probability, showing that the Ni-atoms in the BCC structure increase the ferromagnetic behavior of the Fe-atoms in this phase. This is consistent with what was reported for Fe-Ni disordered alloys [15]. The HMFD with lowers fields, related to the FCC1 ferromagnetic phase, shows two peaks with smaller probabilities than those of the BCC phase. When the temperature decreases the peak with the highest fields slightly shifts to higher fields increasing its probability whereas the one with smaller fields disappears. When the temperature decreases the paramagnetic Fe-atoms in the FCC1 phase, surrounded by Ni-atoms, tend to align with the magnetic field of the ferromagnetic matrix of their neighborhood. This makes those Fe-atoms become ferromagnetic. This tendency increases with decreasing temperature.

The mean hyperfine fields (MHF) of the BCC and FCC1 phases are represented in Fig. 4. The BCC mean hyperfine field at room temperature is higher than that of pure Fe, showing that Ni atoms increase the ferromagnetic behavior of the Fe-

Fig. 4 Mean hyperfine fields of BCC phase and FCC1 phase vs temperature. Line is a guide for the eyes

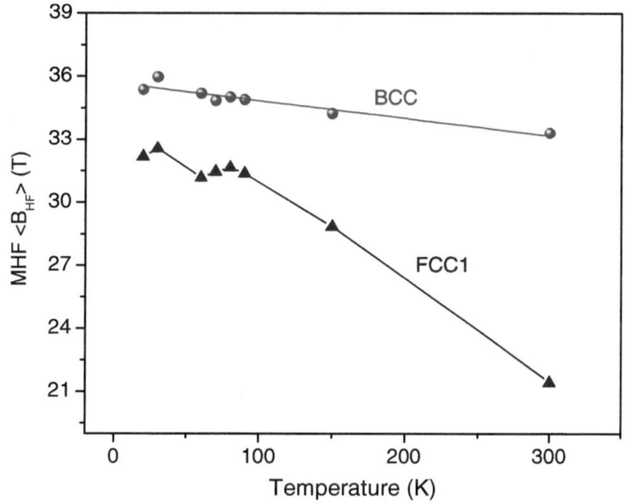

Fig. 5 Relative areas of the singlet and of the HMFD of FCC1 phase vs. temperature. Lines are guides for the eyes

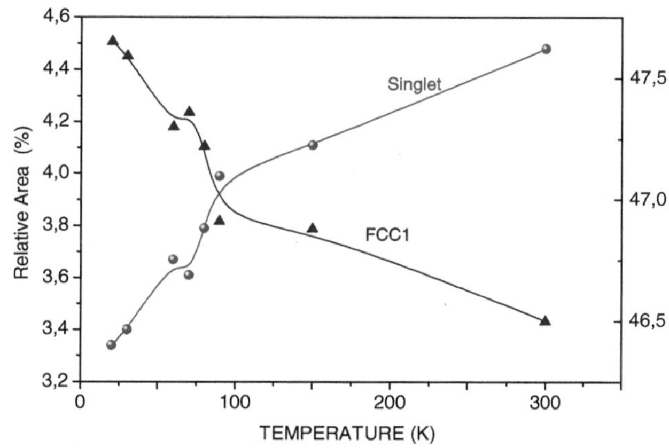

atoms. Also its MHF increases as temperature decreases. The MHF vs. T curve for the FCC1 phase (Fig. 4) is not smooth and shows a clear kink near 60 K (T_k) and a change in slope at about 90 K. When the temperature decreases some spins begin to align due the existence of ferromagnetic exchange interactions. Their number increases as temperature decreases. Due the existence of disorder and competitive interactions which arise from the ferromagnetic exchange between Fe and Ni atoms and the paramagnetic character of Fe-atoms in the FCC structure, some spins tend to remain in the paramagnetic state. For $T > T_k$, these spins are frustrated and show a paramagnetic behavior. Below T_k, these spins are frozen but tend to align with the magnetic field of the ferromagnetic matrix, thereby contributing to the observed average field. Hence, at a given temperature between T_k and 0 K (the re-entrant spin glass region), the average field would be higher than that expected if only ferromagnetic spins were present.

Finally Fig. 5 represents the spectral areas of the paramagnetic site (FCC2) and of the ferromagnetic area of the FCC1 phase, as a function of T. It can be observed that the paramagnetic area decreases with T, and for T = 60 K a kink appears. The FCC2 spectral area exhibits the opposite behavior, showing that the increase of this spectral area is not only due to the normal increase due to the decrease of T, but also to those paramagnetic sites of the FCC2 phase which become ferromagnetic.

4 Conclusions

$Fe_{65}Ni_{35}$ alloys were prepared by the mechanical alloying technique in a high-energy planetary ball mill. The as-milled samples, characterized by X-ray diffraction, show the coexistence of three phases BCC, FCC1 and FCC2. The last phase is related to the paramagnetic phase obtained by Mössbauer spectrometry. The FCC2 phase is related to the antitaenite phase proposed by Scorzelli. As temperature decreases, the MHF of FCC1 phase increases, and shows a kink near 60 K, which can be attributed to a re-entrant spin-glass phase.

Acknowledgement The authors would like to thank Colciencias, Universidad Autónoma de Occidente and Universidad del Valle for their support.

References

1. Tcherdyntsev, V.V., Kaloshkin, S.D., Tomilin, I.A., Shelekhov, E.V., Baldokhin, Yu.V.: Nanostruct. Mater. **12**, 139–142 (1999)
2. Hamzaoui, R., Elkedim, O., Fenineche, N., Gaffet, E., Craven, J.: Mater. Sci. Eng. **A360**, 299–305 (2003)
3. Hellstern, E., Schultz, L.: J. Appl. Phys. **63**, 1408 (1988)
4. Suryanarayana, C.: Prog. Mater. Sci. **46**, 1–184 (2001)
5. Hellstern, E., Fecht, H.J., Fu, Z., Johnson, W.L.: J. Appl. Phys. **65**, 305 (1989)
6. Djekoun, A., Otmani, A., Bouzabata, B., Bechiri, L., Randrianantoandro, N., Greneche, J.M.: Catal. Today **113**, 235–239 (2006)
7. Valderruten, J.F., Pérez Alcázar, G.A., Greneche, J.M.: J. Phys. Condens. Matter **20**, 485204 (2008)
8. Scorzelli, R.B.: Hyperfine Interact. **110**, 143–150 (1997)
9. Valderruten, J.F., Pérez Alcázar, G.A., Greneche, J.M.: Hyperfine Interact. **195**, 219–226 (2010)
10. Rancourt, D.G., Scorzelli, R.B.: J. Magn. Magn. Mater. **150**, 30 (1995)
11. Larson, A.C., Von Dreele, R.B.: General structure analysis system (GSAS). Los Alamos Natl. Lab. Rep. No. LAUR 86–748 (2004)
12. Teillet, J., Varret, F.: Mosfit Programm, University du Maine, France (unpublished)
13. Petrov, Yu.I., Shafranovsky, E.A., Baldokhin, Yu.V., Kochetov, G.A.: J. Appl. Phys. **86**(12), 7001 (1999)
14. Rancourt, D.G., Lagarec, K., Densmore, A., Dunlap, R.A., Goldstein, J.I., Reisener, R.J., Scorzelli, R.B.: J. Magn. Magn. Mater. **191**, L255–L260 (1999)
15. Restrepo, J., Pérez Alcázar, G.A., Bohórquez, A.: J. Appl. Phys. **81**(8), 4101 (1997)

DOI 10.1007/978-94-007-6482-8
Reprinted from *Hyperfine Interactions*

Author Index to Volume 224

MIX
Papier aus verantwortungsvollen Quellen
Paper from responsible sources
FSC® C105338
FSC
www.fsc.org

If you have any concerns about our products,
you can contact us on
ProductSafety@springernature.com

In case Publisher is established outside the EU,
the EU authorized representative is:
Springer Nature Customer Service Center GmbH
Europaplatz 3, 69115 Heidelberg, Germany

Printed by Libri Plureos GmbH
in Hamburg, Germany